Calculate the Orbit of Mars!

Jane Clark

Calculate the Orbit of Mars!

An Observing Challenge and Historical Adventure

 Springer

Jane Clark
Cardiff Astronomical Society
Risca, UK

ISBN 978-3-030-78266-5   ISBN 978-3-030-78267-2   (eBook)
https://doi.org/10.1007/978-3-030-78267-2

This Springer imprint is published by the registered company Springer Nature Switzerland AG
The registered company address is: Gewerbestrasse 11, 6330 Cham, Switzerland

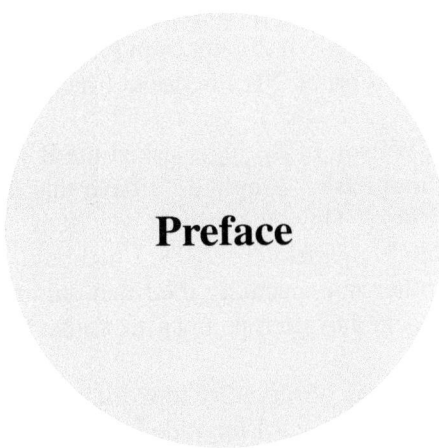

Preface

Twelve years ago, I wrote a book [1] in Springer's Patrick Moore's Practical Astronomy Series entitled *Measure Solar System Objects and their Movements for Yourself!*, in which I worked out some planetary orbits assuming that these orbits, and that of the Earth, were coplanar circles. I deliberately made that restriction to keep the mathematical difficulty to approximately that of college freshman mathematics for science and engineering students. I didn't use much calculus and used almost no vectors at all.

Of course, I was well aware that this simplifying assumption had its limits. In particular, I was not satisfied with my orbital parameters for Mars, because the Red Planet has quite a noticeably elliptical orbit with an eccentricity of almost 0.1. I now believe I got lucky: my simple method didn't work for data from later years.

I quickly discovered methods for deducing the orbits of comets and asteroids, which work so long as they have a decent inclination to the ecliptic, the plane of the orbit of the Earth. Every single one of these methods fails when the celestial body orbits in almost the same plane as the Earth, as the planets do.

So, I hit a brick wall.

Meanwhile I was slowly gathering photographic position data on planets

Life also got in the way. For me 2010 was one of those disastrous years when my business and my marriage both hit the rocks. I ended up moving From King's Lynn in Norfolk, England, to Bristol on the other side of the country to begin a new job. Difficulty selling a house in the recession, plus the need for some major surgery, meant that I was in a rented house in Bristol for 5 years. In late 2015, I bought a house 34 miles from my Bristol office in Risca, across the border in Wales. The house was what real estate agents politely call a "project": it did not even have a kitchen. I gathered no data for the Mars opposition of 2016 because I had to get my house fixed up.

At this point John Watson of Springer asked me if I had any ideas for a book. "Funny you should ask," I replied, "I have this idea for a project to analyse the orbit of Mars." Thus, the project was born.

By the time of Mars' apparition of 2018, I had built an observatory, but unfortunately my mother unexpectedly died that summer after a short illness, and I was too shocked to attempt any astronomy. So, I got no data then either.

In 2019 I began to get my observatory to work well. The money from my parents' estate did not go amiss – I upgraded to an 11" SCT telescope and bought a Celestron CGX mount, which proved to be a massive improvement on its predecessor. Thus, by 2020, I was well positioned to collect another dataset for Mars' apparition.

But what about that brick wall I had hit with the analysis? I made pretty good progress in 2019, but still didn't have a method. The next bit of life that got in the way was the Covid pandemic. My pandemic fortunes were mixed. On the one hand I managed to pick up long Covid, and was plagued by fatigue. Sometimes I still am. But two pieces of good luck came my way. First, I now work from home, so I'm not losing 2 hours a day to commuting. Second, what else was there to do? Astronomy has been the perfect lockdown hobby.

While dealing with the data analysis problem, I coined a word, for an activity which produced a combination of fascination and exasperation: I called it exasprinating. Eventually my persistence paid off, and I got a method to work. Fortunately for you, dear reader, explaining the method is much, much easier than finding it and making it work with no help was.

It is my strict policy to spell out all my mathematical working and to try to leave nothing as an exercise for the reader. That way, I hope you will experience more fascination and less exasperation than I had to put up with.

It proved to be impossible to make all the chapters self-contained. You will need Chap. 4 for all the least-squares fits I do, unless this is already a very familiar subject to you. You will also need to have read the mathematical parts of Chaps. 2 and 3 to read Chap. 7. You also need to read a small section of Chap. 7 while reading Chap. 6 if you have forgotten, or never knew, about rotation matrices.

I have assumed that the reader is a scientist, engineer or mathematician, and knows something about calculus and vectors and trigonometry No doubt you may have forgotten much of it. Recalling half-forgotten material may cause some short-term pain. Please don't expect to pick this book up and read it like a novel. You will need a pencil and paper. Please expect to have to put it down and think from time to time. In emergencies, there are plenty of inexpensive mathematics textbooks from which to revise.

The history of the orbit of Mars is worth telling because in order to provide convincing solutions, the natural philosophers of the Renaissance had to invent both physics and calculus. An English newspaper editor once said that while comment is free, facts are sacred [2]. I have tried to stick to this principle. For example, as a teeneager, I was much impressed by Arthur Koestler's book [3] covering much the same history. Even at that tender eage, I had an uneasy feeling that, for this author, facts might be subordinate to the quality of the yarn he spun. I have tried to provide a reference for every factual claim I make.

The rate-determining step in astronomical progress has always been our ability to observe, not our cleverness at theorizing. A combination of Tycho's positional data and Galileo's telescopes did for the ancients' view of the universe. The same thing happened with Hubble's observations of the distances to galaxies and the redshift. Without those insights, the theorists could have debated until the cows come home.

I would like to thank my editors at Springer, whose chief service was to keep chasing me. This includes Maury Solomon and John Watson, both retired, who encouraged me to start, and Hannah Kaufman and Clement Wilson Kamalesh, whose encouragement kept me going. I would also like to thank my many astronomer friends, who have taught me more than they know. Finally, I would like to thank my best friend Pauline Thomas for her encouragement and friendship.

No doubt there are better ways to do what I did. If anyone finds them, finds a historical nugget or two, or manages to do some other interesting piece of celestial mechanics, please share them with me at jane.clark@ finerandd.com. Where next for me? One of my ambitions is to devise a method for measuring the astronomical unit in miles, based on inverting Rømer's discovery [4], while obsering the Moons of Jupiter, that light has a finite speed. Stay tuned.

Risca, Wales, UK   Jane Clark
April 2021

Acknowledgements

I would like to thank my editors at Springer, whose chief service was to keep chasing me. This includes Maury Solomon and John Watson, both retired, who encouraged me to start; and Hannah Kaufman and Clement Wilson Kamalesh, whose encouragement kept me going. I would also like to thank my many astronomer friends, who have taught me more than they know.

I also express my thanks to Dr. Jeannette M. Fine for permission to use a piece of software and Pauline Thomas for permission to use a photograph.

Finally, I would like to thank my best friend Pauline Thomas for her encouragement and friendship.

Contents

1 In the Beginning.. 1

2 Read My Ellipses.. 45

3 The Behaviour of Two Bodies in Orbit............................... 59

4 Least Squares Fit to Sets of Equations.............................107

5 The Orbit of the Earth...113

6 Here's Looking at You, Mars!..151

7 First Shot at the Orbit of Mars.......................................201

8 Refining the Preliminary Orbit 269

Appendix 1: Parabolas and Hyperbolas 293

Appendix 2: Multiplying Three Vectors................................299

Appendix 3: Proof of Trigonometric Addition Formulae for All Angles....303

References...305

Index 311

About the Author

Jane Clark is a British amateur astronomer who earns her living as an engineer. She has a Ph.D. in physics and an MBA from Warwick University. She completed 2 years of postdoctoral training at Case Western Reserve University in Ohio before returning to England to begin an industrial career. She became interested in both astronomy and photography as a teenager in the 1970s, photography much more seriously, although as her career progressed and family commitments increased, both interests lapsed. She acquired a telescope in 2006, shortly after completing her MBA, and quickly became hooked on observing. This experience made her realize that astronomy is a lot more fun than business administration. In 2017 she achieved her ambition of having an observatory in her back yard. She is a member of Bristol, Cardiff and Newtown Astronomical Societies; and was a founder member of West Norfolk Astronomy Society. Jane gives talks on the Solar System to astronomy clubs, and other societies as diverse as the cub scouts, the University of the Third Age, and church wives' groups.

Chapter 1

In the Beginning

Most, if not all, of the early civilizations developed their astronomy. If nothing else, it gave them a way to measure and predict the seasons. Evidence from Australia, where Europeans discovered a native culture whose land management was there, but could hardly have been called agricultural in a Western sense, is that the first Australians had good naked-eye astronomical knowledge and were well aware of the planets. Europeans found that the Maori in New Zealand used astronomy to predict seasons [5], even though they had no writing before they encountered Europeans. In other words, humans must have been astronomers for a very long time.

Not all the early civilizations had writing as we know it. This invention does not seem to have made it to Central America, for example. Yet we now know that, when they wanted to, the Mesoamerican civilizations could orient their buildings to indicate astronomical events. Therefore, something must have been recorded. The Ancient Egyptians used the disappearance of the star Sirius into the sunrise to know when the Nile would flood. They were exceptionally dependent on this, as rainfall is very rare indeed in much of Egypt. My father served as a draftee in the British Army along the Suez Canal. After it rained, he told that the locals debated whether or not the previous rainfall had been twenty years before. The first indication that the floods were coming was astronomical: there simply was no rainy season.

Civilizations on the Indus, and the Tigris and Euphrates rivers, were among those that kept records. One remarkably successful Macedonian king and general, Alexander the Great (356–323 BCE), conquered lands

J. Clark, *Calculate the Orbit of Mars!*,
https://doi.org/10.1007/978-3-030-78267-2_1,
© Springer Nature Switzerland AG 2021

from Greece to the Indus River. Although he died young and his empire did not last, one side effect was the diffusion of astronomical knowledge from the Indian and Babylonian civilizations to the Middle East.

(CE, or the Common Era, is a way of dating. It is exactly the same as the system BC and AD and is simply a more neutral name that does not imply either recognition or rejection of a particular religious event. BCE is "Before the Common Era". One of the more curious characteristics of this dating system is that it has no year zero. It went from 1 BC/BCE straight to 1 AD/CE.)

In the city of Alexandria, Egypt, one of many founded by Alexander the Great as he conquered his way across the Middle East, there grew up a great library and a research institution called the Musaeum, said to have housed over a thousand scholars at any time. The heyday of the Musaeum was from roughly 300–145 BCE.

In that time, it hosted Archimedes, Euclid, Eratosthenes and Aristarchus of Samos.

Eratosthenes measured the circumference of the Earth. It's hard to know whether his answer was a good or brilliant first effort, because that depends on how we interpret the distance unit he used. He used the stadion, which might be anywhere from 159 m to 209 m [6]. If he used the lower end of the range, his Earth radius is almost smack on the modern value. He is popularly supposed to have done this by comparing the angles of the Sun at a mineshaft at Syene, modern Aswan, and in Alexandria. In fact, he was more careful than that, but the details are lost [7]. His own book has not been preserved.

We know of Aristarchus of Samos directly and indirectly. Directly we know of his book "On the sizes and distances of the Sun and Moon" [8]. Indirectly we know from a book by Archimedes, the Sand Reckoner, that

"Aristarchus has brought out a book consisting of certain hypotheses, wherein it appears, as a consequence of the assumptions made, that the universe is many times greater than the 'universe' just mentioned. His hypotheses are that the fixed stars and the sun remain unmoved, that the earth revolves about the sun on the circumference of a circle, the sun lying in the middle of the orbit, and that the sphere of the fixed stars, situated about the same center as the sun, is so great that the circle in which he supposes the earth to revolve bears such a proportion to the distance of the fixed stars as the center of the sphere bears to its surface" [9]. There is no evidence that Aristarchus ever made his hypotheses quantitative [10], although I find it hard to believe that his idea was purely qualitative.

Archimedes goes on to dismiss Aristarchus' idea. Unfortunately for posterity, so did just about everyone else.

What the world got instead was a textbook by Claudius Ptolemy, known as Ptolemy. Ptolemy's view was firmly geocentric. Why was this? It was partly down to the baleful influence of Aristotle.

Aristotle lived from 384 to 322 BCE. I am clearly not qualified to judge how good a philosopher Aristotle was. But I am suitably qualified and experienced to judge his physics. I cannot say it loudly enough: he was a lousy physicist. "In reality [Aristotle] does little but analyze the meanings of every-day experience and words in order thereby to solve the problems of nature." So says J. L. E. Dreyer in his "History of Astronomy from Thales to Kepler" [11]. Because he had such a following for his other endeavours, people unfortunately assumed that he know what the heck he was talking about. Well, he didn't. His idea that everything was made of four elements, Earth, Air, Fire and Water, was hogwash. His ideas on motion misled people for centuries. The likes of Huyghens, Galileo and Newton took a careful, experimental and quantitative view of motion, unlike Aristotle, and got much nearer to what we now believe to be the truth. Aristotle, on the other hand, tried to argue his physics from philosophical reasoning. He also postulated an idea totally untainted by evidence either for or against: that the heavens above the earth were made of a fifth element, which was "perfect". This frankly bizarre idea took a lot of dislodging by means of evidence.

One of Aristotle's fetishes was "prefect circular motion", that is, uniform circular motion about a point. There *is* a useful characteristic of circular motion: the centripetal force required to keep the motion of a point particle circular is always perpendicular to the distance moved. In an infinitesimal time dt, the work done is

$$dT = \mathbf{F}_{\text{centripetal}} \bullet \mathbf{dx} = \mathbf{0} \tag{1.1}$$

because $\mathbf{F}$ and $\mathbf{dx}$ are always perpendicular. In other words, no work is done to maintain circular motion. The extension to finite bodies is obvious. You simply integrate over the volume. This no energy input is required to keep a planet moving in a circular orbit in a vacuum. It also requires no energy input to keep a solid planet rotating about its axis. Of course, Aristotle did not have the modern vocabulary to describe the phenomenon of circular motion not changing kinetic energy. Nevertheless, he overdid it. As we shall see, it took the world millennia to move on from the fetishization of perfect circular motion.

He acquired over the centuries a tremendous following to the point where it became dangerous to question his claims. We shall see this later in our story.

Ptolemy was strongly influenced by Aristotle, although it would be somewhat unfair to accuse him of being a slavish follower. He modified Aristotle's idea that uniform circular motion was some kind of perfect state.

Ptolemy's Model: Salient Features

Ptolemy lived in Alexandria, rather later than the heyday of the Musaeum, from about 100 to 170 CE.

His was one of the first mathematical models of the universe and one of the earliest mathematical models of a natural phenomenon of any kind. To that extent it was a remarkable achievement. He based it on an earlier, but simpler model, due to Hipparchus. The model was published in a textbook, which we now know by its Arabic name, "the Almagest" [12]. This was partly because some parts survived in the original Greek, while other parts only survived in Arabic translation.

A discussion of the Almagest appears in the self-published online book by Fitzpatrick [13].

It has been superseded by better models. Nowadays, we think that that's how scientific theory advances. New observations and measurements come along which no longer fit the old paradigm, so a new paradigm is required.

That concept came along much later than Ptolemy's model. Even on its own terms, it only approximately matched the then known data. Nowadays we have statistical tools to handle approximation, but that was not true until well after the time of Newton, let alone the time of Ptolemy.

Without tools to help you choose which data to accept, and which to reject, and without much prior experience to go on, Ptolemy was on his own and had to do the best he could. Many scholars have tried to take him down by claiming that he made "convenient" choices of data. Owen Gingerich [14], in an essay entitled "Was Ptolemy a fraud?", argues that that's being a little harsh and judging him by modern standards.

The model he developed had the Earth "not moving", with everything else moving around it. Ancient Greek concepts of space and of motion and stationariness were, of course, pre-Newtonian and very different from ours.

As we now know, the problem the planetary astronomers had to solve was how to model and predict the motion of planets, including the one upon which we live, as they follow elliptical paths. Not only did Ptolemy not know that, but the data available to him were not accurate enough to distinguish elliptical paths from plausible alternatives.

The path of a superior planet (Mars, Jupiter or Saturn) in Ptolemy's model is shown in Fig. 1.1. The Earth is at a fixed point. The sold circle is

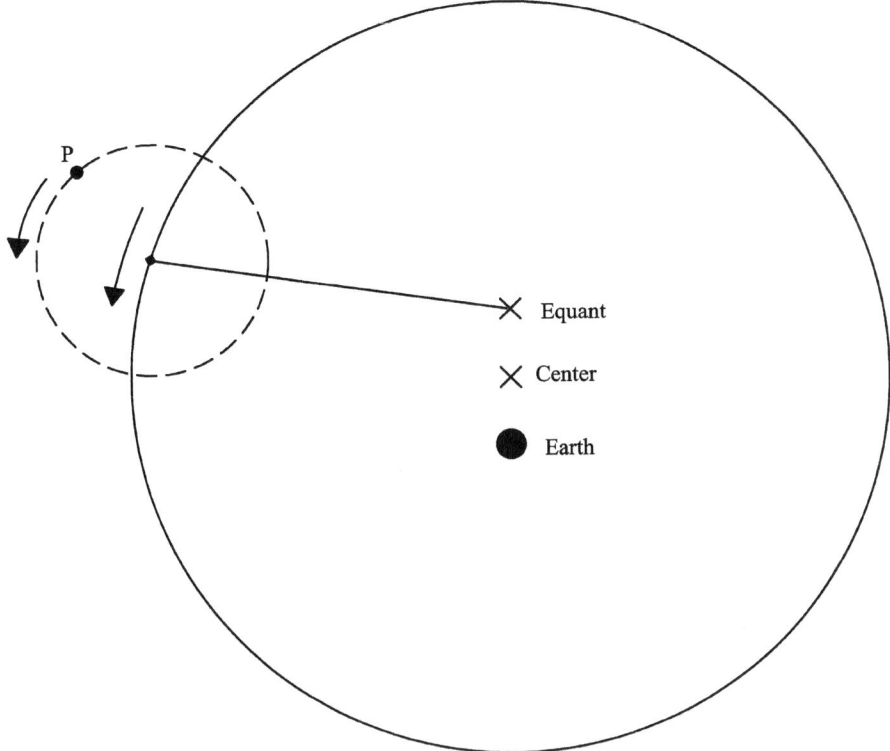

Fig. 1.1 Ptolemy's model of the path of a superior planet, that is, Mars, Jupiter or Saturn. The Earth is at a fixed point. The sold circle is called the "deferent" of the planet. Note that its centre is not at the Earth. Another circle, the "epicycle", shown dashed, rotates around the centre at a variable rate such that an observer at the Equant would see a uniform rate of rotation. The planet itself rotates around the epicycle. The plane of this arrangement is that of the Ecliptic, the apparent orbit of the Sun. (Image: Author)

called the "deferent" of the planet. Note that its centre is not at the Earth. Another circle, the "epicycle", shown dashed, rotates around the centre at a variable rate such that an observer at the Equant would see a uniform rate of rotation. The planet itself rotates around the epicycle.

The time to go around the deferent is what we would now call the orbital period of the superior planet. The time to go around the epicycle is roughly an Earth year.

The plane of this arrangement is that of the Ecliptic, the apparent orbit of the Sun.

The sun in Ptolemy's model also rotates about the Earth in a similar manner, except that the Equant is at the centre of the deferent, and there is no epicycle. The epicycle is not needed because the Sun does not undergo retrograde motion when observed from the Earth. This model is credited to Hipparchus of Nicaea (c. 190–120 BCE) although this author's writings do not survive. Hipparchus is also credited with being the "father" of trigonometry and with discovering the precession of the equinoxes by comparing his data against that from Babylon [15].

Through modern eyes, we might say that the deferent represents the planet's orbit and the epicycle represents the effect of the orbit of the Earth. Of course, we cannot expect Ptolemy to have thought like that.

A quantitative assessment of the calculating capability of this model will be made a little later in our story.

The Copernican Revolution: The Pun That Keeps oving

No doubt this pun has elevated Copernicus' role beyond the admittedly very high level it deserves (Fig. 1.2). It is a myth that he put the Sun at the centre of his system. He almost, but not quite, did this. But he did have the Earth moving around the Sun.

In fact, he was a dedicated Aristotelian, who was offended by the way that Ptolemy's system did not employ uniform motion about the centers of circles. He embarked upon his study with a view to rectifying this situation. In the process, he noticed that some of his epicycles would cancel out if he had the Earth go round the Sun, rather than vice versa.

His model for a heliocentric planet looks like Fig. 1.3.

Part of the conceptual leap that Copernicus had to make was to consider whether the Earth is a planet. He certainly discussed this question [16]. He also worried about the implications of his theory for gravity. If the Earth was but a planet, could it be the sole source of gravity? Did the Sun also exert gravitational pull? What about other bodies? [16]

This model actually made less accurate predictions of planetary motions than Ptolemy's model.

Before Galileo and Newton, particularly Newton, the concepts of moving and stationary were not what they are now. Nowadays, we believe that the universe has no centre and that a body, if it obeys classical mechanics to a good approximation, experiences force but is not affected by translational movement. Rotation is another matter: the General Theory of Relativity is required to explain this. Such explanation is outside the scope of this book. Any interested reader is referred to the section on Mach's principle in

Fig. 1.2 Nicolaus Copernicus portrait from Town Hall in Thorn/Toruń – 1580. (Image courtesy of http://en.wikipedia.org/wiki/Copernicus#mediaviewer/File:Nikolaus_Kopernikus.jpg). Public domain

Misner, Thorne and Wheeler [17], but is warned that General Relativity is not for the faint-hearted. To understand what little I myself know, I had to sign up for a distance-learning undergraduate course, over 40 years after I graduated, having totally failed to teach myself the subject.

The next significant character in this story, Tycho Brahe, got rather hung up on this question, as we will see.

Tycho Brahe, the Greatest Pre-telescope Observer

Tycho (Fig. 1.4), who lived from 1546 to 1601 [18], did not quite live to see the telescope applied to astronomy. In this, he was a little unlucky: he died of an illness so sudden that rumours abounded that he was poisoned. These were eventually laid to rest in 2010 after his body was exhumed and tested [19]. It seems as though he died of an advanced bladder infection. This post-mortem did indicate that the best known legend about Tycho, that he had a

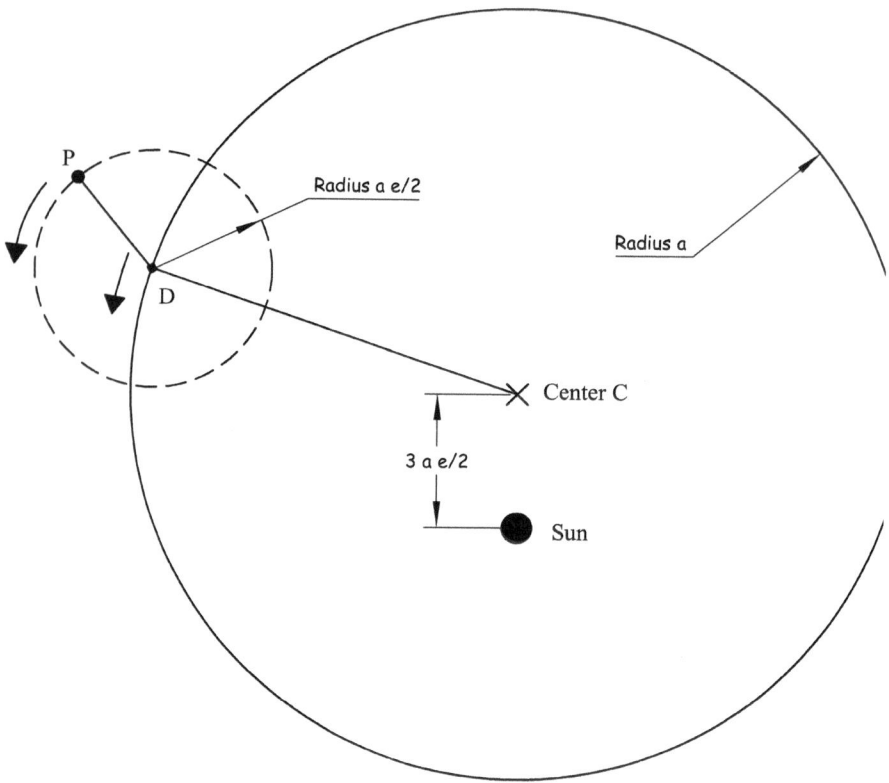

Fig. 1.3 A heliocentric orbit according to Copernicus. The lines CD and DP rotate uniformly about C and D, respectively, such that DP rotates twice as fast as CD. The perihelion is at the top and the aphelion is at the bottom. (Image: Author)

metal prosthetic nose because part of his was cut off in a duel, was true. The post-mortem showed traces of brass around the nose wound.

This was simply another example of what an extreme character Tycho was. Everything about him was outsized. He was a high-ranking, very wealthy Danish aristocrat. Confusingly, the land where he was born and mostly lived was in what is now south-western Sweden, but back then it was Danish. Unusually for an aristocrat, he married for love and produced thirteen children. He was a prodigiously determined observational astronomer. He had his own magnificent observatory on an Island, together with a paper mill and printing press. Over the years, he employed perhaps sixty people and drove them hard [20]. He even had his own court jester [21]. He was a great correspondent and author [22]. This

Fig. 1.4 Tycho Brahe. (Image courtesy of http://cache.eb.com/eb/image?id=83677&
rendTypeId=4). Public domain

enabled him to disseminate his results and, in effect, to shout louder than
any of his contemporaries. He was certainly competitive and keen to
establish his own legend. But the thing is this: he had the talent to back
this up. His measurements were an order of magnitude better than anything that went before.

Well educated in the Universities of Germany and Scandinavia [21], he
discovered early in his adulthood that he could measure astronomical phenomena better than the ancient Greeks could. He would not have known
about the astronomers of Islam. They were not widely known in Europe
during the Renaissance [23].

Tycho and Mathematics

Geometry was an essential tool to convert Tycho's measurements into celestial positions. Living in the computer age as I do, I have to admit that I never gave this a second thought until I realized that he worked before logarithms were invented. Multiplication and division were therefore much more severe and labour-intensive problems than they were once logarithms became available.

The key art was spherical trigonometry, which is a shorthand name for the analysis of triangles on the surfaces of spheres. Of course, the sphere the early scientists thought of was a faraway sphere on whose inside surface were the "fixed" stars.

The parameters governing such a triangle are built up in Figs. 1.5, 1.6, 1.7, and 1.8.

Thus, our triangle on the surface of a sphere can be uniquely defined entirely in terms of angles, provided that its sides all lie on great circles.

Now let me define some planar triangles to help me analyse the spherical triangle (Fig. 1.9). I do this by drawing lines tangential to the two great circles that meet at A.

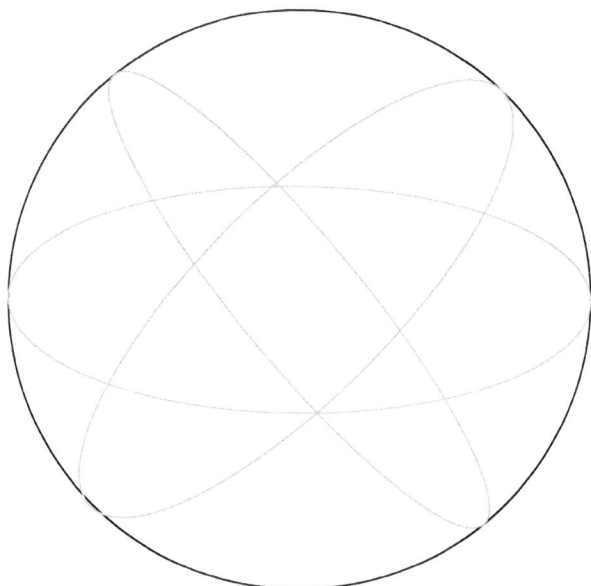

Fig. 1.5 Imagine a sphere onto whose surface three great circles are drawn. A great circle is a circle with the same radius as the sphere, whose centre is the centre of the sphere. (Image: Author)

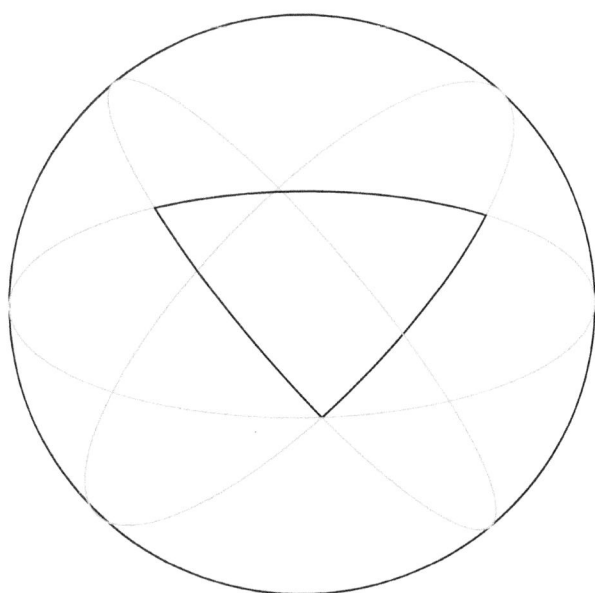

Fig. 1.6 These three great circles form the sides of a triangle on the sphere. They actually host another triangle on the other side of the sphere, but we will not consider that for now. (Image: Author)

Then by the cosine rule, these two equations are true.

$$DE^2 = AD^2 + AE^2 - 2(AD)(AE)\cos A;$$
$$DE^2 = OD^2 + OE^2 - 2(OD)(OE)\cos a. \tag{1.2}$$

Because we chose tangents, the angles $\angle OAD$ and $\angle OAE$ are right angles. Thus,

$$OD^2 = OA^2 + AD^2;$$
$$OE^2 = OA^2 + AE^2. \tag{1.3}$$

Substitute the equations of Eq. (1.3) into Eq. (1.2):

$$DE^2 = AD^2 + AE^2 - 2(AD)(AE)\cos A;$$
$$DE^2 = OA^2 + AD^2 + OA^2 + AE^2 - 2(OD)(OE)\cos a. \tag{1.4}$$

Now subtract the upper from the lower of Eq. (1.4).

$$0 = 2(OA^2) + 2(AD)(AE)\cos A - 2(OD)(OE)\cos a. \tag{1.5}$$

Rearranging Eq. (1.5) gives

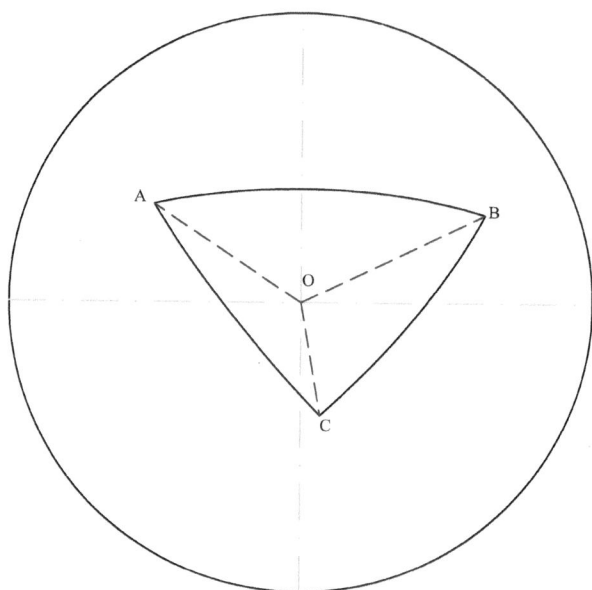

Fig. 1.7 Let this triangle have vertices A, B and C, all in upper case. Let the centre of the sphere be at O. (Image: Author)

$$\cos a \quad = \quad \frac{OA}{OE}\frac{OA}{OD} + \frac{AE}{OE}\frac{AD}{OD}\cos A \qquad (1.6)$$

$$\text{so } \cos a \quad = \quad \cos b \cos c + \sin b \sin c \cos A$$

using the definitions of sine as opposite/hypotenuse and cosine as adjacent/hypotenuse. Rearranging Eq. (1.6) gives

$$\cos A = \frac{\cos a - \cos b \cos c}{\sin b \sin c} . \qquad (1.7)$$

In the pre-logarithmic age, the very useful Eq. (1.7) would have been messy to manipulate. It involves two multiplications and a division.

It is therefore worth some effort to see if we can turn the multiplications into additions or subtractions.

We will only deal with the cases where angles are right angles or less because Tycho's measurements did not use angles >90°. By following the logic in Fig. 1.10, it can be seen that

$$\sin(g+h) = \sin g \cos h + \cos g \sin h \qquad (1.8)$$

and

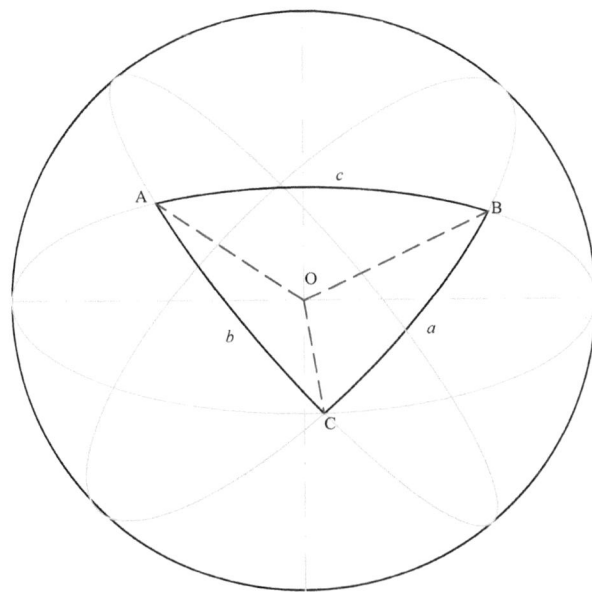

Fig. 1.8 Let the side opposite A has length a, let that opposite B has length b and let that opposite C has length c. A better way to analyse the triangle is to regard a, b and c as angles. a is the angle BOC; b is the angle COA and c is the angle BOA. (Image: Author)

$$\cos(g+h) = \cos g \cos h - \sin g \sin h \tag{1.9}$$

Proofs of these formulae for all angles, whether greater than 90° or not, are given in Appendix 3.

Eq. (1.8) can be used like this:

$$\sin g \sin h = \frac{1}{2}\left[\cos(f-g) - \cos(f+g)\right], \tag{1.10}$$

and Eq. (1.9) can be used like this:

$$\cos g \cos h = \frac{1}{2}\left[\cos(f-g) + \cos(f+g)\right], \tag{1.11}$$

These relations can be used in Eq. (1.7) to eliminate the multiplications but not the division:

$$\cos A = \frac{2\cos a - \left[\cos(b-c) + \cos(b+c)\right]}{\cos(b-c) - \cos(b+c)}. \tag{1.12}$$

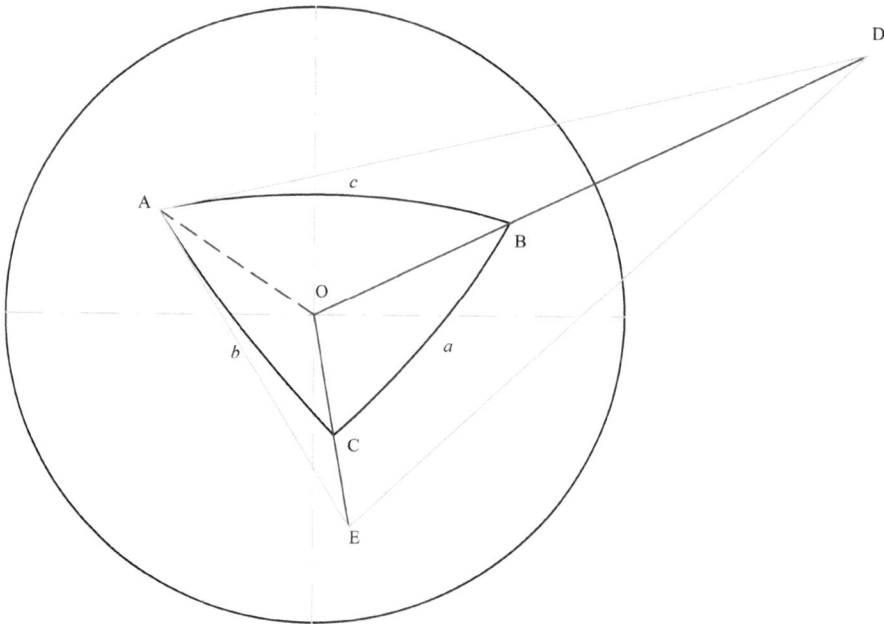

Fig. 1.9 Two lines are drawn tangent to the two great circles that cross at A. The lines OB and OC are projected until they meet these tangent lines at D and E, respectively. (Image: Author)

This elimination of multiplication was well worth the bother when you could not use logarithms. Indeed, the process was given a name: *prosthaphaeresis*. This compound word was built from the Greek words *prosthesis* and *aphaeresis* (subtract).

In those days, Eqs. (1.2)–(1.12) were not written out as formulae. They were written in words [24].

Tycho claimed priority for this technique. For this he has been in hot water with historians ever since. It's the perfect storm in a teacup. There does not appear to be enough evidence to settle the matter decisively [24], but it seems likely that Tycho was introduced to the formula by an assistant, Johannes Wittich [25]. Thoren [25] reports that Wittich brought the method to Tycho's observatory, having developed it beforehand. Tycho, who was nothing if not competitive, seems to have viewed prosthaphaeresis as something of a trade secret. He was not best pleased when he found out that, after leaving his employment, Wittich published the technique [24]. Such a driven man as Tycho venting his anger may well have been an impressive sight.

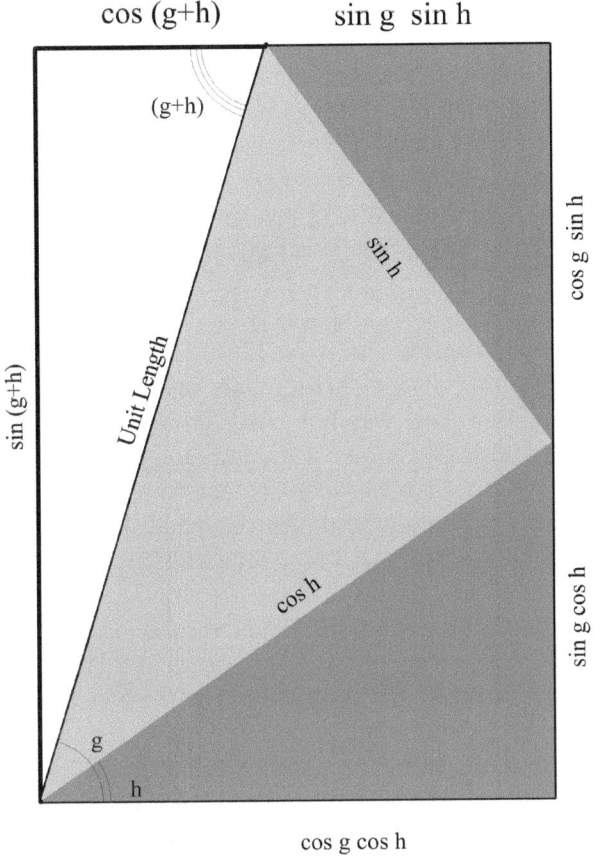

Fig. 1.10 Diagram showing how to manipulate sines and cosines of sums of angles

Tycho's Instruments

Tycho was no lone genius. In fact, you could even make a case for saying that his operation at his observatory in Uraniborg, Hveen, the Danish island now part of Sweden and known as Ven, was the first example of "big science". By the standards of today's big science, where the biggest projects employ thousands, Tycho's operation was small, yet his grant from the King of Denmark was 1% of the total royal expenditure. That was not to be sneezed at.

His first big challenge was to observe a supernova in 1572. He was careful to measure the position of the supernova relative to those of the other stars in Cassiopeia, within which constellation it had appeared. He looked very hard for evidence of parallax. His method was to compare what happened when the supernova first appeared in November and Cassiopeia was very high in the sky with what happened later when the circumpolar Cassiopeia was low in the sky. Cassiopeia is circumpolar for latitudes above 34°N: his latitude was around 56°N. He was looking for parallaxes of a few degrees for a nearby object. He found no parallax, and even checked that over its 18-month apparition, it kept pace with the precession of the equinoxes. He was in fact unable to detect whether it did or not, the effect in question being about 20 arcseconds. He reckoned that the supernova was more distant than the planets and was probably on the sphere on which people then believed the stars to exist [26]. The appearance of this very bright star cast doubt on the Aristotelian idea that the heavens were unchanging. That, however, was not what Tycho wrote about. He wrote about the astrological implications of the new star [26].

His instrument was a large wooden sextant on a stand, basically a huge protractor. When Cassiopeia was almost overhead, he was unable to make measurements. A later version of Tycho's instrument is shown in Fig. 1.11.

His next famous observation was of the great comet of 1577. This was by no means the only comet that he observed over the years. He again took an interest in parallax measurement and demonstrated that the comet was not in the Earth's atmosphere at all, but was beyond the Moon. If the supernova made him doubt the Greek belief in heavenly spheres to carry the planets, the comet blew his faith in this idea out of the water. He observed that the angular velocity of the comet decreased and deduced from this that it was travelling *through* these spheres. This made him doubt their very existence. This in turn drove him, over a period of years, to grope towards a cosmological model of his own [28, 29]. He had, however, another fixation to deal with. He could never shake off his belief that the Earth could not be moving. So he worked his way towards a system, without epicycles, in which the Earth was fixed, the Moon and Sun go around the Earth and the planets go around the Sun. The resultant system is shown in Fig. 1.12. It was published in 1588 [30].

For our story, the significance is that Fig. 1.12 appears to be the first diagram showing the Solar System without planets travelling in epicycles. The system became less elegant the more quantitatively it was analysed [28], and never caught on, despite the best efforts of Tycho, and later his heirs, to foist it on the world.

Over thirty years, he built better and better instruments as he sought the elusive prize of accuracy. He began with conventional enough instruments, but developed them. Over time they got bigger because it was then possible to make the scales finer. He had to contend with typical problems as you

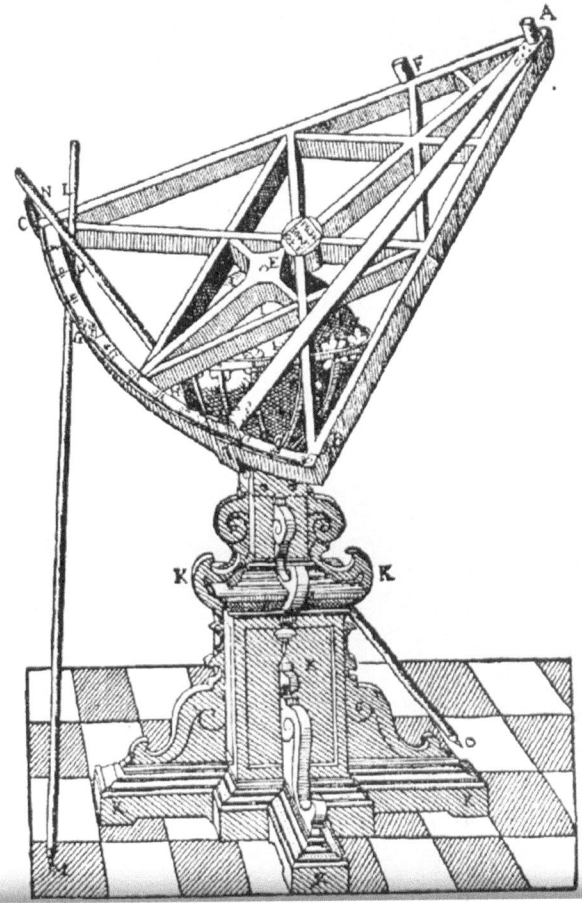

Fig. 1.11 A sextant built by Tycho in 1582, ten years after the supernova. He says in the text of his book on instruments that he built three of these instruments over the years. (Source: *Astronomiæ Instauratæ Mecanica* by Tcho Brahe, 1598 [27]). Public domain

make things bigger and bigger. Maintaining the rigidity of the equipment was a challenge. So was operating them in the wind. In the end, he built a separate observatory next door, in which the instruments were basically below ground.

Another of Tycho's instruments was the great equatorial armillary, a device capable of measuring both right ascension and declination. Despite owning the best clocks then available, Tycho was not able to measure time reliably. But he could use the equatorial armillary to get the declination of a star and then compare its right ascension with that of a known and trusted

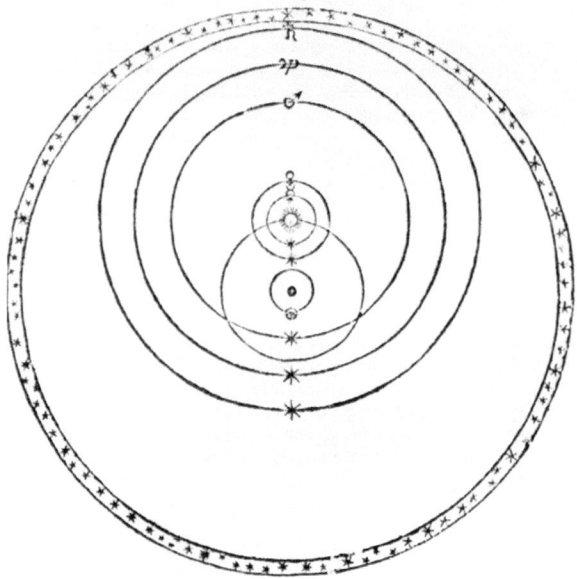

Nova Mvndani Systematis Hypotyposis ab Authore nuper adinuenta, qua tum vetus illa Ptolemaica redundantia & inconcinnitas, tum etiam recens Coperniana in motu Terræ Phyfica abfurditas, excluduntur, omniaq́, Apparentiis Cælestibus aptißime correspondent.

Fig. 1.12 The Tycho model of the world. The Earth is fixed, the Moon and Sun go around the Earth and the planets go around the Sun. My translation of the Latin inscription reads "New Hypotheses of the system of the World proposed recently by the author, by which the redundancy & inelegance of the Ptolemaic, and the physical absurdity of the Copernican, are both excluded, in very apt agreement with the appearance of the heavens". (Source: Brahe's *De Mundi Aetherei Recentioribus Phaenomenis*, 1588). Public domain

reference star. Armillaries were not new, but they were either too small to be accurate or very big and unwieldy: in particular they would flex. The word "armillary" is derived from the Latin word "armilla" meaning "bracelet". Tycho reduced this device to the bare minimum. Originally armillaries were much more complex than Tycho's eventual design. The one in Fig. 1.13 shows that they would typically have great circle rings for each of the equator, both tropics, the Arctic and Antarctic circles and the ecliptic.

The principle of Tycho's armillary is illustrated in Fig. 1.14.

His actual Great Equatorial Armillary is illustrated in Fig. 1.15. It was graduated to ¼-arcminute intervals [31].

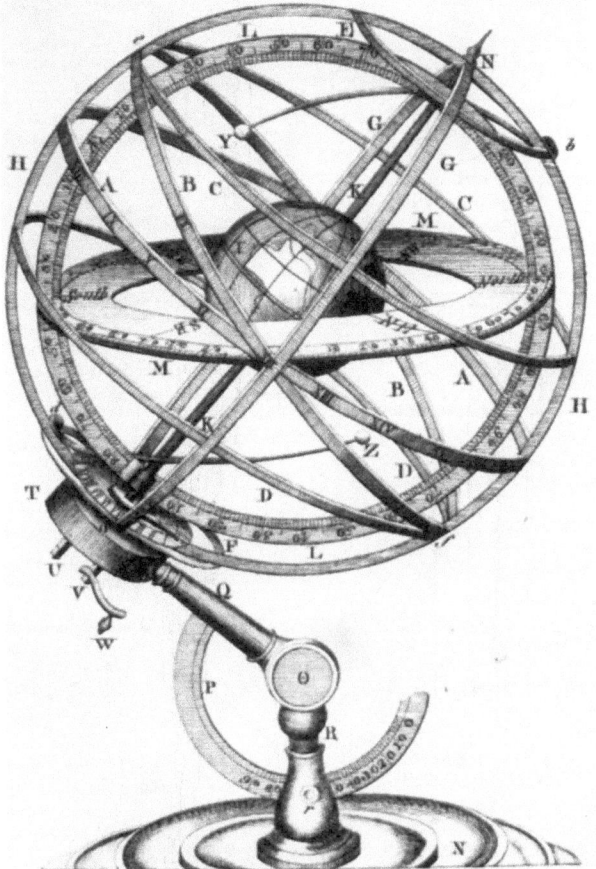

Fig. 1.13 An armillary from the 1771 edition of the Encyclopedia Britannica. Public domain

One of the uses to which he put this armillary was to measure the phenomenon of refraction by the atmosphere. He and his team would simultaneously use this instrument and a quadrant to measure the right ascension and declination of a star and measure its altitude, thereby mapping the phenomenon of atmospheric refraction as a star travelled across the sky.

The other big instrument of Tycho's was the Mural Quadrant, an enormous 90-degree protractor, shown in Fig. 1.16. This instrument was graduated to $1/6$-arcminute intervals [31].

Tycho put immense effort into calibrating the Mural Quadrant and the Great Equatorial Armillary.

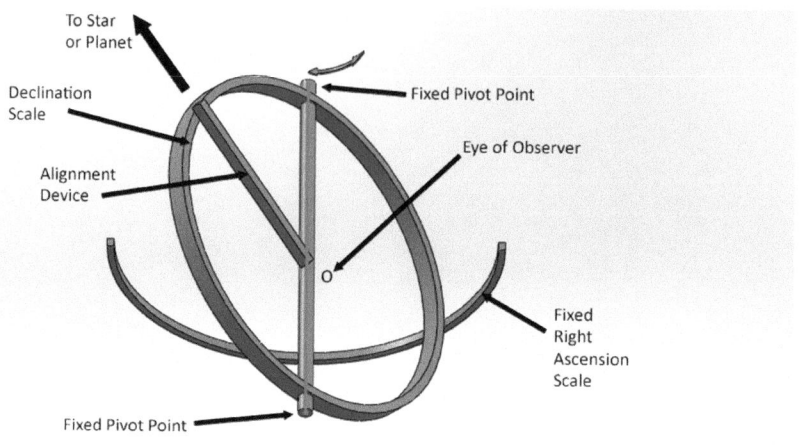

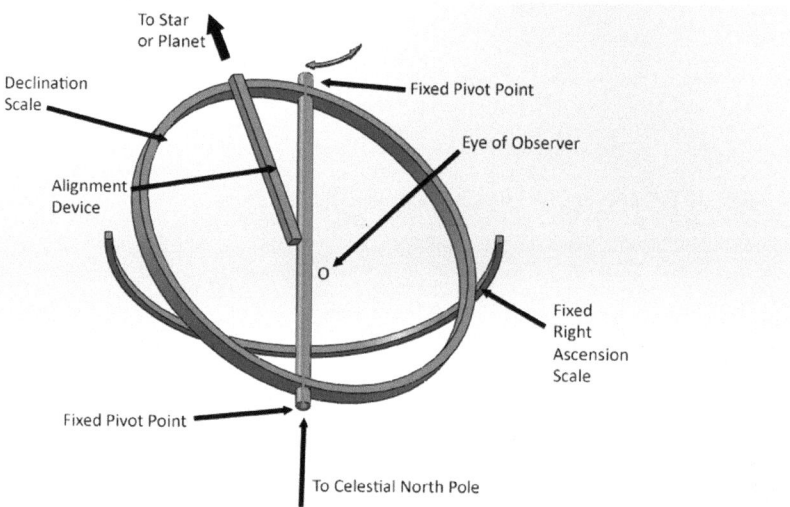

Fig. 1.14 The principle of Tycho's great Equatorial Armillary. The diameter of the ring was around nine feet [31]. The axis of the ring is allowed to rotate but fixed against translation at both ends and points towards the North Celestial Pole. Using the alignment device, the observer rotates the ring until they are looking at the star and reads off the declension on the ring. The fixed semi-circular scale enables the right ascension to be read. The observer would then quickly move to a reference star and do the same. The difference in the two right ascensions is then read directly without the need for trigonometric calculation

Fig. 1.15 Tycho's Great Equatorial Armillary as realized in practice. (Source: Tycho's *Astronomiæ Instauratæ Mecanica* [27]). Recreated as a copperplate engraving by Tycho's former assistant Willem Janszoon Blaeu (c. 1570–1638), who later acquired a reputation as a cartographer. The image was published posthumously by his son in *Novus Atlas* [32]. This image brings out the scale markings better than Tycho's original. Public domain

He had two more tricks up his sleeve to boost accuracy. The first is what we would now call by its Arabic-derived name, an *alidade*. Tycho would not have used the word "alidade".

(You can often spot Arabic-derived words because of the al- or Al-prefix. "Al" is Arabic for "the". It is not a separate word as in English, but a prefix. Some, but not all, people insert a hyphen between "al" and the rest of the word when transliterating from the Arabic alphabet. The most common place where this occurs in astronomy is in star names, a striking number of which begin with "Al".)

Even in Latin, he used the Greek term "dioptra", rather than a Latin-derived name. Little things like this give clues about which ancient cultures Tycho looked to. His alidade is shown in Fig. 1.17.

The principle of Tycho's alidade is that it can be used by two observers, one to the left and one to the right. Tycho says that he would raise or lower it until the star can be seen through slit AD and in the slit EH at the other end. If it can also be seen through slits BC and FG, the star is aligned accurately. Slits CD and DH can also be used to align in the opposite direction. For solar observations, the hole in the middle of EFGH is projected onto a circle behind ABCD that we cannot see in Fig. 1.17. The widths of the slits at the near end can be adjusted. Tycho remarks [27] that he is surprised that

Fig. 1.16 Tycho's Mural Quadrant. (Source: Tycho's *Astronomiæ Instauratæ Mecanica* [27]). Notice the top right sub-illustration showing how two people would use the sextant shown in Fig. 1.11. Public domain

Fig. 1.17 Tycho's alidade or dioptera. (Source: Tycho's *Astronomiæ Instauratæ Mecanica* [27]). Public domain

nobody else complained about the extreme difficulty of seeing stars through pinholes, which was evidently the favoured method.

The last account of Tycho's instrumentation is of his method of reading fine detail from his scales, shown in Fig. 1.18.

This method appears to have been invented some centuries before by a rabbi named Levi ben Gershon (1288–1344), who lived in what is now the south of France [33]. This is the first notable Jewish contribution to astronomy. The next significant Jewish contribution came several centuries later when A. A. Michelson (1852–1931), the first Jew and the first American to win a Nobel Prize, made important discoveries about the speed of light.

The principle is that by spreading the dots over a greater distance by angling them steeply, it is possible to magnify the scale by a factor of $1/\tan\theta$, where θ is the angle YAE in Fig. 1.18, and thereby make the scale easier to mark and read. The instrument is thereby made more accurate. The

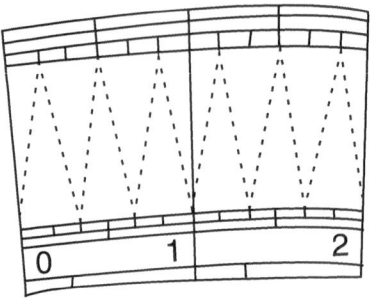

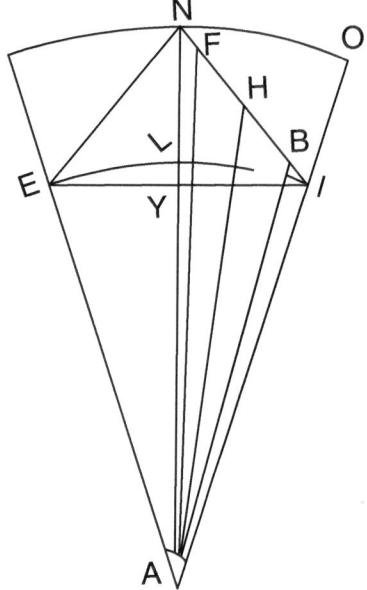

Fig. 1.18 The method of subdivision by transversal points, known as a Jacob's Ladder [33]. (Source: Tycho's *Astronomiæ Instauratæ Mecanica* [27]). Public domain

care of Tycho's calibration of his Jacob's Ladder scale is a wonder to behold. In his book of his instruments [27], he reports having measured calibration corrections as low as 3‴, where 1‴ is 1/60 arcsecond and is known as one "third".

So what did Tycho measure about Mars? He measured declinations and angles to stars (see Figs. 1.19 and 1.22). Figure 1.19 shows a typical observation from the Julian date 31 January 1596. This is a diary-like entry,

DIE 31 JANUARIJ.
H. 6 M. 15⅓ Tranſiuit ♂ per Meridi-
anum, cum haberet Altitudinem
per Volub.   55 23
Fuit tunc ſiniſter Humerus Orion.
orient. 29 33   Declin. ♂ 21 18
                              21 17⅓

Pro examinandis Quadrantibus &
præcipue Chalybeo, cuius pinnacidium hodie reparatum ac correctum
eſt, capiebatur Altitudo Meridiana
H. 6 57⅓ Aldeboræ
    per Chalyb.   49 43⅓
    per Volub.   49 44
    Declinatio   15 38
                        15 37⅓
Hic corrigebatur horologium re-
troponendo indicem in H. 6 M. 37.

For the Day of 31 January
H. 6 M. 15¼ ♂ [Mars] Transited by the Meridian with maximum Altitude
By the Revolv[ing Azimuthal Quadrant]: 55°23′
The Humerus of Orion [γ Orionis, Bellatrix] was then 29°33′ to the left.
Declination ♂ 21°18′
21°17½′
For examining the scale of the steel quadrant, repaired today, the altitude of
Aldebraran at the Meridian at [time] 6 H. 57½.
By Steel [Quadrant]   49°43½′
By Reolv[ing Quadrant] 49°44′
Declination 15°38′
15°37½′
This is corrected by putting the clock back to the index H. 6 M. 37.

Fig. 1.19 A Typical Tycho observation, from 31 January 1596 in the Julian calendar, that is, 10 February 1596 in the Gregorian calendar (Denmark switched in 1700). In the lower image, the Latin is translated by the author. Tycho's text is in the public domain

observation notes made as Mars was observed. It records the data faithfully enough, but it is also quite chatty, talking about the problems of calibrating a repaired instrument and of calibrating his clock. Tycho's Latin is also a little idiosyncratic. The word pinnacidium appears nowhere in dictionaries. For a discussion, see Michael Covington's blog [34].

Reading Tycho's Latin requires knowledge of astrological symbols for the constellations of the Zodiac and of the astrological character used for oppositions ($\mathcal{S}$). He uses these symbols freely. It is also beneficial to read his instruments book [27] to pick up his Latin nomenclature for his instruments.

His star names add to the fun. Most of the common star names that we use nowadays have their origins in Arabic or Greek – and they aren't that similar to the original words. For example Denebola (β Leonis) was originally Dhanab al-Assad (ذيل الأسد), Tail of the Lion. Tycho calls it Cauda Leonis, which is Latin for Tail of the Lion. He often abbreviates these names, just to make sure that this isn't getting too easy. My Arabic is alas minimal: I once did an introductory course of 30 lessons, but where I then lived, no follow-up courses were available. But I did have an illustrated child's book of the first 1000 Arabic words [35] open while reading Tycho. Very helpful it was too, when it came to guessing what Tycho's star names mean.

Let's have a look and see what a modern planetarium software application, Cartes du Ciel v. 4.0 [36], thinks was going on that evening. I had to feed in the Gregorian date, 10 February 1596. I also inserted modern longitude and latitude coordinates for Tycho's observatory, which I took from Wikipedia.

From Figs. 1.20 and 1.21, it can be seen that *Cartes du Ciel* agrees with Tycho's measured declinations to within 2 arcminutes for Mars and 1 arcminute for Aldebaran. *Cartes du Ciel* predicts an altitude of Mars at transit of 55°22′16″, within ¾ arcminute of Tycho's value. It predicts an altitude of Aldebaran at transit altitude of 49°44′, which is the value measured by Tycho.

While I have no way to check the algorithm used by *Cartes du Ciel* to calculate the orbit of Mars, I can't help wondering if Tycho's measurement is more accurate than his spherical trigonometric calculation to obtain a declination.

Cartes du Ciel tells me that the angle between Mars and Bellatrix that night, mentioned by Tycho, is 26°20′, not the 29°33′ measured by Tycho. I don't think I have mistranslated "Humerus Orionis", although that is a documented alternative name for Bellatrix [37], as is "Sinister Humerus Orionis". "Left" and "right" in Latin are "sinister" and "dexter", respectively.

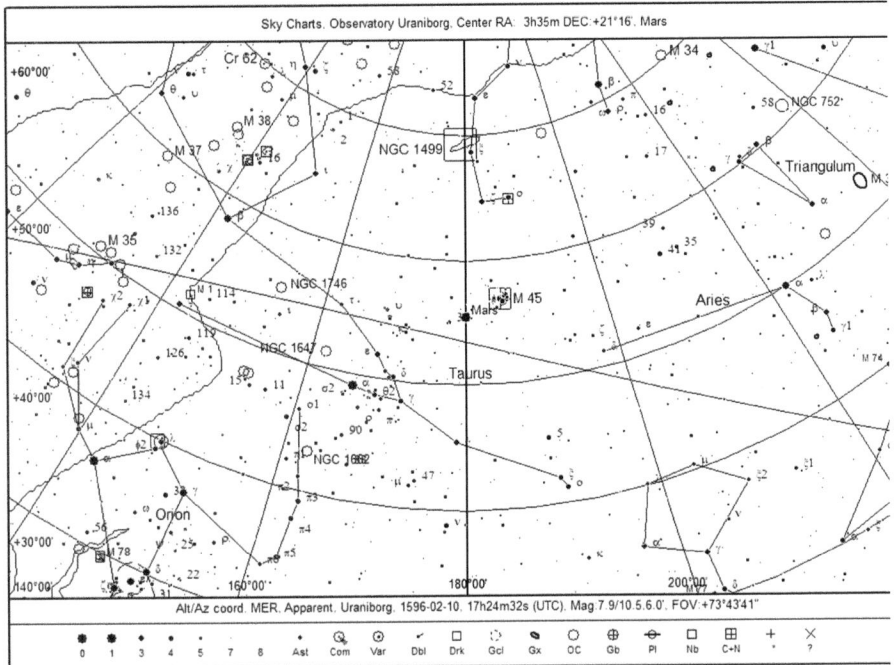

Fig. 1.20 Mars at transit on the night of 31 January 1596 (Julian), that is, 10 February 1596 (Gregorian), according to Cartes du Ciel. The software inserts an entirely anachronistic time of 17:24:32 UTC. Nobody thought of time that way in those days. Tycho was using a local time. The declination of Mars is calculated at +21°16′, which is within 2 arcmin of Tycho's measured value

It is clear that time measurement was a serious problem for Tycho. For one so dedicated to accuracy, this must have been quite a source of frustration.

Now, an important point to note is that Tycho gives a declination for Mars, but not a right ascension. Oh no. No, no, no, no, no. That would be too easy. Way too easy. Instead he gives us a riddle: we know the declination and the angle between Mars and another star. Let me ignore for a moment the apparent mismeasurement of the angular separation to Bellatrix and take Tycho's record at face value. Let

$$\beta = 26°20′ = 0.4596 \text{ rad} \qquad (1.13)$$

be the angle to the target star Bellatrix. Also let me work in a coordinate system such that the right ascension of Bellatrix is zero. This turns out to be a significant simplification. In this coordinate system, the RA and Dec of Bellatrix are, according to Cartes du Ciel,

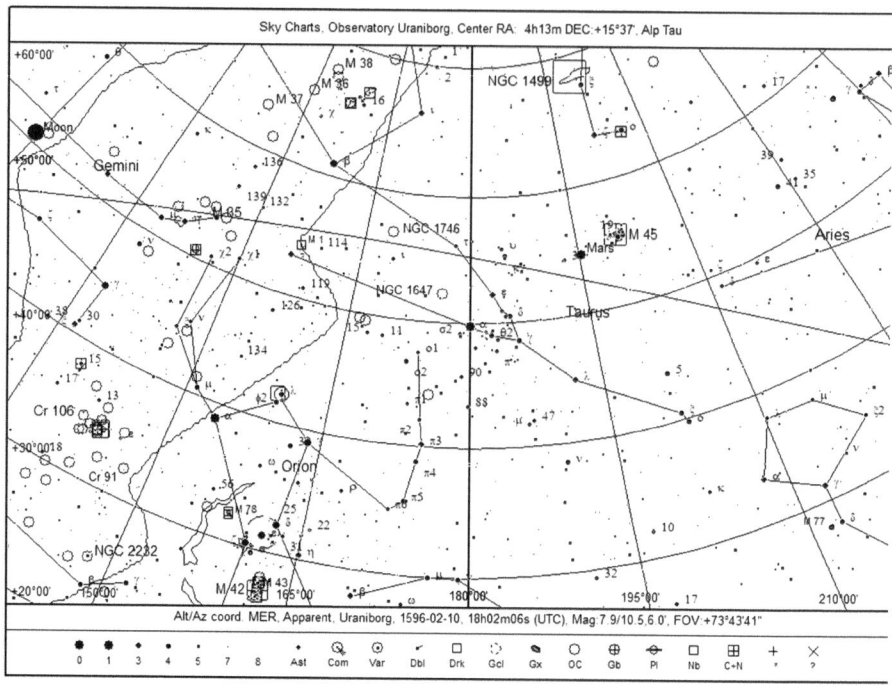

Sky Charts, Observatory Uraniborg, Center RA: 4h13m DEC:+15°37', Alp Tau

Alt/Az coord. MER, Apparent, Uraniborg, 1596-02-10, 18h02m06s (UTC), Mag:7.9/10.5,6.0', FOV:+73°43'41"

Fig. 1.21 Aldebaran at transit on the night of 31 January 1596 (Julian), that is, 10 February 1596 (Gregorian), according to Cartes du Ciel. The time is now noted at 18:02:06 UTC. The difference between that time and the time of the Mars transit illustrated in Fig. 1.20 is 37 m 34 s, cf. Tycho's claim of 41 m 10 s. The declination calculated is 15°37', within an arc minute of Tycho's measured value

$$\alpha_1 = 0$$
$$\alpha_{1.\text{conventional}} = 5\text{H } 04' = -1.310 \text{ rad} \tag{1.14}$$
$$\delta_1 = 5°54' = 0.103 \text{ rad}$$

The radians are made negative to avoid making a mirror-image of the heavens. In modern terms, the RA and declination coordinate system is a left-handed coordinate system if you transform the angles on a sphere of any given radius into Cartesian coordinates. Making the RAs negative switches us to a right-handed coordinate system, which is how most mathematics is set up.

We are also given the declination of Mars

$$\delta_M = 12°18' = 0.372 \text{ rad.} \tag{1.15}$$

Now we can write down the dot product of the unit vector to Mars and to Bellatrix, bearing in mind that the unit vector of a point (α, δ) in the sky is

$$\hat{\mathbf{r}} = (\cos\alpha\cos\delta, \ \sin\alpha\cos\delta, \ \sin\delta), \qquad (1.16)$$

$$\cos\beta = (\cos 0 \cos\delta_1, \sin 0 \cos\delta_1, \sin\delta_1)\cdot(\cos\alpha_M\cos\delta_M, \sin\alpha_M\cos\delta_M, \sin\delta_M)$$

$$= (\cos\delta_1, 0, \sin\delta_1)\cdot(\cos\alpha_M\cos\delta_M, \sin\alpha_M\cos\delta_M, \sin\delta_M)$$

$$= \cos\delta_1\cos\alpha_M\cos\delta_M + \sin\delta_1\sin\delta_M$$

$$\qquad\qquad\qquad\qquad\qquad\qquad\qquad\qquad\qquad\qquad (1.17)$$

$$\cos\alpha_M\cos\delta_1\cos\delta_M = \cos\beta - \sin\delta_1\sin\delta_M$$

$$\cos\alpha_M = \frac{\cos\beta - \sin\delta_1\sin\delta_M}{\cos\delta_1\cos\delta_M}.$$

In our case, in radians

$$\cos\alpha_M = \frac{\cos 0.4596 - (\sin 0.103 \times \sin 0.372)}{\cos 0.130 \times \cos 0.372}$$

$$\qquad\qquad\qquad\qquad\qquad\qquad\qquad\qquad (1.18)$$

$$= 0.930$$

$$\text{so } \alpha_M = 0.377 \text{ rad.}$$

Adding this to the conventional right ascension of Bellatrix gives us

$$\alpha_{Mars.conventional} = \pm\alpha_{Mars} + \alpha_{1.conventional}$$

$$= 0.377 \text{ rad} - 1.310 \text{ rad}$$

$$= -0.933 \text{ rad} \qquad\qquad (1.19)$$

$$= 3 \text{ H } 34'.$$

Tycho's riddle is thus solved. We can get both right ascension and declination data for Mars from his measurements.

The ± sign in Eq. (1.19) arises because if Mars is to the left of the reference star, the negative option must be used; and if it is to the right, the positive option must be used. (By left, of course I mean having a higher RA, and by right I mean having a lower RA than the reference star.)

By the way, I checked: the apparent error I reported above in the angle to Bellatrix makes a 1-arcminute difference to the calculated right ascension of Mars.

Tycho's Mars data for the years 1582–9 have been collected and are shown in Figs. 1.22 and 1.23. Not every one of his data points is shown. I selected those cases where he quoted a declination and rejected one or two

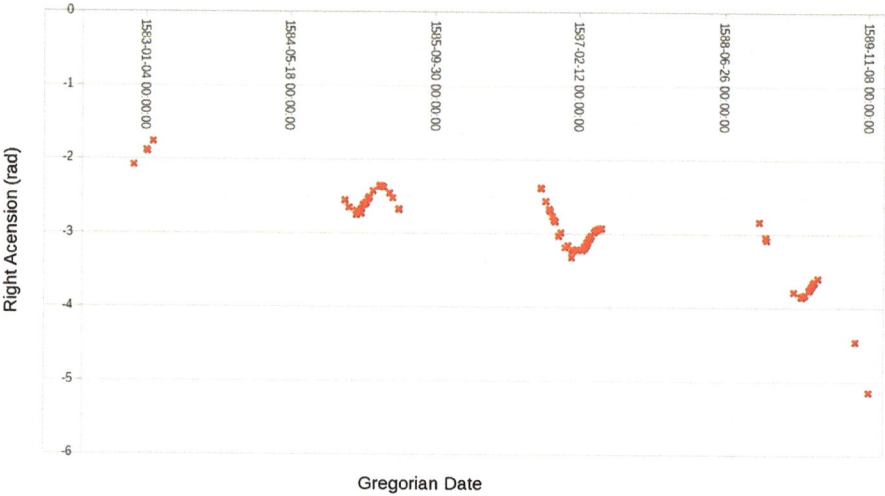

Fig. 1.22 Tycho's measurements of the right ascensions of Mars over several apparitions. (Source: Tycho's Complete Works [39]. Image from data calculated by the author using Eq. (1.19))

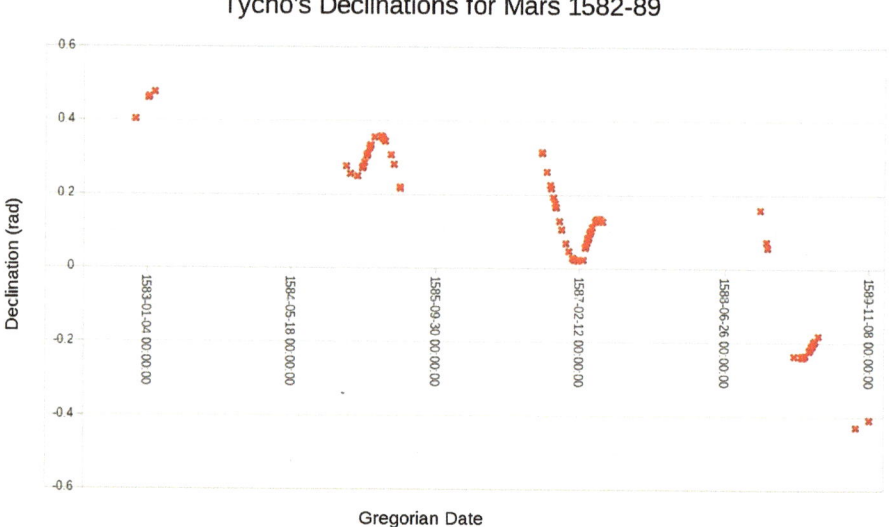

Fig. 1.23 Tycho's measurements of the declinations of Mars over several apparitions. (Source: Tycho's Complete Works [39]). Public domain

cases where I suspected errors. There are other cases where he quotes angular separations from more than one star, but did not quote a declination. These cases could no doubt be analysed and added to the datasets shown in Figs. 1.22 and 1.23.

Incidentally, I used LibreOffice Calc™, Version 6.2.7.1, to plot Tycho's data, because Microsoft Excel™ cannot handle dates before 1 January 1900. LibreOffice Calc has a very nice date facility, which even knows when to convert from Julian to Gregorian dates. In Roman Catholic jurisdictions Thursday, 4 October 1582, was followed by the first day of the Gregorian calendar, Friday, 15 October 1582. In Protestant jurisdictions, the date change was later. In Denmark, it happened in 1700. In the British Empire, which then included what is now the United States, it happened in 1752. The dates recorded by Tycho would therefore have been Julian [38]. I have changed them to Gregorian dates to simplify any later calculations.

Tycho's data were accurate to about 4 arcminutes, which was enough to cause the models of Ptolemy and Copernicus a problem: they no longer adequately agreed with observations.

Something was going to have to be done about that.

Galileo – Telescopes and Free Fall

Galileo di Vincenzo Bonaiuti de' Galilei (1564–1642) (Fig. 1.24), universally known as Galileo, was as great an experimenter and observer as Tycho and a more successful theoretician. One of his theories has stood the test of time to this day: he discovered the phenomenon of free fall [40].

He was the son of an impoverished gentleman, that is, someone with a coat of arms who believed himself to be a cut above the ordinary folk. His father therefore had to work for a living. He was a musician – a lutenist. As a young man, Galileo would sometimes accompany his father to gigs as a lute player. Galileo was a competent musician, so he would have developed a good sense of timing. We will see how this contributed to his science.

His early career is not suggestive of the kind of focus that drove Tycho, whose only major interest was astronomy. A college dropout, Galileo dabbled in all sorts of activities. He wrote. He painted. Having failed to graduate from university, he studied mathematics privately, with a view to becoming a mathematics professor. He kept the company of literary friends. He tried poetry, not very successfully. But he also wrote prose and had some success as a literary critic. Indeed, some of his literary criticism has been used in Italian schools. All this practice at writing made him, according to some, the greatest Italian prose writer since Machiavelli, if not the best ever [41]. Whether this opinion is universal or not, he was a great writer. That

Fig. 1.24 Galileo in a portrait showing him younger than most such portraits. In the Royal Museums at Greenwich, probably painted by Francesco Apollodoro of Padua, c. 1531–1612. Public domain

fact is central to understanding him: first because he communicated his scientific discoveries with crystal clarity, and second because in his great clash with the Inquisition, of which more later, his enemies knew quite well what he thought.

Galileo's first university teaching position was in his native city of Pisa, to which he was appointed in 1589. At least one of his colleagues, Professor Borro, who may have been one of his undergraduate teachers, was imprisoned several times by the Inquisition [41]. The primary tool of the Inquisition in guarding the doctrines of Rome was terror. In 1600, they tied the Dominican friar Giordano Bruno to a stake upside down and burned him [42]. He held numerous unorthodox views, including the idea that the stars were other suns, with planets inhabited by living beings. The behaviour of the Inquisition reminds me of other reigns of terror, such as those of Stalin in the USSR and Robespierre in France. It must have been horrible to have to live with them as a background menace. It was during this time that he was alleged to have dropped cannonballs of various weights from the leaning Tower of Pisa to show that they landed simultaneously [43], whereas Aristotle had implied that this would not be the case. This may or may not be a tall tale [44]. As a story to illustrate a point, it is excellent. It is comprehensible to everyone, scientist or not.

Galileo did not stay long in Pisa. After three years he tripled his salary by moving to a university in Padua, near Venice. Venice's duke ("doge" in Italian) was not a fan of the Inquisition and did not encourage its thought police. Venice and Padua were therefore oases of freedom in religiously difficult times – up to a point. The Jesuits were run out of Venetian Territory in 1606. And the Friar Bruno mentioned above was extradited to Rome from Venice in 1593.

It would be quite wrong to think of the Italy of those days as a united country. It wasn't. It was a collection of small, independent states, often separately dominated by foreign powers, which were not united until the mid-1800s. Even then, the process was not completed until the end of World War I in 1919 [45]. The Italian states relevant to Galileo are shown in Fig. 1.25.

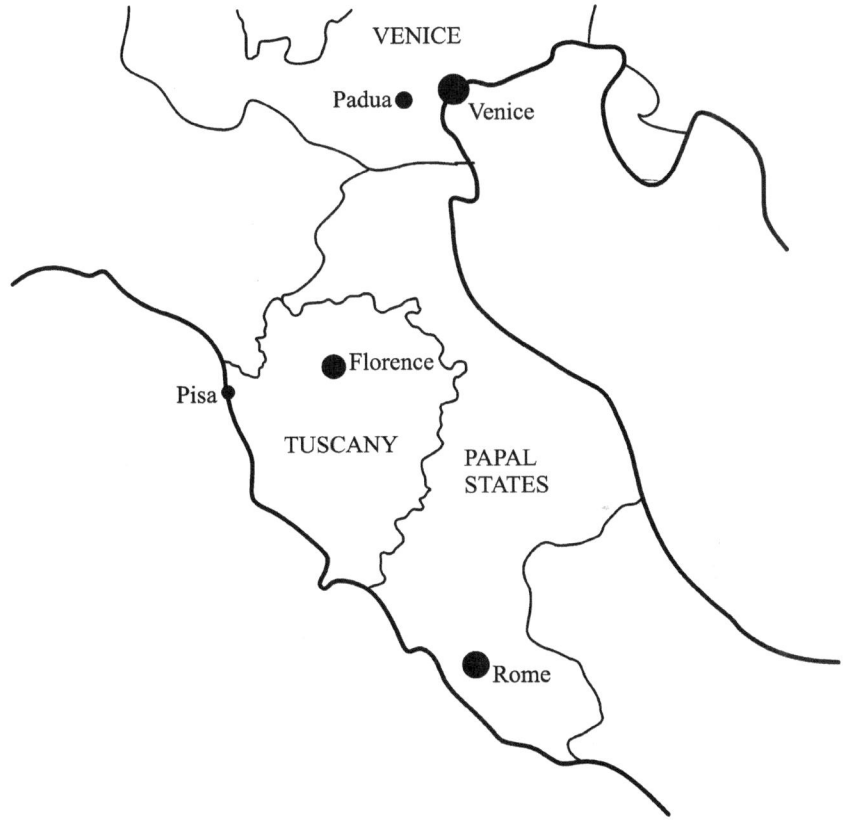

Fig. 1.25 Galileo's Italy. (Image: Author)

At first, Galileo set himself up with a basic lifestyle in Padua. He acquired a house, evidently a large one for it contained student lodgings and a workshop with a live-in coppersmith, so it was as much a place of business as a home [41]. He obtained a patent on a horse-driven pump to irrigate elevated land, which gave him a monopoly to manufacture it [41]. He also set up a mistress in a nearby house, with whom he had two daughters and a son. They did not live with him [41]. He abandoned the mistress when he later took a fancy job in Florence [46]. He also manufactured a military compass (Fig. 1.26), a kind of primitive slide rule. He is credited, possibly falsely, with its invention [47]. At any rate, his workshop manufactured and sold over 100 of them. So he was a businessman as well as an academic.

Free Fall and Motion

Galileo did his basic research into this subject in the years before his attention was diverted to telescopic astronomy [48]. He particularly studied free fall, motion on inclined planes and pendula.

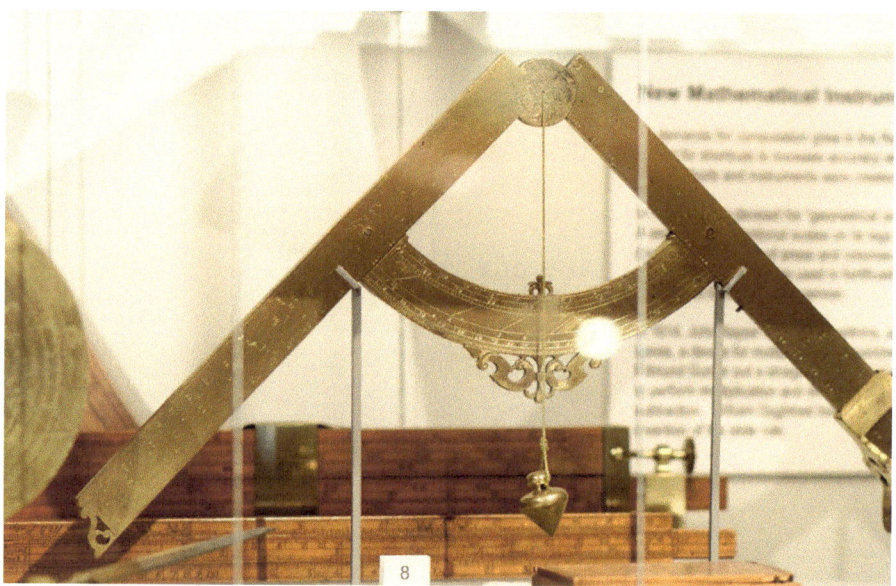

Fig.   1.26 Galileo's   compass.   (Source: https://en.wikipedia.org/wiki/File:Galileo%27s_geometrical_and_military_compass_in_Putnam_Gallery,_2009-11-24.jpg, Credit: Sage Ross, GNU Free Documentation License, Version 1.2)

In order to get any meaningful concept of speed, he had to have a measure of the passage of time. This was before the era of decent clocks. This is where he had a brainwave of world class brilliance. As his balls rolled on inclined planes, he placed strings of gut, such as would be used in a stringed instrument like a lute, at such intervals that the ball passing over them would beat time. He would adjust the distances between the strings until the ball rolling over them would beat time regularly. Stillman Drake, who described this method in *Galileo at Work*, reckoned that anyone can spot errors in musical rhythm to $1/32$ second, and a competent musician could do at least twice that well [48]. So Galileo could time to better than a sixtieth of a second without a clock.

I did not find a diagram showing this apparatus, so I sketched one myself, shown in Fig. 1.27. It shows the principle, but is missing quite a few details, including the mechanism for keeping the gut strings taut and for moving and holding them along the groove. In the example shown I have made the string distances go as the square of the distance from the start, knowing with my modern knowledge that the distance s travelled would be

$$s = \frac{1}{2}\left(g\cos\theta\right) t^2 \tag{1.20}$$

where g is the gravitational acceleration and θ is the slope of the incline. The clicks as each gut string is crossed would be at equal time intervals.

Galileo did not know this and worked it out the hard way. By this and similar experiments, he worked out the law of freefall, that bodies of different masses are *accelerated* equally. That was a key break from Aristotle, who asserted otherwise, without troubling himself about evidence. Thanks to Galileo, we talk of the acceleration due to gravity.

Galileo was not the first person to invent the concept of speed, despite frequent claims to the contrary. This concept was used by scholars at Merton College, Oxford University, who in the 1300s devised the Merton rule of uniform acceleration

$$s = \frac{1}{2}\left(v_i + v_f\right) t, \tag{1.21}$$

where s is distance travelled, v_i is initial velocity, v_f is final velocity and t is the elapsed time [49]. This rule was graphically proved by scholars in France and Italy during the same century. Galileo may or may not have known of these developments. His original contribution was to demonstrate speed-related phenomena experimentally because he had a way to measure time, which the earlier scholars did not.

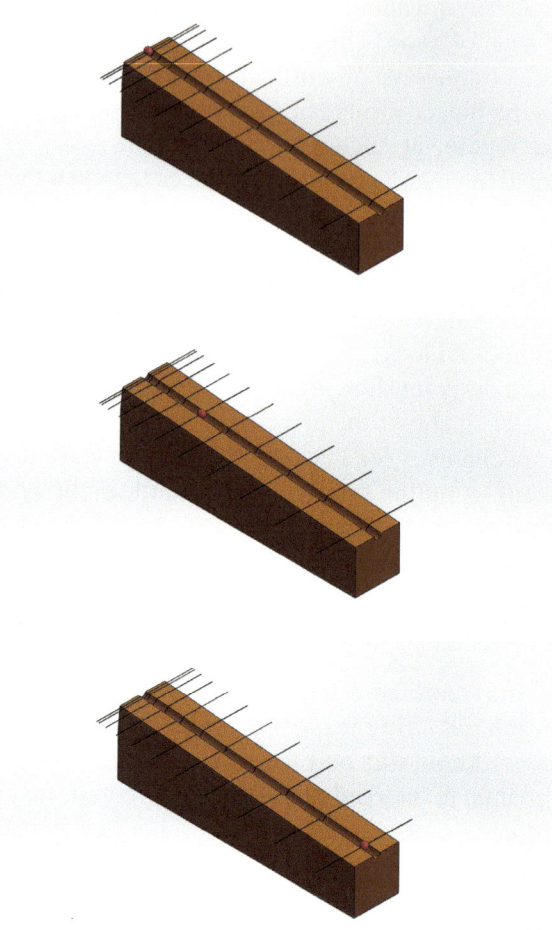

Fig. 1.27 Principle of Galileo's inclined plane experiments. The ball makes a click-ing sound every time it crosses one of the strings. Hence, the experimenter can hear the time interval between clicks. We do not know exactly how Galileo constructed this apparatus. (Image: Author)

He also discovered that the period of one swing of a given pendulum does not depend on the amplitude of the swing (Fig. 1.28). Late in life, he came close to inventing the pendulum clock, but by then he was too blind and infirm to carry the idea forward [50]. The first pendulum clock was built by Christiaan Huyghens seventeen years after Galileo died. Galileo did his experiments on a roughly five-feet-long pendulum, measuring time by the flow of grains of material through an orifice [48].

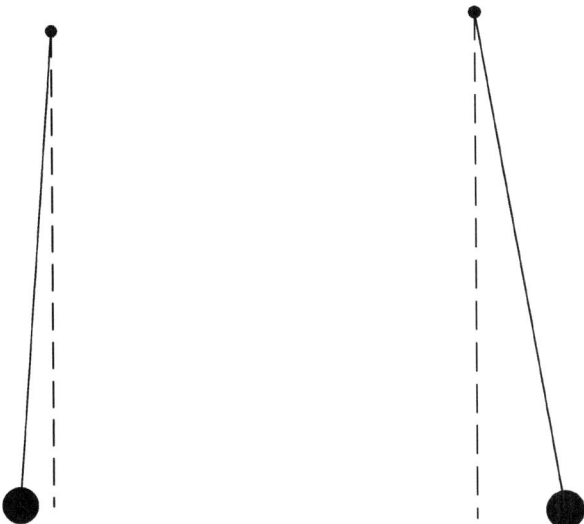

Fig. 1.28 Two pendulums of the same length, but different amplitudes, have the same swing period. This is true for swing angles up to about 10° and is true regardless of the mass of the swinging pendulum bob. (Image: Author)

The other great contribution of Galileo to movement was to discover what we now call Newton's first law of motion, namely, that a body continues to move in a straight line unless it is acted on by an external force. This is in flat contradiction to what Aristotle said and therefore controversial in those Aristotle-revering times. Armed with this discovery, he was able to analyse the motion of a projectile in the presence of minimal air resistance and show that it would fly in a parabola, because the horizontal speed is constant, while the vertical speed is determined by gravity [51]. In modern terms, the position of a projectile fired at a speed u at an angle θ to the horizontal is given by

$$
\begin{aligned}
s_{\text{horizontal}} &= u\cos\theta\; t \\
s_{\text{vertical}} &= u\sin\theta\; t - \frac{1}{2}\, g\; t^2
\end{aligned}
\tag{1.22}
$$

This is illustrated in Fig. 1.29.

According to Drake [40], most of this work was accomplished during Galileo's time at Padua.

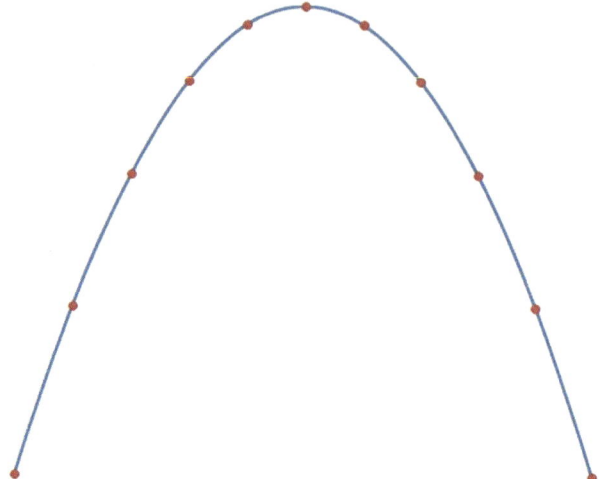

Fig. 1.29 Parabolic flight of a projectile when the air resistance is negligible. (Image: Author)

The Telescope Bursts onto the Scene

Towards the end of Galileo's time at Padua, a patent application was filed in the low countries by one Hans Lippershey, for a device that could make distant objects look bigger. The patent was rejected, and news of this invention quickly leaked. It reached more than one astronomer, but Galileo was the one who got into the history books. There are several reasons for this. First, according to his own account, he had to figure out how the telescope worked, given an incomplete story. He had the brains to do this. Second, the workshop in his house was doing a side line in spectacles for sale, so he had a stock of lenses to play with and help him work out what he needed [41] (Fig. 1-30). Third, he was a skilled artist who could draw and paint what he saw. Fourth, his high order of writing skills meant that his write-up was the one that got the attention, especially as his account was systematic.

One of the hypotheses that he demolished was the Aristotelian idea that the heavens were "perfect".

One of his discoveries was of the Moons of Jupiter. This was an earth-shattering discovery, because it was an Aristotle-shattering discovery. Aristotle's claim that everything went around the Earth was disproved. Here were some objects that went around Jupiter. It took Galileo a while to figure out what was going on, as these moons came into and out of view.

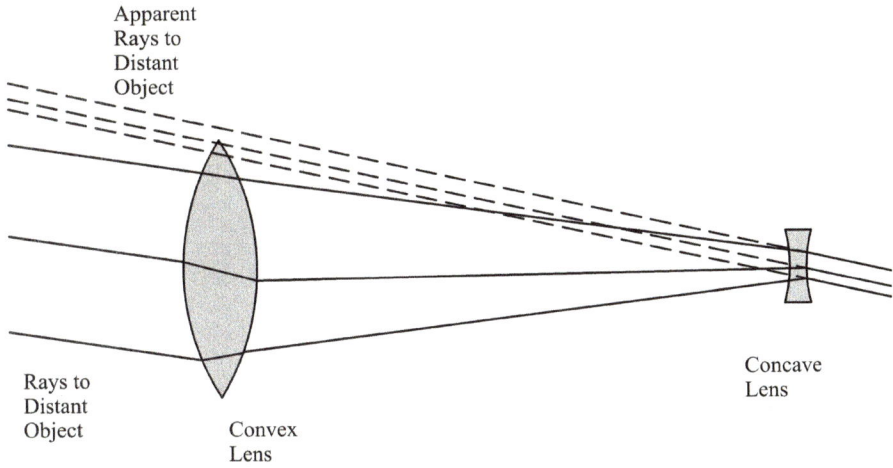

Fig. 1.30 Principle of the Galilean Telescope. It consisted of a convex objective lens and a concave eyepiece and gave upright images. (Image: Author)

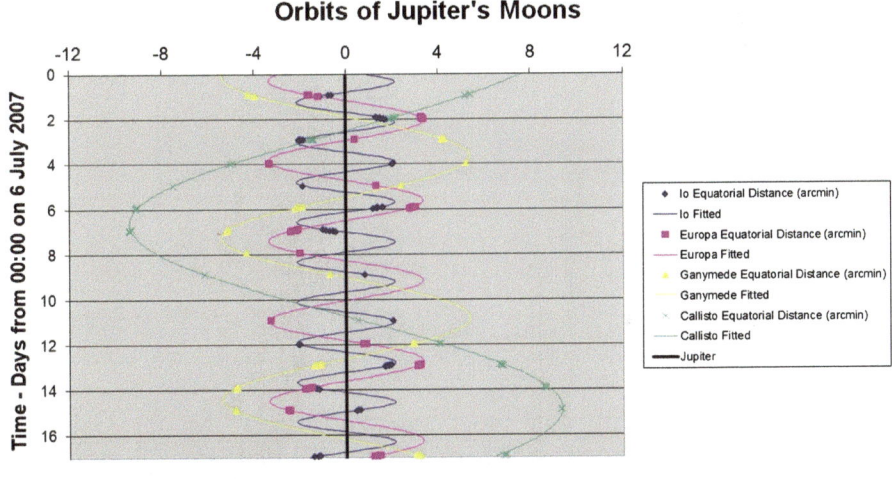

Fig. 1.31 My own observations of Jupiter's moons over one orbit of Callisto, the outermost. The fits are to models of circular orbits. (Image: Author)

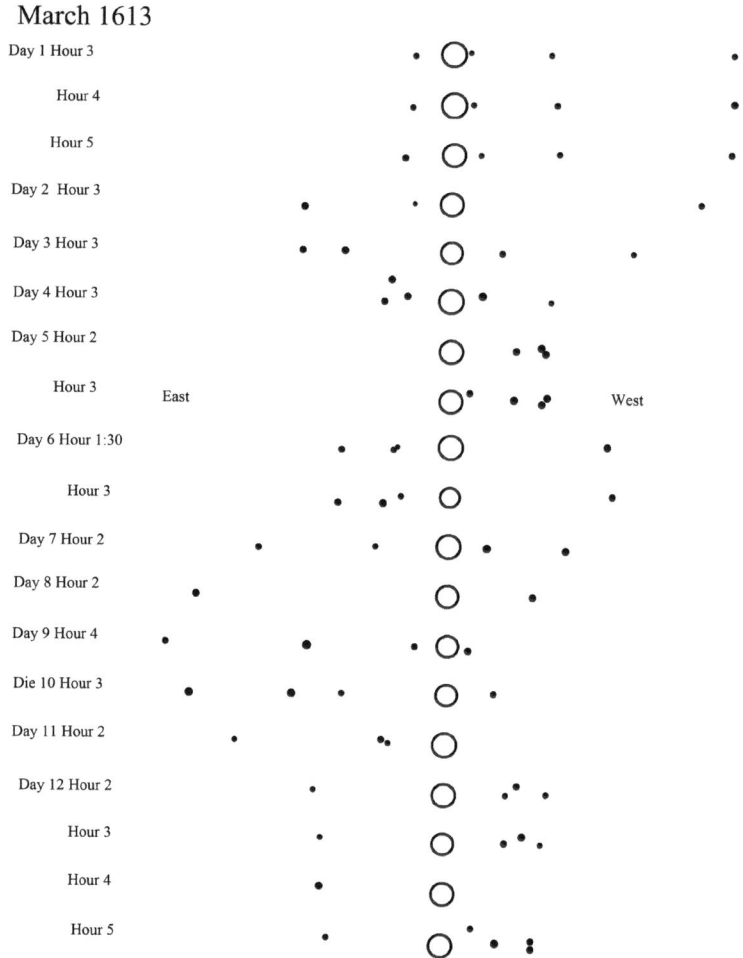

March 1613

Fig. 1.32 One of Galileo's sets of drawings of Jupiter's Moons, traced by the author from the original [52]. (Copyright Expired)

In Fig. 1.31, I show my own observations of Jupiter's Moons. Galileo's observations in Fig. 1.32 are noticeably similar. With his small, primitive telescope, he saw pretty much what I saw with an 8″ Newtonian and a webcam. That tells me that his observation skills were of a very high order.

Galileo also observed that the surface of the Earth's Moon was not "perfect". His drawings look very credible when compared to modern photographs (Figs. 1.33 and 1.34). The crater at the bottom is exaggerated, but otherwise the main features are caught. The more mountainous regions in the Southern Hemisphere are recorded. It should not be forgotten that these images represent the first time anyone ever observed the moon through a

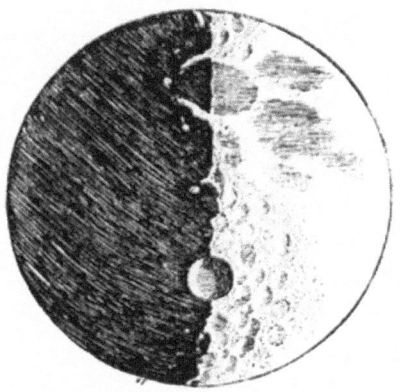

Fig. 1.33 Comparison of Galileo's Moon at first quarter with a montage of photos taken with an 8″ telescope and a webcam. (Left Image: Galileo [53] (Copyright Expired). Right Image: © Author)

Fig. 1.34 Comparison of Galileo's Moon at third quarter with a montage of photos taken with an 8″ telescope and a webcam. (Left Image: Galileo [53] (Copyright Expired). Right Image: © Author)

telescope, which was a rather primitive instrument by today's standards. Therefore, what strikes me is how well Galileo observed.

The discovery of the "imperfection" of the Moon was yet another blow to Aristotle's physics.

The final blow that Galileo struck against the geocentric view of the cosmos was to discover the phases of Venus. His drawings, made during the

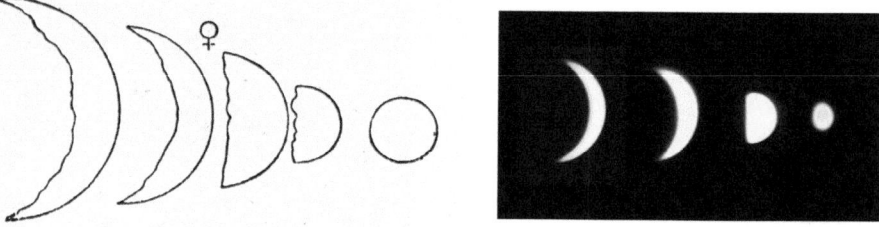

Fig. 1.35 Galileo's drawings of the Phases of Venus [54] (Left) compared to those photographed by author during the 2008–9 apparition (Right). (Left Image: Public domain. Right Image: Author)

Venus apparition of 1610–11, are compared in Fig. 1.35 to a series of photos by the author with an 8″ telescope and a webcam.

The point is that the apparent size of the planet varies, such that it is gibbous when apparently small, and therefore far away, and crescent when apparently large and near. This can only happen if Venus is illuminated by the Sun and orbits the Sun. Were both Venus and the Sun to orbit the Earth, then the phases would be either always gibbous or always crescent. I am once again very impressed by Galileo's observational skill.

Encounter with the Inquisition

Galileo was not slow to turn his good fortune with the telescope into a career boost. He went to Venice to demonstrate his telescope. The Venetian government gave him a significant salary boost and made it lifelong. This, however, was not where Galileo's ambition lay. He persuaded the Medicis, the ruling family in Florence, to equal the new salary and make him their chief philosopher with no teaching duties. Part of the persuasion was to name the Jovian moons the Medici Stars. This didn't go down well with the Venetians, who had just handsomely promoted him, but Galileo left for his new job in Florence anyway [41].

He did not take his mistress or his son. He took the daughters and promptly enrolled them in a convent, believing that he could not afford dowries for them [46]. He had to go into quite a lot of debt to fund dowries for his sisters, and one of his brothers-in-law was quite litigious about trying to collect his dowry, so there was a justification of sorts for this [41]. Nevertheless, Galileo's behaviour at this point does come across as selfish.

There is no shortage of evidence that Galileo comes across as waspish in print: one only needs to read his writings to see this. He often wrote in the

Table 1.1 Willings' categorisation of people and creativity

Not creative	Creative
Able to appreciate creative people	Able to appreciate creative people
Not creative	Creative
Does not appreciate creative people	Does not appreciate creative people

form of dialogues, and the character supporting the view he is trying to knock down gets called names like Simplicio [51]. In fact, I am tempted to think of him in the form suggested to me by a psychologist friend, the late David Willings [55]. Willings was an expert on creativity in people [56]. Willings' model is shown in Table 1.1. Willings said that those who do not appreciate creativity in others are profoundly incapable of the required insight. Galileo seems to me to fit this bill. Creative he most certainly was. But he was very insensitive to the feelings or achievements of others.

What's worse, the leaders of the Inquisition in Rome fit into the bottom left category of this 2 × 2 matrix. They weren't creative and neither appreciated nor liked people who, in their view, rocked the boat.

Part of Galileo's motive for moving to Florence was to take the intellectual struggle for his ideas to the Jesuits [41]. In this he was profoundly unsuccessful. Despite his great gifts as a writer, he could not persuade them. Eventually, the Inquisition put him on trial and sentenced him to lifelong house arrest for refusing to accept that the Earth is immovable. Given the nastiness of some of the punishments the Inquisition had meted out, for Galileo to take this stance took immense courage.

Equally bravely, Galileo worked out ways to circumvent the prohibition imposed by the Inquisition. He continued to publish by writing long letters. He was allowed to receive visitors. His last great book, Two New Sciences, was snuck out and published in the low countries [41].

In the end, the moral victory was Galileo's. History has not been kind to those small-minded Inquisitors who tried to snuff out his scientific ideas.

Chapter 2

Read My Ellipses

Kepler

The great contribution of Kepler (Fig. 2.1) was his three planetary laws:

- A planet orbits the Sun in an ellipse with the Sun at one focus.
- The area swept out by the radius vector from the Sun of a given planet is equal over equal time intervals.
- The semimajor axis a of the orbit is proportional to $T^{2/3}$, where T is the period of the orbit.

In order to follow Kepler to these conclusions, we had better learn how to handle ellipses.

First, I want to tell you a little bit about what Johannes Kepler (1571–1630) was like. The authoritative biography of him is by Max Caspar [57]. He was born near Stuttgart in South-West Germany, the son of a mercenary, who probably died fighting when Johann was a small boy. His mother was a healer and herbalist. He was not robust as either a child or an adult, but was very obviously a mathematical prodigy from an early age. He was a Protestant, a fact which mattered in those religiously dangerous times. He went to university to train to become a minister and became a convinced Copernican when he learned of this theory. He was persuaded to forego the ministry and take a post at the Protestant School in Graz, now in Austria, teaching mathematics and astronomy. Nevertheless, he was a profoundly religious man, a fact which shines through in his astronomical writings.

J. Clark, *Calculate the Orbit of Mars!*,
https://doi.org/10.1007/978-3-030-78267-2_2,
© Springer Nature Switzerland AG 2021

Fig. 2.1 Johannes Kepler, 1610, artist unknown. (Image courtesy of http://en.wikipedia.org/wiki/Kepler). Public domain

While there, he conceived a theory we now regard as misplaced, that the orbits of the planets could be fitted into the five regular spheres. He published this idea in a book *Mysterium Cosmographicum* [58]. He never dropped this idea, which seems to have been quite a driving force for him. He also married a prosperous woman and started a family.

In 1599, Graz became an unsafe place for a Protestant. He tapped Tycho up for a job and left for Prague on 1 January 1600, where Tycho had moved, in the hopes of being offered work. Fortunately for him, and for posterity, he was. (Tycho had decamped there partly because his Danish royal funding proved to be a problem, whereas Rudolph II, King of Bohemia and Holy Roman Emperor, was happy to employ him, and partly because it enabled him to leave his estate to his lower class wife and children, which he could not have done in Denmark.)

Tycho set Kepler to work on his Mars data. He certainly took to this problem with relish. After Tycho's death a few months later, Kepler got his job and his data. The former was at a lower salary than Tycho, and the latter caused some friction and legal wrangling with Tycho's heirs.

The main part of the job was to be an astrological adviser to the Emperor. Nevertheless, Kepler kept at Mars like a dog with a bone. As if that were not enough, he also produced a seminal work on optics in 1604 [59]. He invented the form of the refracting telescope nowadays used for astronomy, with positive lenses for both the objective and the eyepiece (Fig. 2.2), and in the context of lenses, he coined the term "focus", which was originally the Latin for "hearth".

Eventually in 1609, he published *Astronomia Nova* (New Astronomy) [60]. This was a very different book form Galileo's popular science books. It is difficult to read and aimed at an academic audience. But then, Kepler was a different character from either Tycho or Galileo: he was agreeable and lacked their arrogance [57].

Trying to follow Kepler's reasoning through the 500 pages (in translation) of *Astronomic Nova* is very difficult. Because he spends so much time setting up alternative theories in order to demolish them, you end up exhausted. Many good people have been misled by the reasoning [61]. In the end, he notes that an ellipse fits the data, and then he stops. He cannot prove that an ellipse is superior to alternatives. He is motivated by a belief that there must be a physical cause to the orbit of Mars and its shape, but doesn't even almost get the cause right. The only idea available to him in the early 1600s for action at a distance was magnetism, and he believed that a tangential force was needed to drive the planets. Newton said in a letter to Halley in 1686 that "Kepler Knew ye Orb to be not circular but oval and guest [i.e. guessed] it to be Elliptical" [62].

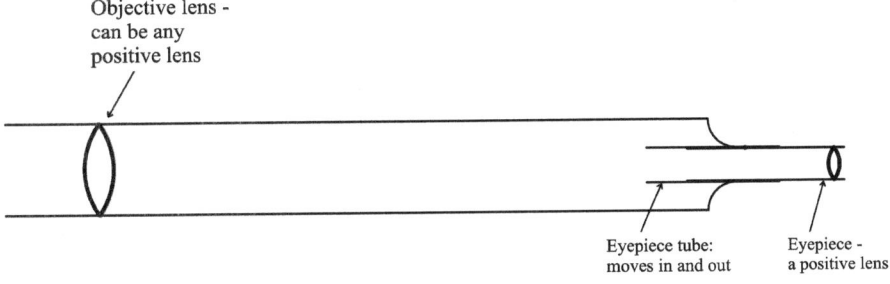

Fig. 2.2 A refractor of the type invented by Kepler

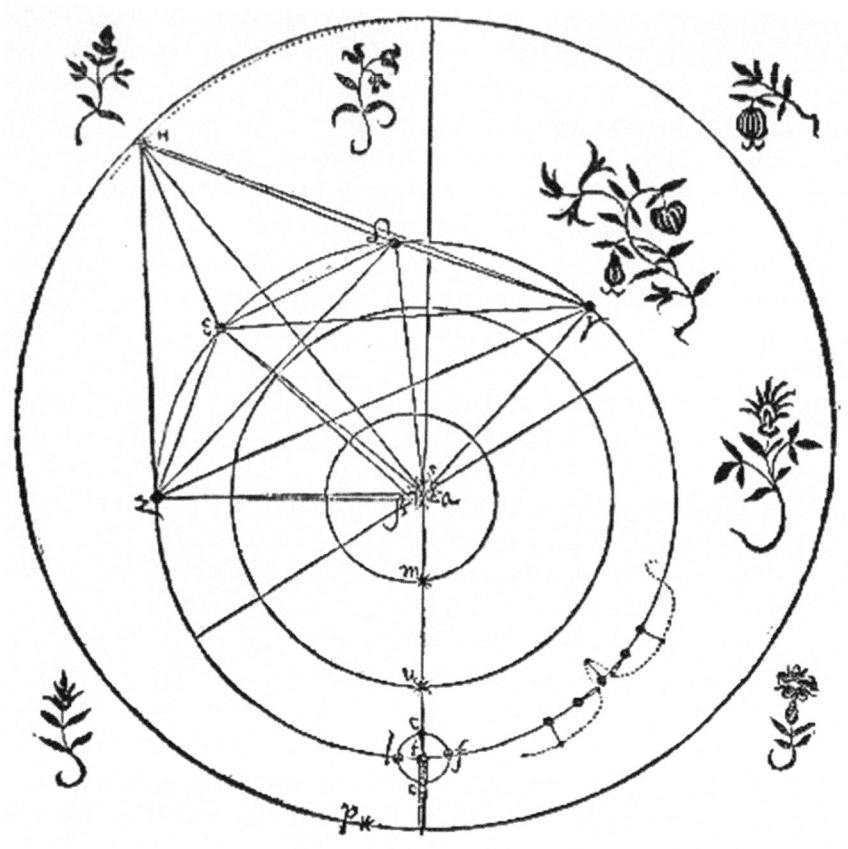

Fig. 2.3 Kepler's diagram from Chap. 27 of *Astronomia Nova*, showing points marked on the orbits of the Earth and Mars. It is interesting to note that he has drawn these orbits as ellipses, even though he does not introduce the concept of an elliptical orbit until much later. Note the ornate style and the difficulty of reading the letters. This is typical of Kepler's style generally: lacking clarity. Public domain

In *Astronomia Nova*, Kepler provides convincing evidence that the orbit is oval, not round and not composed of rotations-within-rotations. He shows that an elliptical orbit fits Tycho's data better than the systems of either Tycho himself or Copernicus.

He also finds a way to show by range-finding that the orbit of the Earth is no different in form to that of Mars. This again differs from the theory of Copernicus, and indeed Ptolemy, who treat it differently. I am impressed by the cunning of his method.

In Fig. 2.4, every angle in each of the triangles SE_1M, SE_2M, SE_3M and SE_4M is known, and hence the triangles are fully defined. Kepler's basic

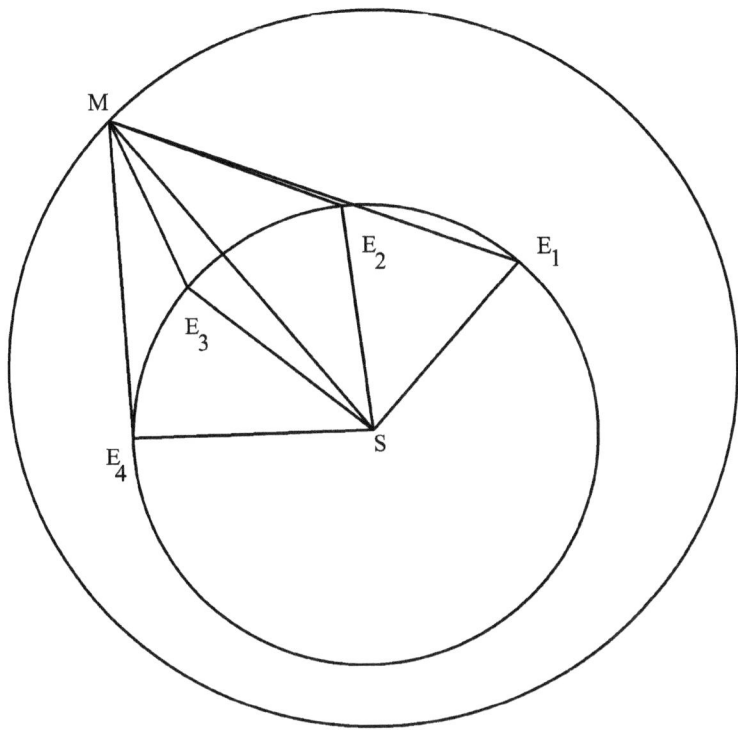

Fig. 2.4 A simplified version of Fig. 2.3, showing the Sun S, Mars M and the Earth at various times, E_1, E_2, E_3 and E_4. These times are at 687-day intervals so that Mars is in the same place, but on successive orbits, at each of these times. (Image: Author)

distance unit tended to be such that the semimajor axis of the Earth's orbit is 100,000 units. Decimal notation, as we know, was introduced by John Napier a little later after the publication of *Astronomia Nova*. Napier was a Scottish aristocrat, who is credited with the invention of logarithms [63]. Using the modern Astronomical Unit would therefore have been very inconvenient for Kepler.

Using this method, he was able to get an eccentricity of 0.0168 for the Earth's orbit (*Astronomia Nova*, Chapters 26 & 27).

He tried the converse method of range-finding to obtain the Sun-Mars distance at various times. Tycho's data were good enough to obtain the location (and therefore time) of Mars' aphelion and perihelion (Fig. 2.5), but they were unfortunately not accurate enough to show any more than that the orbit had some kind of oval shape [61, 62].

He eventually guessed this, not by range-finding but by fine-tuning his ovals to reduce errors compared to Tycho's data. This he did by constructing

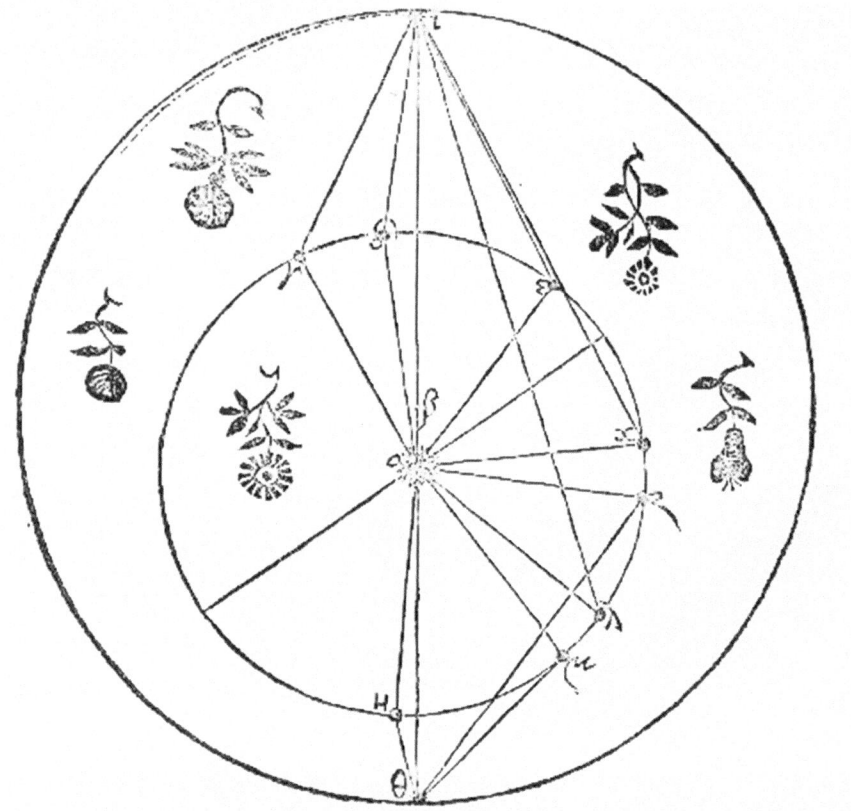

Fig. 2.5 Kepler's diagram showing how he measured the position of the aphelion (top) and perihelion (bottom) of the orbit of Mars (*Astronomia Nova*, Chap. 42). Public domain

multiple circles with rotation at set rates to obtain a net shape, shown exaggerated in Fig. 2.6. Kepler had secondary circles moving round circles to create his ovals. By manipulating the parameters in these shapes, he showed that the best fit to Tycho's data was an elliptical orbit of Mars.

Kepler published *Harmonices Mundi*, the Harmonies of the World, in 1619 [64]. Here, he gives what we now call his third law, that is,

$$a^3 \propto T^2 \tag{2.1}$$

where a is the semimajor axis of the orbit and T is the orbital period (Tables 2.1 and 2.2).

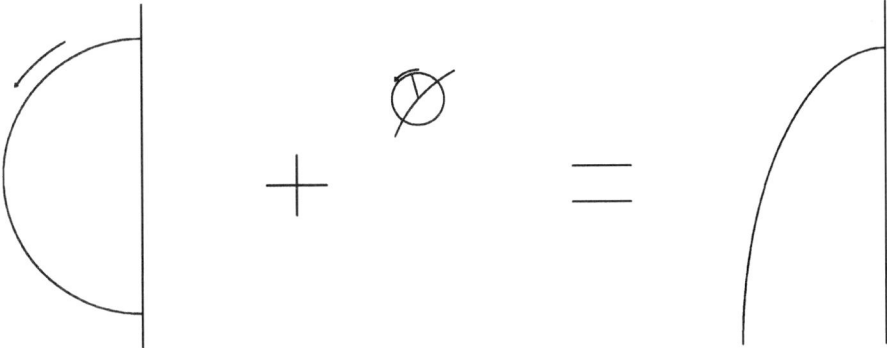

Fig. 2.6 Kepler had secondary circles moving round circles to create his ovals. By manipulating the parameters in these shapes, he showed that the best fit to Tycho's data was an elliptical orbit of Mars. (Drawn by the Author, based loosely on a figure in Chap. 61 of *Astronomia Nova*)

Table 2.1 Kepler's data from *Harmonices Mundi* demonstrating his third law

	Period (days)	Aphelion (AU)	Perihelion (AU)	Semimajor axis (AU)	R^3/T^2 (10^{-6} AU^3/ day^2)	Eccentricity e
Saturn	10,759	10.118	8.994	9.556	7.54	0.059
Jupiter	4332	5.464	4.948	5.206	7.52	0.050
Mars	686	1.661	1.384	1.523	7.50	0.091
Earth	365	1.017	0.983	1.000	7.51	0.017
Venus	224	0.726	0.716	0.721	7.47	0.007
Mercury	87	0.476	0.308	0.392	7.96	0.214

Eccentricities calculated from the other data in the table

(continued)

Table 2.2 Modern data demonstrating Kepler's third law

	Period (days)	Aphelion (AU)	Perihelion (AU)	Semimajor axis (AU)	R^3/T^2 (10^{-6} AU^3/ day^2)	Eccentricity e
Saturn	10,759	10.052	9.022	9.537	7.49	0.054
Jupiter	4332.7	5.453	4.953	5.203	7.50	0.048
Mars	687.0	1.666	1.382	1.524	7.50	0.093
Earth	365.2	1.017	0.983	1.000	7.50	0.017
Venus	224.7	0.728	0.718	0.723	7.49	0.007

(continued)

Table 2.2 (continued)

	Period (days)	Aphelion (AU)	Perihelion (AU)	Semimajor axis (AU)	R^3/T^2 (10^{-6} AU^3/ day^2)	Eccentricity e
Mercury	88.0	0.467	0.307	0.387	7.49	0.206

Sources: Bakich [65] (periods), Murray & Dermott [66] (other data)
Aphelions and perihelions calculated from other data in the table

As Kepler stated, his third law does not work so well with his Mercury data. This does not surprise me. Mercury is hard to spot from my own location at 51½°N, let alone at Tycho's location at nearly 56°N. Additionally, it is always close to the horizon, where refraction alters the data, and it is only visible in the dawn or dusk.

Although you can find Kepler's first two laws in *Astronomia Nova*, they are easier to find in a more popular account he later wrote, *Epitome of Copernican Astronomy* [67].

Kepler lived on the fault-line where Protestantism met Catholicism. He had to move a couple of times because his Protestantism became dangerous. One, possibly unintended, consequence of the Reformation was an increase in witch-burning [68]. Kepler had to spend a couple of years defending his own mother from an accusation of witchcraft. Fortunately for her, he had enough prestige to make his influence felt.

For all that one can fault Kepler for his lack of clarity, he was without doubt a very great man, whose ideas overthrew a lot of ancient and mediaeval baggage. For the first time, he produced a *Solar System*, in which the Sun really was the dominant body.

Ellipses

Having established that planets orbit in ellipses, it is time to learn something about them. I went all the way from kindergarten to physics Ph.D. without learning very much about them. Everything I learnt was written on five pages of high school notes. I still have them. My working assumption, dear reader, is that you have no more background in ellipses than me, and maybe less. So, like Maria in *the Sound of Music*, I'll start from the very beginning.

I have no intention of using the mathematical methods of the early scientists. To me, this is as clumsy and absurd as travelling to visit the Tycho Brahe Museum in Scandinavia on horseback just because that was the only way Tycho could get there. We have better mathematical methods nowadays, which are both more powerful and easier to understand.

We also have ready access, in our homes, to amounts of computer power that used to cost millions of dollars when I was young. In my day job, I occasionally get into hot water because I advocate using computers to mechanize calculations that some of my colleagues would prefer to do by hand. If that is your preference, please be my guest.

I am not an advocate of doing manually what machines can do better. I drive a car with an automatic transmission and find some people's enthusiasm for stick-shift cars incomprehensible. I never wash pots and dishes by hand. I have a machine in my kitchen that gets them cleaner than I ever could. Similarly, I hate mathematical drudgery and avoid it as far as I can.

Back to ellipses. Kepler relied heavily on, and quoted widely from [69], the works of Apollonius of Perga on ellipses. He himself was no slouch when it came to developing the theory of ellipses. We will eventually come to the equation named after him.

But let's start simple. Mostly I taught myself the basics from the excellent textbook of George B Thomas Jr., titled *Calculus and Analytic Geometry*, aimed at college freshmen and women [70]. Recent editions of this book are still available.

My immediate aim is to describe an ellipse by an equation in polar coordinates. Since Kepler showed us that the planets orbit along ellipses with the Sun at one focus, not at the centre, there isn't a lot of point in obtaining an equation in a coordinate system with its origin at the centre. We want our origin to be at one focus.

One way to define an ellipse is to imagine two points F_1 and F_2 and to choose a convenient coordinate system such that they lie on the x-axis, equidistant from the origin, at the points $(c,0)$ and $(-c,0)$ (Fig. 2.7).

Let us further define a line by all those points where the sum of the distances from F_1 and F_2 is constant. Let P be a point on this line. Then

$$2a = \sqrt{(x+c)^2 + y^2} + \sqrt{(x-c)^2 + y^2}. \tag{2.2}$$

This can be seen by dropping a perpendicular from P to the x-axis (Fig. 2.7) and using Pythagoras' theorem. That the sum of the distances is indeed $2a$ can be seen by considering the cases where $y = 0$.

We now perform some operations on this equation. In particular, I want to tease out Eq. (2.3), which is probably the best-known equation describing an ellipse. It is certainly the only thing I learned about ellipses in high school. Then we'll move on to polar coordinates.

$$2a - \sqrt{(x-c)^2 + y^2} = \sqrt{(x+c)^2 + y^2}.$$

$$4a^2 + (x-c)^2 + y^2 - 4a\sqrt{(x-c)^2 + y^2} = (x+c)^2 + y^2$$

$$4a^2 + x^2 + c^2 - 2xc + y^2 - 4a\sqrt{x^2 + c^2 - 2xc + y^2} = x^2 + c^2 + 2xc + y^2$$

$$4a^2 - 4a\sqrt{x^2 + c^2 - 2xc + y^2} = 4xc$$

$$-4a\sqrt{x^2 + c^2 - 2xc + y^2} = 4xc - 4a^2$$

$$a\sqrt{x^2 + c^2 - 2xc + y^2} = -xc + a^2$$

$$\sqrt{x^2 + c^2 - 2xc + y^2} = a - \frac{xc}{a}$$

$$x^2 + c^2 - 2xc + y^2 = a^2 + \left(\frac{xc}{a}\right)^2 - \frac{2axc}{a}$$

$$x^2 + c^2 - 2xc + y^2 = a^2 + x^2\left(\frac{c}{a}\right)^2 - 2xc$$

$$x^2\left(\frac{a^2 - c^2}{a^2}\right) + y^2 = a^2 - c^2$$

$$\frac{x^2}{a^2} + \frac{y^2}{a^2 - c^2} = 1$$

$$\frac{x^2}{a^2} + \frac{y^2}{b^2} = 1. \qquad (2.3)$$

Here $b = \sqrt{a^2 - c^2}$. In the case where $a = b = r$ this reduces to

$$a^2 + b^2 = r^2, \qquad (2.4)$$

which is the equation of a circle. We can thus also think of an ellipse as a stretched circle. By setting $x = 0$, we can also see that the line described by Eq. (2.3) passes through $(x, y) = (0, \pm b)$.

If we also let

$$c = ae \qquad (2.5)$$

Then
$$\begin{aligned} b &= \sqrt{a^2 - c^2} \\ &= \sqrt{a^2 - a^2 e^2} \\ &= a\sqrt{1 - e^2}. \end{aligned} \qquad (2.6)$$

We have the equation of an ellipse (Eq. (2.3)) in Cartesian coordinates. Our aim was to write an equation in polar coordinates. For orbital mechanics, the centre of the polar coordinates needs to be F_1 or F_2, i.e.

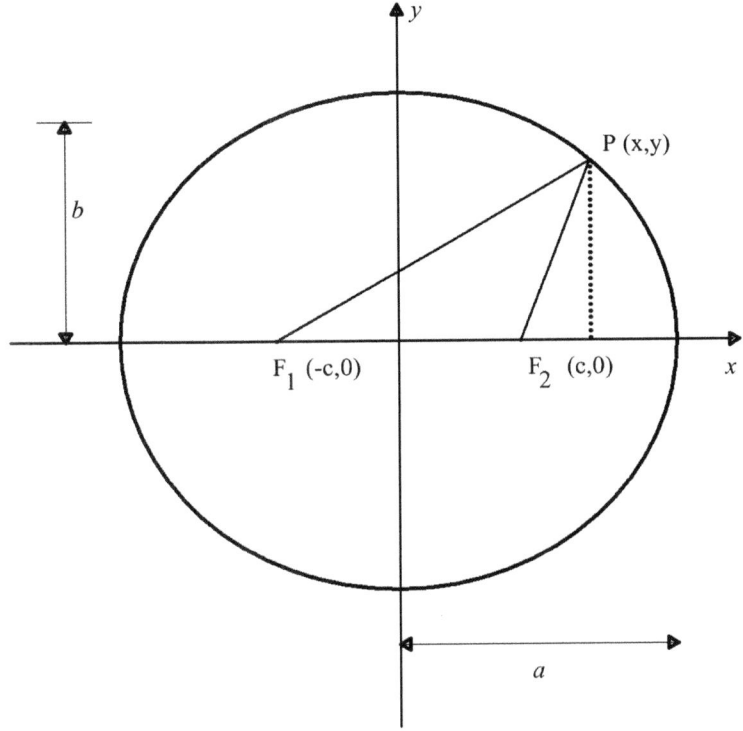

Fig. 2.7 An ellipse as the locus of points (i.e. all points), the sum of whose distances from F_1 and F_2 is constant

displaced by $(\pm c, 0)$ from the origin we have hitherto used, which is at the centre of the ellipse. These points are known as *focuses* or *foci*.

To find the foci, we use the fact, shown in Fig. 2.8, that

$$
\begin{aligned}
BF_1 + BF_2 &= 2a \\
BF_1 &= BF_2 \\
\therefore BF_1 &= BF_2 = a.
\end{aligned}
\tag{2.7}
$$

Thus, a circle of radius a centred at B cuts the x-axis at the foci $(\pm c, 0)$.

Now, let us use Fig. 2.9 to define a line F_2P, of length r, which goes from one focus to the ellipse.

$$
\begin{aligned}
x &= c + r\cos f \\
y &= r\sin f \\
\frac{(c + r\cos f)^2}{a^2} + \frac{r^2 \sin^2 f}{b^2} &= 1 \tag{2.8} \\
b^2\left(c^2 + 2cr\cos f + r^2 \cos^2 f\right) + a^2 r^2 \sin^2 f &= a^2 b^2 \\
b^2 c^2 + 2rcb^2 \cos f + b^2 r^2 \cos^2 f + a^2 r^2 - a^2 r^2 \cos^2 f &= a^2 b^2
\end{aligned}
$$

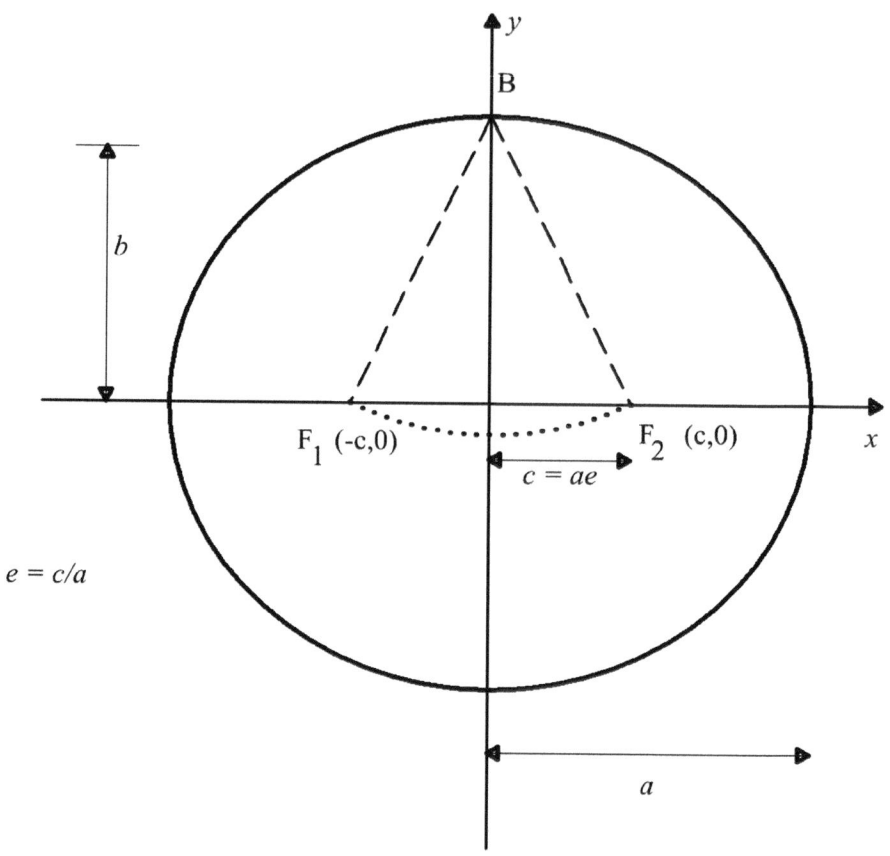

Fig. 2.8 To find the foci, use Eq. (2.7)

since

$$\sin^2 f = 1 - \cos^2 f. \tag{2.9}$$

Now substitute from Eqs. (2.5) and (2.6) into Eq. (2.8).

The quantity e is called the "eccentricity" of the ellipse. It can take values such that

$$0 \le e < 1. \tag{2.10}$$

If $e = 1$, the shape is not an ellipse but a parabola. If $e > 1$ the shape is a hyperbola. These results, not directly relevant to our little tale, are given in Appendix A.

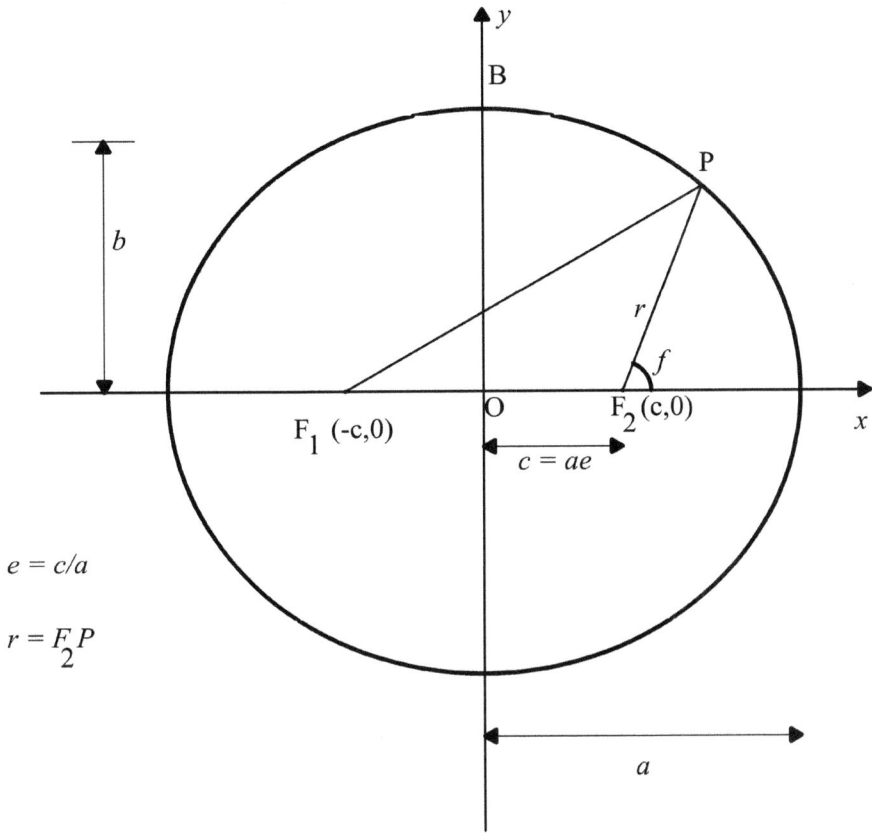

Fig. 2.9 Defining the line $r = F_2P$

$$a^2\left(1-e^2\right)a^2e^2 + 2aea^2\left(1-e^2\right)r\cos f + a^2\left(1-e^2\right)r^2\cos^2 f + a^2r^2 - a^2r^2\cos^2 f \quad = \quad a^2a^2\left(1-e^2\right)$$

$$a^2\left(1-e^2\right)a^2e^2 + 2aea^2\left(1-e^2\right)r\cos f + a^2\left(1-e^2\right)r^2\cos^2 f + a^2r^2 - a^2r^2\cos^2 f - a^2a^2\left(1-e^2\right) = \quad 0$$

$$\left(1-e^2\right)a^2e^2 + 2ae\left(1-e^2\right)r\cos f + \left(1-e^2\right)r^2\cos^2 f + r^2 - r^2\cos^2 f - a^2\left(1-e^2\right) \quad = \quad 0$$

$$\left(1-e^2\right)a^2e^2 + 2ae\left(1-e^2\right)r\cos f + \left(1-e^2\right)r^2\cos^2 f + r^2 - r^2\cos^2 f - a^2\left(1-e^2\right) \quad = \quad 0$$

$$r^2 - e^2r^2\cos^2 f - a^2\left(1-e^2\right)^2 + 2a\left(1-e^2\right)er\cos f \quad = \quad 0$$

$$-r^2 + e^2r^2\cos^2 f + a^2\left(1-e^2\right)^2 - 2a\left(1-e^2\right)er\cos f \quad = \quad 0$$

$$-r^2 + \left[er\cos f - a\left(1-e^2\right)\right]^2 \quad = \quad 0$$

$$\pm\left[er\cos f - a\left(1-e^2\right)\right] \quad = \quad r$$

$$r \quad = \quad er\cos f - a\left(1-e^2\right) > 0$$

$$r\left(1 + e\cos f\right) \quad = \quad a\left(1-e^2\right)$$

$$r \quad = \quad \frac{a\left(1-e^2\right)}{\left(1+e\cos f\right)}$$

$$(2.11)$$

This gives us the equation of an ellipse in polar coordinates about a focus. The fact that $r > 0$ has been used to eliminate the negative possibility when taking square roots.

The angle f has a rather strange name. It is called the "true anomaly".

The area of an ellipse is straightforward to deduce. It is well known that the area of a circle of radius a is

$$A_{circ} = \pi \, a^2 = \pi \, a \, a. \tag{2.12}$$

An ellipse is made by "stretching" one axis of a circle by factor b/a. We can see this because if we take the last line of Eq. (2.3) for the case $b = a$, then

$$
\begin{aligned}
1 &= \frac{x^2}{a^2} + \frac{y^2}{b^2} \\
1 &= \frac{x^2}{a^2} + \frac{y^2}{a^2} \quad \text{if } b = a \\
a^2 &= x^2 + y^2 \quad \text{if } b = a.
\end{aligned}
\tag{2.13}
$$

The last line of Eq. (2.13) is the equation of a circle.

Hence, the area of an ellipse is obtained by reversing the $b \to a$ transformation and changing one of the a's in Eq. (2.12) back to b.

$$
\begin{aligned}
A_{ell} &= \pi \, a \, \frac{b}{a} \, a \\
&= \pi \, a \, b .
\end{aligned}
\tag{2.14}
$$

Unfortunately, you can't play the same trick to obtain the circumference. There is no simple formula for the circumference of an ellipse.

Chapter 3

The Behaviour of Two Bodies in Orbit

Christiaan Huyghens (1629–1695)

Huyghens (Fig. 3.1) was a Dutch scientist who worked in both his native land and Paris. Armed conflict meant that he had to leave Paris at one point.

His greatest contribution to mechanics was to understand centripetal force [71], except that he actually identified centrifugal force. In modern notation, if we have a body rotating about a centre distant from it, its equation of motion can be written in the following form:

$$\mathbf{r} = r_0 \left[\hat{\mathbf{x}} \cos\left(\omega t + \phi\right) + \hat{\mathbf{y}} \sin\left(\omega t + \phi\right) \right]. \tag{3.1}$$

Here, r_0 is the magnitude of the radius. We can differentiate this twice to obtain the acceleration.

$$\frac{d\mathbf{r}}{dt} = \dot{\mathbf{r}} = r_0 \left[-\omega\hat{\mathbf{x}} \sin\left(\omega t + \phi\right) + \omega\hat{\mathbf{y}} \cos\left(\omega t + \phi\right) \right]$$

$$\frac{d^2\mathbf{r}}{dt^2} = \ddot{\mathbf{r}} = r_0 \left[-\omega^2\hat{\mathbf{x}} \cos\left(\omega t + \phi\right) - \omega^2\hat{\mathbf{y}} \sin\left(\omega t + \phi\right) \right] \tag{3.2}$$

$$= \quad -\omega^2\mathbf{r}$$

Later on, he and two others (Wallace and Wren) co-discovered the law of conservation of linear momentum [72]. Of these co-discoverers, Sir

J. Clark, *Calculate the Orbit of Mars!*,
https://doi.org/10.1007/978-3-030-78267-2_3,
© Springer Nature Switzerland AG 2021

Fig. 3.1 Christiaan Huyghens painted by Casper Netscher, c. 1639–84. Public domain

Christopher Wren was an interesting character. He is a household name in the UK, not for his scientific achievements, but for his achievements as an architect. As the Surveyor of the King's Buildings, he designed the cathedral and many churches after the 1666 Great Fire of London. These churches all have a characteristic style, and St. Paul's Cathedral is a great work of art [73]. Less well known is that his previous day job was as Savilian Professor of Astronomy at Oxford University. One of his main contributions there was to develop micrometre-containing telescopes to facilitate astrometry [74]. In his architectural activities, Wren's chief sidekick was Robert Hooke, who did most of the negotiating with builders and other suppliers [73]. Hooke was well known [75] as a scientist in his own right, having developed the law of elasticity named after him, being one among many who thought in terms of an inverse square law of gravity, being a pioneer of microscopy. Wren and Hooke got to hear of the inventions of telescope cross-hairs (sometimes called reticles) through John Flamsteed, the then Astronomer Royal, who was in correspondence with a young astronomer in the north of England named William Gascoigne, who had invented these devices. Gascoigne, despite his isolation from the academic centres, was very well read. Great things may have come of him, but unfortunately, he was killed in battle in the English Civil War of 1642–1651 [76].

Sir Isaac Newton (1642–1727): The Great Synthesizer

It is commonly but erroneously supposed that Newton (Fig. 3.2) was the Great Do-It-Yourselfer who developed his ideas from scratch and took the world by storm with them. Don't get me wrong: as great scientists go, he was in the very top rank.

Fig. 3.2 Newton at the height of his powers, aged 46. Portrait by Godfrey Kneller (1646–1723). (Image courtesy of http://commons.wikimedia.org/wiki/Isaac_Newton#mediaviewer/File:GodfreyKneller-IsaacNewton-1689.jpg). Public domain

But here's the thing. The things he was most credited with were somewhat "in the air" at the time. His mentor at Cambridge, Isaac Barrow, had already developed what gets called the "fundamental theorem of calculus" [77]. Others, notably Robert Hooke [77], not only had guessed the inverse square law of gravity but also had even worked out that it gave rise to elliptical orbits of the type identified by Kepler. Newton's laws of motion were mainly established by others, notably Galileo.

Newton resynthesized these ideas and systematized them. In the process, he developed them and made them much, much more robust.

Despite these fantastic contributions to mechanics, most of his experimental work was in other fields, alchemy and optics. It is well known to amateur astronomers that he made the first reflecting telescope, whose design is named after him. Newtonian telescopes are still popular today. I myself have owned two over the years, with which I took more than 2400 photographs.

Newton was never the easiest of people. His profound insecurity may have stemmed from the fact that his father died before he was born, and his mother remarried when he was a toddler, leaving young Isaac in the care of his grandparents. His reputation at Cambridge University was that of a hermit who did not socialize much outside his college rooms [77].

His insecurity also showed up in his extreme reluctance to publish and in the priority disputes in which he fought tooth and nail. Sometimes, the root cause of these disputes was his slowness to publish [77].

These irksome personal traits may have had a scientific benefit. His main work on mechanics and astronomy, *Mathematical Principles of Natural Philosophy* [79], usually known by the first word of its original Latin name, *Principia*, is remarkably thorough and rigorous in making sure that every possible objection is anticipated and addressed. This thoroughness is in good part what distinguishes Newton from contemporaries like Hooke [78]. Edmund Halley, of comet fame, undertook the task of being Newton's publisher of the *Principia*. This required formidable powers of persuasion to get a reluctant Newton to write the book [77].

Newton's stupendous genius, plus the fact that he put a good deal of effort into "spin-doctoring" his own reputation [77], means that he became the stuff of legend, so much that it is hard to separate fact from fiction in the legend of Isaac Newton. Patricia Fara has written a whole book identifying and debunking myths about him [80].

Two books helped me to find my way through the *Principia*. One was Harper's *Isaac Newton's Scientific Method* [81] and the other was Chandrasekhar's *Newton's Principia for the Common Reader* [82]. The former is a philosophy of science book. In the latter, Nobel Laureate

S. Chandrasekhar takes us through the *Principia* using modern mathematical notation. I have to say that his estimate of the capabilities of a common reader is, shall we just say, a high estimate. I found it very tough going. Nevertheless, the mathematical language of the *Principia* is so old fashioned and such hard going that I doubt if I would have persevered without Chandrasekhar's work. If it took a Nobel Laureate 5 years to work his way through the *Principia* and write his findings up, I dread to think how long it would have taken for me.

In what follows, I will develop enough of the story to get me to my own measurement of the orbit of Mars, but will also digress a little at times to show how thorough Newton was.

Universal Gravitation

I assume two bodies, treated as points. The vector form of Newton's law of gravity is that the force exerted by body 2 on body 1 is

$$F_{21} = -\frac{Gm_1 m_2}{r_{12}^3} r_{12}, \tag{3.3}$$

Where G is a constant, m_1 and m_2 are masses of the two bodies attracting each other, and r_{12} is their separation. The equal and opposite reaction is the force exerted on body 1 by body 2

$$\begin{aligned} F_{12} &= -\frac{Gm_1 m_2}{r_{21}^3} r_{21} \\ &= \frac{Gm_1 m_2}{r_{12}^3} r_{12}, \end{aligned} \tag{3.4}$$

since $r12 = -r21$. Appling Newton's second law of motion to Eqs. (3.3) and (3.4) gives

$$\begin{aligned} F_{12} &= -\frac{Gm_1 m_2}{r_{21}^3} r_{21} = m_1 \ddot{r}_1 \\ F_{21} &= -\frac{Gm_1 m_2}{r_{12}^3} r_{12} = m_2 \ddot{r}_2 \end{aligned} \tag{3.5}$$

In Eq. (3.5), r_1 and r_2 are the positions of bodies 1 and 2 respectively with respect to an arbitrary origin.

We neglect any other forces that may be exerted on this pair of bodies. In other words, this is a closed system so that

$$m_1\ddot{r}_1 + m_2\ddot{r}_2 = 0. \tag{3.6}$$

Integrating Eq. (3.6) twice with respect to time gives

$$\begin{aligned}
m_1\dot{r}_1 + m_2\dot{r}_2 &= a, \\
m_1 r_1 + m_2 r_2 &= at + b.
\end{aligned} \tag{3.7}$$

We can move our arbitrary origin to the centre of mass (CoM) of the two bodies. The CoM is at the point

$$R = \frac{m_1 r_1 + m_2 r_2}{m_1 + m_2}. \tag{3.8}$$

Combining Eqs. (3.7) and (3.8) gives

$$\begin{aligned}
\dot{R} &= \frac{a}{m_1 + m_2}; \\
R &= \frac{at + b}{m_1 + m_2}.
\end{aligned} \tag{3.9}$$

As we might expect for a closed system, the CoM is moving at a constant velocity (which could be zero if we were to choose the CoM as our frame of reference).

Taking Eqs. (3.95), we can rearrange them to obtain

$$\begin{aligned}
-\frac{Gm_2}{r_{21}^3} r_{21} &= \ddot{r}_1 \\
-\frac{Gm_1}{r_{12}^3} r_{12} &= \ddot{r}_2 \\
-\frac{G(m_1 + m_2)}{r_{12}^3} r_{12} &= \ddot{r}_2 - \ddot{r}_1 \\
-\frac{G(m_1 + m_2)}{r_{12}^3} r_{12} &= \ddot{r}_{12} \\
\ddot{r}_{12} + \frac{G(m_1 + m_2)}{r_{12}^3} r_{12} &= 0.
\end{aligned} \tag{3.10}$$

c 12.
We need the second time derivative of **r** with respect to time.

$$\begin{pmatrix} \hat{r} \\ \hat{\theta} \end{pmatrix} = \begin{pmatrix} \cos\theta & \sin\theta \\ -\sin\theta & \cos\theta \end{pmatrix}\begin{pmatrix} \hat{x} \\ \hat{y} \end{pmatrix}.$$

$$\begin{pmatrix} \dfrac{d\hat{r}}{dt} \\[3mm] \dfrac{d\hat{\theta}}{dt} \end{pmatrix} = \begin{pmatrix} \dfrac{d\hat{r}}{d\theta}\dfrac{d\theta}{dt} \\[3mm] \dfrac{d\hat{\theta}}{d\theta}\dfrac{d\theta}{dt} \end{pmatrix}$$

$$= \begin{pmatrix} -\sin\theta & \cos\theta \\ -\cos\theta & -\sin\theta \end{pmatrix}\begin{pmatrix} \hat{x} \\ \hat{y} \end{pmatrix}\dfrac{d\theta}{dt}$$

$$\begin{pmatrix} \dfrac{d\hat{r}}{dt} \\[3mm] \dfrac{d\hat{\theta}}{dt} \end{pmatrix} = \begin{pmatrix} \hat{\theta} \\ -\hat{r} \end{pmatrix}\dfrac{d\theta}{dt}$$

(3.11)

$$\frac{dr}{dt} = \frac{dr}{dt}\hat{r} + r\hat{\theta}\frac{d\theta}{dt}. \tag{3.12}$$

$$\begin{aligned}
\frac{d^2\mathbf{r}}{dt^2} &= \frac{d}{dt}\left(\frac{dr}{dt}\hat{r}\right) + \frac{d}{dt}\left(r\hat{\theta}\frac{d\theta}{dt}\right) \\[2mm]
&= \frac{d^2r}{dt^2}\hat{r} + \frac{dr}{dt}\frac{d\hat{r}}{dt} + r\hat{\theta}\frac{d^2\theta}{dt^2} + r\frac{d\hat{\theta}}{dt}\frac{d\theta}{dt} + \frac{dr}{dt}\hat{\theta}\frac{d\theta}{dt} \\[2mm]
&= \frac{d^2r}{dt^2}\hat{r} + \frac{dr}{dt}\hat{\theta}\frac{d\theta}{dt} + r\hat{\theta}\frac{d^2\theta}{dt^2} - r\hat{r}\frac{d\theta}{dt}\frac{d\theta}{dt} + \frac{dr}{dt}\hat{\theta}\frac{d\theta}{dt} \\[2mm]
&= \frac{d^2r}{dt^2}\hat{r} - r\hat{r}\frac{d\theta}{dt}\frac{d\theta}{dt} + \frac{dr}{dt}\hat{\theta}\frac{d\theta}{dt} + r\hat{\theta}\frac{d^2\theta}{dt^2} + \frac{dr}{dt}\hat{\theta}\frac{d\theta}{dt} \\[2mm]
&= \left(\frac{d^2r}{dt^2} - r\frac{d\theta}{dt}\frac{d\theta}{dt}\right)\hat{r} + \left(\frac{dr}{dt}\frac{d\theta}{dt} + r\frac{d^2\theta}{dt^2} + \frac{dr}{dt}\frac{d\theta}{dt}\right)\hat{\theta} \\[2mm]
&= \left(\frac{d^2r}{dt^2} - r\left(\frac{d\theta}{dt}\right)^2\right)\hat{r} + \left(\frac{1}{r}\left(r\frac{dr}{dt}\frac{d\theta}{dt} + r^2\frac{d^2\theta}{dt^2} + r\frac{dr}{dt}\frac{d\theta}{dt}\right)\right)\hat{\theta} \\[2mm]
&= \left(\frac{d^2r}{dt^2} - r\left(\frac{d\theta}{dt}\right)^2\right)\hat{r} + \left(\frac{1}{r}\left(2r\frac{dr}{dt}\frac{d\theta}{dt} + r^2\frac{d^2\theta}{dt^2}\right)\right)\hat{\theta} \\[2mm]
&= \left(\frac{d^2r}{dt^2} - r\left(\frac{d\theta}{dt}\right)^2\right)\hat{r} + \left(\frac{1}{r}\left(\frac{dr^2}{dr}\frac{dr}{dt}\frac{d\theta}{dt} + r^2\frac{d^2\theta}{dt^2}\right)\right)\hat{\theta} \\[2mm]
&= \left(\frac{d^2r}{dt^2} - r\left(\frac{d\theta}{dt}\right)^2\right)\hat{r} + \left(\frac{1}{r}\left(\frac{dr^2}{dt}\frac{d\theta}{dt} + r^2\frac{d^2\theta}{dt^2}\right)\right)\hat{\theta} \\[2mm]
&= \left(\frac{d^2r}{dt^2} - r\left(\frac{d\theta}{dt}\right)^2\right)\hat{r} + \left(\frac{1}{r}\left(\frac{d}{dt}\left(r^2\frac{d\theta}{dt}\right)\right)\right)\hat{\theta}.
\end{aligned}$$

(3.13)

Pick up and rewrite Equation (3.10)

$$\frac{d^2 r_{12}}{dt^2} + \frac{G(m_1 + m_2)}{r_{12}^3} r_{12} = 0$$

$$\frac{d^2 r}{dt^2} + \frac{\mu}{r^3} r = 0, \tag{3.14}$$

where

$$\mu = G(m_1 + m_2) \tag{3.15}$$

and the "12" subscript has been dropped from r.

Take the vector cross product of r with both sides of Eq. (3.14):

$$r \times \frac{d^2 r}{dt^2} + \frac{\mu}{r^3} r \times r = 0$$

$$r \times \frac{d^2 r}{dt^2} + \frac{\mu}{r^3} 0 = 0 \tag{3.16}$$

$$r \times \frac{d^2 r}{dt^2} = 0.$$

We use the fact that

$$\frac{d(r \times \dot{r})}{dt} = (r \times \ddot{r}) + (\dot{r} \times \dot{r})$$

$$= (r \times \ddot{r}) + 0 \tag{3.17}$$

$$= (r \times \ddot{r})$$

to integrate Eq. (3.16) to give

$$r \times \dot{r} = \mathbf{h} = \text{constant}, \tag{3.18}$$

so that

$$r \cdot (r \times \dot{r}) = \begin{vmatrix} r_x & r_y & r_z \\ r_x & r_y & r_z \\ \dot{r}_x & \dot{r}_y & \dot{r}_z \end{vmatrix} = 0 = r \cdot \mathbf{h}, \tag{3.19}$$

where the zero follows because any determinant with two identical rows is equal to zero. Since Eq. (3.19) is true for all r, it follows that the motion is always perpendicular to the constant $\boldsymbol{h}$. In other words, the motion is in a plane. The large black dot denotes the scalar product operator for two vectors.

Equation (3.18) can be multiplied by m_2 to tell us that the angular momentum of the orbiting body 2 is

$$\boldsymbol{r} \times m_2 \dot{\boldsymbol{r}} = m_2 \boldsymbol{h} = \text{constant} \tag{3.20}$$

if

$$m_1 \gg m_2. \tag{3.21}$$

$$\therefore \ G(m_1 + m_2) \approx Gm_1. \tag{3.22}$$

If the masses are similar, as in a mutually orbiting binary star pair, the story is slightly more complex. But the mass of the Sun is something like three million times that of Mars [83], so this is not going to be an issue to us: Eq. (3.21) is well satisfied. This, incidentally, was a point at which Newton showed that Kepler's laws could only be approximately correct, although of course an error of one part in three million would not have been detectable by Tycho or Kepler.

The idea of gravity acting at a distance was enough to make some people, not least Huyghens, doubt it [77]. Newton himself was worried. His famous quote from an essay, "General Scholium", which was appended to the second (1713) edition of *Principia*, "Hypotheses non fingo," is often translated as, "I make no hypotheses." I beg slightly to dissent from this translation. "Fingo" can also mean "I deceive" or "I make a pretence of". The English word we have inherited from "fingo" is "to finagle". That suggests to me that Sir Isaac was very wary of trying to explain rather than describe the phenomenon of gravity.

I often wonder what he would make of modern theories of gravity.

Gravity Due to Point Masses and Spheres

The Sun and the planets are not points occupying no volume. They are roughly spherical. Any law of gravity needs to take that into account. There is no evidence that anyone else who thought of either the inverse-square law (r^{-2}) dependence or elliptical orbits did this, but Newton did.

Consider a hollow sphere of radius R, which is infinitesimally thin, but has mass. It is attracting a test mass m, i.e. a mass so small that its gravity does not significantly affect the sphere, e.g. by distorting it. The sphere has mass dM. There is a ring of this sphere with mass dM_{ring}, subtending an infinitesimal angle $d\theta$. From the mass m, this ring subtends an angle ϕ, as shown in Fig. 3.3.

The volume of the ring

$$dV_{ring} = (\text{circumference})(\text{height})(\text{thickness})$$
$$= (2\pi R \sin\theta)(R\ d\theta)(dR). \tag{3.23}$$

Therefore, if the ring has uniform density ρ, then its mass

$$dM_{ring} = 2\pi\rho R^2 \sin\theta\ d\theta\ dR. \tag{3.24}$$

The mass of the sphere

$$dM = 4\pi R^2 dR, \tag{3.25}$$

since the surface area of the shell is $4\pi R^2$ and its thickness is dR. Therefore,

$$dM_{ring} = \frac{1}{2} dM \sin\theta d\theta. \tag{3.26}$$

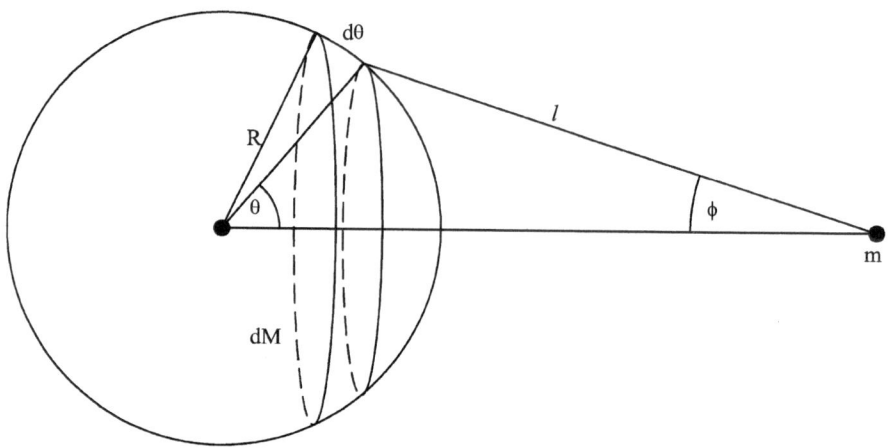

Fig. 3.3 A hollow sphere of radius R, which is infinitesimally thin but has mass. It is attracting a test mass m, i.e. a mass so small that its gravity does not significantly affect the sphere, e.g. by distorting it. The sphere has mass dM. There is a ring of this sphere with mass dM_{ring}, subtending an infinitesimal angle $d\theta$. (Image: Author)

The net force exerted by this ring normal to the line between m and the centre of the sphere will be zero by symmetry. The force along the line between m and the centre of the sphere is as follows:

$$
\begin{aligned}
dF &= -\frac{GmdM_{ring}}{l^2} \\
&= -\frac{GmdM \sin\theta d\theta}{2l^2} \cos\phi.
\end{aligned}
\tag{3.27}
$$

To get the gravitational force due to the whole thin shell shown in Fig. 3.3, we need to integrate Eq. (3.27) over all values of θ. This is not quite as easy as it looks because l and ϕ also depend on θ. This little difficulty is overcome by two applications of the cosine rule. First

$$
\cos\theta = \frac{r^2 + R^2 - l^2}{2Rr}
\tag{3.28}
$$

and then

$$
\cos\phi = \frac{r^2 + R^2 - l^2}{2Rl}.
\tag{3.29}
$$

From Equation (3.27)

$$
\begin{aligned}
\frac{d\cos\theta}{dl} &= \frac{d\cos\theta}{d\theta}\frac{d\theta}{dl} \\
&= \frac{-2l}{2Rr}.
\end{aligned}
\tag{3.30}
$$

Therefore,

$$
\sin\theta d\theta = \frac{ldl}{Rr}.
\tag{3.31}
$$

Hence,

$$dF = \frac{GmdM\sin(\theta)d\theta}{2l^2 Rr}\cos(\phi)$$

$$= \frac{GmdM\sin(\theta)d\theta}{2l^2 Rr}\left[\frac{r^2 + l^2 - R^2}{2rl}\right]$$

$$= \frac{GmdMldl}{2l^2 Rr}\left[\frac{r^2 + l^2 - R^2}{2rl}\right] \qquad (3.32)$$

$$= \frac{GmdM\,dl}{4r^2 R}\left[\frac{r^2 + l^2 - R^2}{l^2}\right]$$

$$= \frac{GmdM\,dl}{4r^2 R}\left[1 + \frac{r^2 - R^2}{l^2}\right].$$

This is integrated to obtain F:

$$F = \int_{r-R}^{r+R} dl(dF)$$

$$= \int_{r-R}^{r+R} dl\left[\frac{G\,m\,dM}{4\,r^2 R}\right]\left[1 + \frac{r^2 - R^2}{l^2}\right]$$

$$= \left[\frac{G\,m\,dM}{4\,r^2 R}\right]\int_{r-R}^{r+R} dl\left[1 + \frac{r^2 - R^2}{l^2}\right]$$

$$= \frac{G\,m\,dM}{4\,r^2 R}\left\{\int_{r-R}^{r+R} dl + \int_{r-R}^{r+R} dl\left[\frac{r^2 - R^2}{l^2}\right]\right\}$$

$$= \frac{G\,m\,dM}{4\,r^2 R}\left\{[l]_{r-R}^{r+R} + \left[-\frac{(r^2 - R^2)}{l}\right]_{r-R}^{r+R}\right\}$$

$$= \frac{G\,m\,dM}{4\,r^2 R}\left\{\{r + R - r + R\} + \left[-\frac{(r + R)(r - R)}{l}\right]_{r-R}^{r+R}\right\}$$

$$= \frac{G\,m\,dM}{4\,r^2 R}\left\{2R - \left[\frac{(r + R)(r - R)}{(r + R)} - \frac{(r + R)(r - R)}{(r - R)}\right]\right\}$$

$$= \frac{G\,m\,dM}{4\,r^2 R}\{2R - [(r - R) - (r + R)]\}$$

$$= \frac{G\,m\,dM}{4\,r^2 R}\{2R - [-2R]\}$$

$$= \frac{G\,m\,dM}{r^2}. \qquad (3.33)$$

Thus, the shell attracts the test mass m as if it were a point at its centre. It further follows that this is true for any shell that is concentric with the one we have considered, even if it has a different density. So, if ρ is a function of the radius R, we can integrate

$$
\begin{aligned}
F_P &= \frac{Gm}{r^2} \int_0^{R_P} dM \\
&= \frac{Gm}{r^2} \int_0^{R_P} 4\pi R^2 \rho(R) dR \\
&= \frac{GmM_P}{r^2},
\end{aligned}
\tag{3.34}
$$

where R_P is the total radius of the planet and M_P is its total mass. Thus, the planet will attract as if it were a point provided that it is spherically symmetric, even if its density varies through its thickness. Given that the self-gravity of a planet or a star is considerable, such density variation is likely to occur.

Angular Momentum and Its Implications

Since the term proportional to μ in Eq. (3.16) vanishes, this constancy of angular momentum only depends on the fact that the force between the two bodies acts along the line joining them. It does not depend on the inverse-square law. To me, this seems eminently reasonable: if there are no tangential forces, the rotation can't speed up or slow down, so the angular momentum won't change.

Because h is a constant vector, its direction, which is always perpendicular to the orbital movement of body 2, does not change. In other words, body 2 has to ply its orbit in a plane.

This constant angular momentum does something else that's interesting. It gives a physical reason for one of Kepler's laws.

Let us remind ourselves how the area of a triangle is calculated. Figure 3.4 shows a triangle with corners A, B and C, and sides of lengths a, b and c. d is the height of the perpendicular from side AB to corner C. The angle at corner A is α.

The area S of the triangle ABC is half the base times the perpendicular height, or

$$
S = \frac{1}{2} cd = \frac{1}{2} cb \sin \alpha.
\tag{3.35}
$$

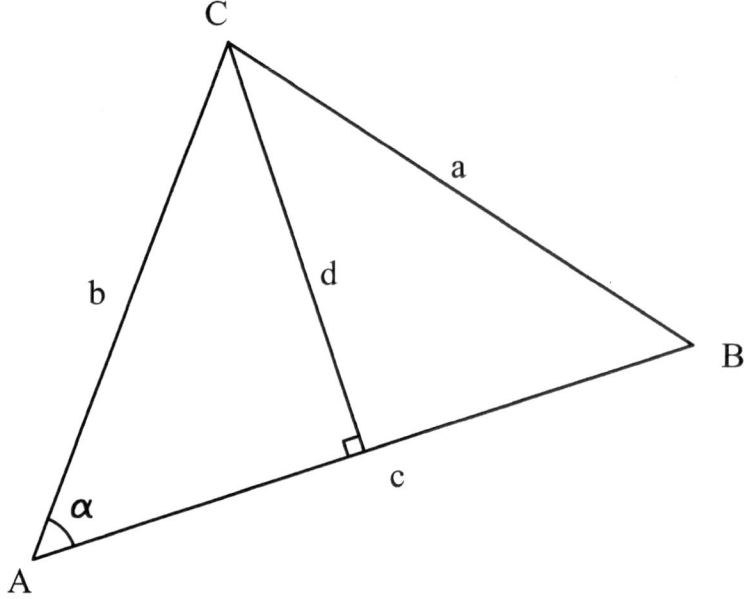

Fig. 3.4 A typical triangle showing variables and we need to calculate its area. a, b and c are the lengths of the sides, and d is the perpendicular height from side AB to corner C. (Image by the author)

We can rewrite Equation (3.35) in vector language as

$$\mathbf{S} = \frac{1}{2}\ \mathbf{c} \times \mathbf{b}, \tag{3.36}$$

where now $\mathbf{c}$ and $\mathbf{b}$ are vectors of magnitude c and b respectively, pointing along the sides AB and AC of our triangle.

Now let us return to our orbiting planet. Equation (3.18) can be applied to the situation shown in Fig. 3.5. Here, the grey area is given by

$$\mathbf{S}_{\text{Grey}} = \frac{1}{2}\ \mathbf{r} \times \dot{\mathbf{r}}\Delta t = \frac{\Delta t}{2}\ \mathbf{r} \times \dot{\mathbf{r}} = \frac{\Delta t}{2}\ \mathbf{h}. \tag{3.37}$$

In other words, the area $\mathbf{S}_{\text{Grey}}$ is proportional to Δt for short time intervals. A larger time interval can always be constructed by adding or, in the limit of small Δt, integrating over many such intervals.

$$\mathbf{S}_{\text{Swept-out}} = \frac{\mathbf{h}}{2} \int_0^t dt' = \frac{\mathbf{h}}{2}\ t. \tag{3.38}$$

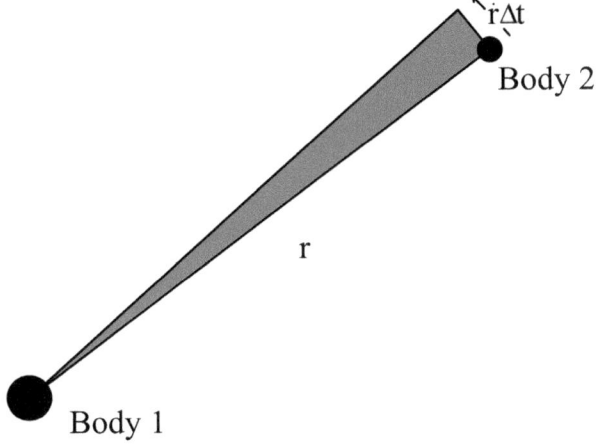

Fig. 3.5 The grey area is the area swept by the orbiting Body 2 in a short time Δt

Thus, we have derived Kepler's second law of planetary motion: Body 2 sweeps out the same area $\mathbf{h}t/2$ in any time interval t. This is true so long as the force between the two bodies acts along the line joining them. Because $\mathbf{h}$ is a constant vector, the areas are all co-planar.

One other result for future use will be noted here. If t is equal to the Period T to complete one orbit, then $S_{Swept\text{-}out}$ is equal to the area of the ellipse describing the orbit. We already obtained the area of the ellipse as Eq. (2.14), where we showed that it is equal to $\pi\,a\,b$, where a and b are the semi-minor and semi-major axes of the ellipse respectively. Thus,

$$\frac{\mathbf{h}}{2}\,T = \pi ab\,\hat{\mathbf{h}}. \tag{3.39}$$

Therefore, ignoring the unit vector,

$$T = \frac{2\pi ab}{h}. \tag{3.40}$$

The Bit Where Conic Sections Come In

Equation (3.14) is still in the form of a differential equation. It would be a lot more useful if we were to solve it, i.e. remove the derivatives.

To recap, Eq. (3.14) is

$$\frac{d^2\mathbf{r}}{dt^2} + \frac{\mu}{r^3}\mathbf{r} = \mathbf{0}.$$

Let us substitute the vector components from Eq. (3.13).

$$\left(\frac{d^2r}{dt^2} - r\left(\frac{d\theta}{dt}\right)^2\right)\hat{\mathbf{r}} + \left(\frac{1}{r}\frac{d}{dt}\left(r^2\frac{d\theta}{dt}\right)\right)\hat{\boldsymbol{\theta}} + \frac{\mu}{r^2} \tag{3.41}$$

The $\hat{\mathbf{r}}$ components and the $\hat{\boldsymbol{\theta}}$ components in Eq. (3.41) must each be equal to zero. Consider the $\hat{\mathbf{r}}$ component.

$$\frac{d^2r}{dt^2} - r\left(\frac{d\theta}{dt}\right)^2 = -\frac{\mu}{r^2}. \tag{3.42}$$

We need a way to remove the term $d\theta/dt$. We can do this by substituting Eq. (3.12) into Eq. (3.18).

$$\begin{aligned}\mathbf{h} &= \mathbf{r} \times \dot{\mathbf{r}} \\ &= \mathbf{r} \times \left(\frac{dr}{dt}\hat{\mathbf{r}} + r\hat{\boldsymbol{\theta}}\frac{d\theta}{dt}\right) \\ &= \mathbf{r} \times r\hat{\boldsymbol{\theta}}\frac{d\theta}{dt} \\ &= r^2\frac{d\theta}{dt}\left(\hat{\mathbf{r}} \times \hat{\boldsymbol{\theta}}\right). \end{aligned} \tag{3.43}$$

But we already know that $\mathbf{h}$ is perpendicular to both $\hat{\mathbf{r}}$ and $\hat{\boldsymbol{\theta}}$. We can write down the scalar equation

$$h = r^2\frac{d\theta}{dt}. \tag{3.44}$$

Equation (3.42) becomes

$$\begin{aligned}\frac{d^2r}{dt^2} - \frac{r}{r^4}h^2 &= -\frac{\mu}{r^2}. \\ \frac{d^2r}{dt^2} - \frac{1}{r^3}h^2 &= -\frac{\mu}{r^2}. \end{aligned} \tag{3.45}$$

Most of the textbooks, which I have looked, simplify this problem by making a substitution

$$u = \frac{1}{r}. \tag{3.46}$$

Then

$$\begin{aligned}
\frac{dr}{dt} &= \frac{dr}{du}\frac{du}{d\theta}\frac{d\theta}{dt} \\
&= \frac{-1}{u^2}\frac{du}{d\theta}\frac{d\theta}{dt} \\
&= -h\frac{du}{d\theta}
\end{aligned}$$
(3.47)

and

$$\begin{aligned}
\frac{d^2r}{dt^2} &= \frac{d}{dt}\left(-h\frac{du}{d\theta}\right) \\
&= -h\frac{d\theta}{dt}\frac{d}{d\theta}\frac{du}{d\theta} \\
&= -h\left(hu^2\right)\frac{d}{d\theta}\frac{du}{d\theta} \\
&= -h^2u^2\frac{d^2u}{d\theta^2}.
\end{aligned}$$
(3.48)

We can now substitute Eq. (3.48) into Eq. (3.45).

$$-h^2u^2\frac{d^2u}{d\theta^2} - u^3h^2 = -u^2\mu.$$
(3.49)

Divide every term by u^2:

$$\frac{d^2u}{d\theta^2} + u = \frac{\mu}{h^2}.$$
(3.50)

Let us try as solution to Eq. (3.50)

$$u = \frac{\mu}{h^2}\left[1 + e\cos(\theta - \varpi)\right].$$
(3.51)

Here, ϖ is a constant angle. At school, I was taught to pronounce this symbol "pomega", to rhyme with "omega", though I think other pronunciations exist. It is supposed to be some ghastly combination of π and ω. Given that there are 22 other perfectly good Greek letters, and 26 English letters to choose from, I cannot for the life of me think why people thought they needed to use pomega, but they did, and we are stuck with it.

Then,

$$
\begin{array}{ccc}
\dfrac{du}{dt} & = & \dfrac{\mu}{h^2}\left[-e\sin(\theta-\varpi)\right] \\[2ex]
\dfrac{d^2u}{dt^2} & = & \dfrac{\mu}{h^2}\left[-e\cos(\theta-\varpi)\right] \\[2ex]
\dfrac{\mu}{h^2}\left[-e\cos(\theta-\varpi)\right]+\dfrac{\mu}{h^2}\left[1+e\cos(\theta-\varpi)\right] & = & \dfrac{\mu}{h^2}\left[1-e\cos(\theta-\varpi)+e\cos(\theta-\varpi)\right] \\[2ex]
& = & \dfrac{\mu}{h^2}. \quad \text{QED.}
\end{array}
$$

$$(3.52)$$

It is legitimate to ask whether Eq. (3.51) is the most general solution to Eq. (3.50); in other words, to ask whether there could be other solutions.

I came to the conclusion that there can't. Here is my reasoning. To satisfy Eq. (3.50), u needs to look something like this:

$$u = \frac{\mu}{h^2}\left[1+f\left(\theta\right)\right], \tag{3.53}$$

such that

$$\frac{d^2 f}{d\theta} = -f\left(\theta\right). \tag{3.54}$$

This can be seen by back-substitution into Eq. (3.50):

$$-f\left(\theta\right)+\frac{\mu}{h^2}\left(1+f\left(\theta\right)\right)=\frac{\mu}{h^2}. \tag{3.55}$$

You can't satisfy this with terms like x, x^2,..., x^n or any combination of them. You can't satisfy it with hyperbolic sines or cosines, because they do not satisfy Eq. (3.54). The only things that do are sines, cosines, and their exponential equivalents [84]

$$
\begin{aligned}
\cos\theta &= \frac{e^{i\theta}+e^{-i\theta}}{2} \\[2ex]
\sin\theta &= \frac{e^{i\theta}-e^{-i\theta}}{2i}
\end{aligned}
\tag{3.56}
$$

If I take the most general formula, I can think of using exponentials of $\pm i\theta$, let me try

$$u = \frac{\mu}{h^2}\left[1 + Ae^{ik(\theta-\alpha)} + Be^{-iq(\theta-\beta)}\right]. \tag{3.57}$$

Then,

$$\frac{du}{d\theta} = \frac{\mu}{h^2}\left[1 + ikAe^{-ik\alpha}e^{ik\theta} - iqBe^{iq\beta}e^{-iq\theta}\right]$$

$$\frac{d^2u}{d\theta^2} = \frac{\mu}{h^2}\left[1 - k^2Ae^{-ik\alpha}e^{ik\theta} + q^2Be^{iq\beta}e^{-iq\theta}\right] \tag{3.58}$$

Equations (3.57) and (3.58) can't satisfy Eq. (3.54) unless $k = q = 1$. Then,

$$\frac{\mu}{h^2} = \frac{\mu}{h^2}\left[1 + Ae^{-i\alpha}e^{i\theta} + Be^{i\beta}e^{-i\theta}\right]. \tag{3.59}$$

The right-hand side of Eq. (3.59) must always be real for all θ. This can only happen if $A = B$ and $\alpha = -\beta$. Setting $A = e$ and $\alpha = \varpi$ gets us back to Eq. (3.51).

Now let us replace u with r.

$$r = \frac{h^2}{\mu\left[1 + e\cos(\theta - \varpi)\right]}. \tag{3.60}$$

This is an important equation. We showed earlier (Eq. (2.11), and in Appendix 1, we will show (Eqs. (A1.10) and (A1.11)) that Eq. (3.60) is the equation of a conic section in polar coordinates with the origin at one focus. When $e < 1$, it is the equation of an ellipse, with $f = \theta - \varpi$.

Thus, Newtonian gravity causes two-body orbits to be conic sections. The importance of this result, which we have now shown to be true, cannot be over-emphasized. Indeed, we now see that Kepler's first law, that the planets orbit in ellipses with the Sun at one focus, is a consequence of Newtonian Gravity and Newton's laws of motion. We will discuss later what Newton did to put this point beyond a reasonable doubt, given the knowledge and observations of the day. We don't quite yet have the basic results we need to address this question.

When $e < 1$, Eq. (3.60) is the equation of an ellipse. When $e = 0$, we have the equation of a circle (radius = constant).

By the way, the quantity h^2/μ is given the name "semilatus rectum" in the literature. I promise you that this does not sound anything like as rude in Latin as it does in English.

It is possible to obtain an expression for h^2/μ in terms of the semi-major axis, using our earlier result Eq. (2.11) that

$$r = \frac{a\left(1-e^2\right)}{1+e\cos\left(\theta-\varpi\right)}. \tag{3.61}$$

We end up with

$$p = \frac{h^2}{\mu} = a\left(1-e^2\right). \tag{3.62}$$

Equation (3.62) defines the quantity p.

Time Evolution

While it was vital to demonstrate that the two-body Newtonian problem gives rise to orbits as conic sections, Eq. (3.60) has a practical limitation: it does not explicitly contain time. We will be watching how Mars moves relative to the heavens over time, so we need to understand how the orbit evolves over time.

A Brief Aside: the Word Anomaly

One of the biggest irritants in the jargon of celestial mechanics is the way that the word anomaly is used. In everyday speech, an anomaly is something that is unusual, unexpected or abnormal.

This is not so in astronomy. Here, an anomaly is an angle. It is a particular angle; the angle between the line joining the focus of a conic section and the periapsis, or occasionally the apoapsis. It can also be an angle defined by a line from the centre of the ellipse. Anomalies are often angles that evolve over time as the body plies its orbit.

(The periapsis is the closest point of the conic section to one of the focuses. The apoapsis is the furthest point on an ellipse from this focus. The plurals of periapsis and apoapsis are periapses and apoapses, respectively. The apoapses of parabolas and hyperbolas are not defined.)

Figure 3.6 shows an angle f, which astronomers would call anomalies. The former is the anomaly with respect to (w.r.t.) the focus; the latter is the anomaly w.r.t. the centre of the ellipse shown.

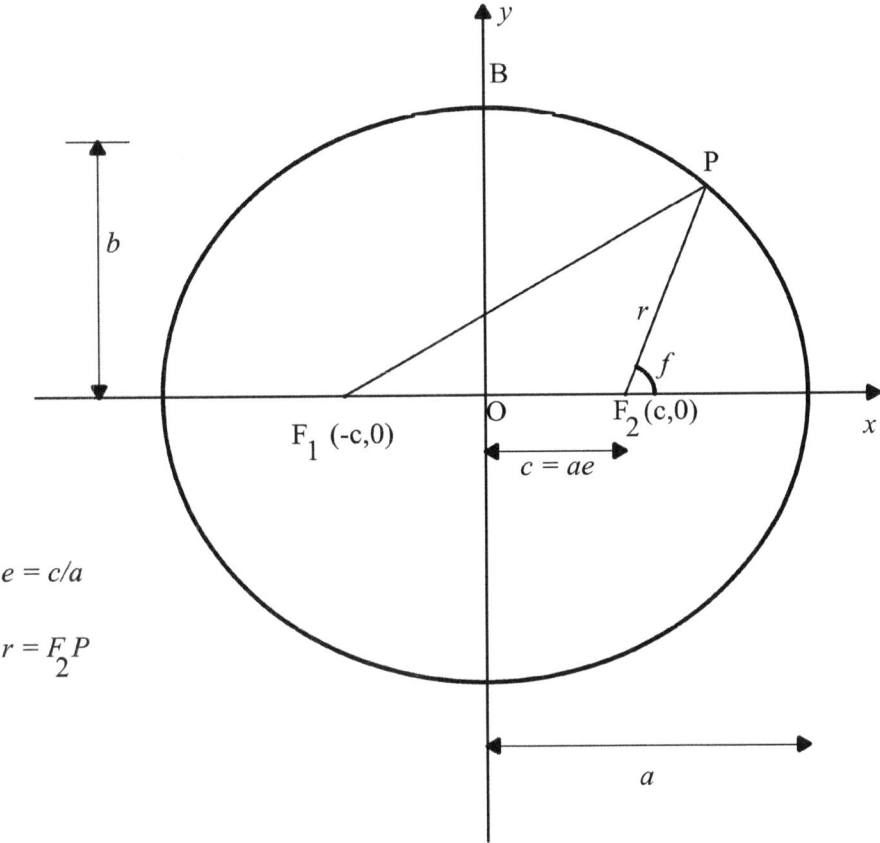

Fig. 3.6 An ellipse, showing an angle that could be called an anomaly. f is the anomaly taken from F_2

Figure 3.7 shows another angle E which is also called an anomaly. This angle is constructed from the ellipse by extending the line F_2P onto the circle shown in Fig. 3.7. This circle passes through the centre of the ellipse and through the periapsis and apoapsis w.r.t. F_2. (Of course this also means that it touches the ellipse at the two points on the major axis of the ellipse, but I digress.)

In Fig. 3.7, the angle $\angle SOQ$ has been called E. Angle E has a special name: it is called the *eccentric anomaly*. Again this name is rather non-obvious to the beginner. It is tempting to suspect that whoever thought up this jargon was using an illegal stimulant. The angle f also has a special name. It is called the *true anomaly* (Fig. 3.8).

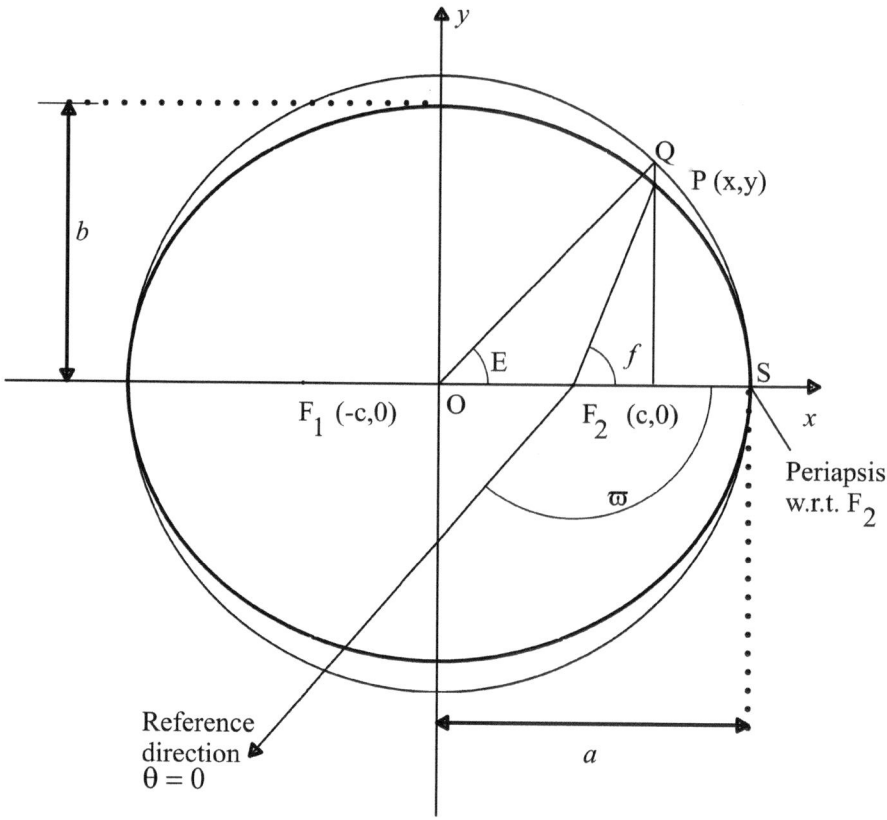

Fig. 3.7 An ellipse, showing two angles that could be called anomalies. The closest point S to the focus $F2$ on the ellipse is called the periapsis with respect to (w.r.t.) this focus. The angle f is the anomaly taken from $F2$. E, taken from O to the circle circumscribing the ellipse, is called the "eccentric anomaly"

The angles f and θ are related by

$$f = \theta - \varpi;$$
$$\theta = f + \varpi.$$
(3.63)

Therefore,

$$r = \frac{a\left(1-e^2\right)}{1+e\cos f}.$$
(3.64)

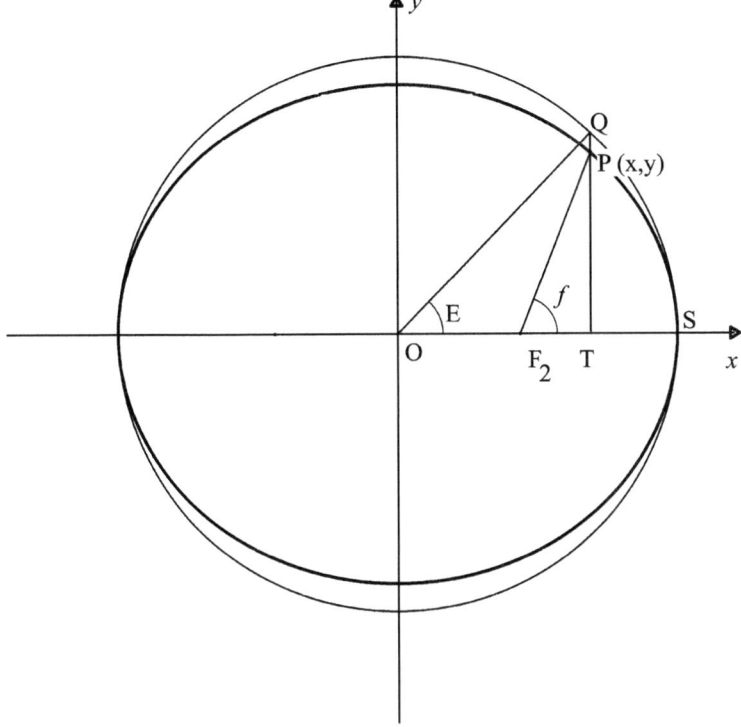

Fig. 3.8 Our ellipse and circumscribing circle again

Why on Earth would we bother with the eccentric anomaly E? It has no physical significance. It does not point to a planet, which would be at the point P. The answer, I am afraid, became horribly clear to me as I progressed through this project. I tried to do everything in terms of the true anomaly f on the grounds that this was the physical quantity. This, dear reader, was a big, big mistake. It is much easier to manipulate the eccentric anomaly E and calculate the true anomaly f from this whenever I needed it.

I will now show you how to convert between anomalies. My destination for the next few minutes will be Eq. (3.72), the formula for converting between f and E.

Remembering that Eq. (2.6) tells us that

$$b = a \sqrt{1-e^2},$$

we can write

$$\frac{b}{a} = \frac{PT}{QT} = \frac{\text{Area}PTS}{\text{Area}QTS}. \tag{3.65}$$

Therefore,

$$r \sin f = b \sin E,$$ (3.66)

$$r \cos f = CT - CS = a \cos E - ae.$$ (3.67)

We can square Eqs. (3.66) and (3.67) and add them to give

$$
\begin{aligned}
r^2 &= & r^2 \left(\cos^2 f + \sin^2 f \right) \\
&= & \left(a \cos E - ae \right)^2 + b^2 \sin^2 E \\
&= & \left(a \cos E - ae \right)^2 + a^2 \left(1 - e^2 \right) \sin^2 E \\
&= & a^2 \left[\left(\cos E - e \right)^2 + \left(1 - e^2 \right) \sin^2 E \right] \\
&= a^2 \left[\cos^2 E + e^2 - 2e \cos E + \sin^2 E - e^2 \sin^2 E \right] \\
&= & a^2 \left[1 + e^2 - 2e \cos E - e^2 \left(1 - \cos^2 E \right) \right] \\
&= & a^2 \left[1 - 2e \cos E + e^2 \cos^2 E \right] \\
&= & a^2 \left[1 - e \cos E \right]^2, \\
\text{so} \quad r &= & a \left(1 - e \cos E \right).
\end{aligned}
$$
(3.68)

Generate some more useful stepping-stone results.

$$
\begin{aligned}
\frac{\sin f}{1 + \cos f} &= \frac{\sin \left(2\dfrac{f}{2} \right)}{1 + \cos \left(2\dfrac{f}{2} \right)} \\
&= \frac{2 \sin \left(\dfrac{f}{2} \right) \cos \left(\dfrac{f}{2} \right)}{1 + 2 \cos^2 \left(\dfrac{f}{2} \right) - 1} \\
&= \frac{2 \sin \left(\dfrac{f}{2} \right) \cos \left(\dfrac{f}{2} \right)}{2 \cos^2 \left(\dfrac{f}{2} \right)} \\
&= \frac{\sin \left(\dfrac{f}{2} \right)}{\cos \left(\dfrac{f}{2} \right)} \\
&= \tan \left(\dfrac{f}{2} \right),
\end{aligned}
$$
(3.69)

In which we have used the formulae

$$\sin(2\theta) = 2\sin\theta\cos\theta, \quad \& $$
$$\cos(2\theta) = \cos^2\theta - \sin^2\theta \qquad (3.70)$$
$$= 2\cos^2\theta - 1.$$

Using Eqs. (3.66), (3.67) and (3.68), Eq. (3.69) becomes

$$\tan\frac{f}{2} = \frac{r\sin f}{r + r\cos f}$$
$$= \frac{b\sin E}{a(1 - e\cos E) + a(\cos E - e)} \qquad (3.71)$$
$$= \frac{b\sin E}{a(1-e)(1+\cos E)}$$
$$= \frac{a\sqrt{1-e^2}\,\sin E}{a(1-e)(1+\cos E)}$$

$$\tan\frac{f}{2} = \sqrt{\frac{1+e}{1-e}}\,\tan\frac{E}{2}. \qquad (3.72)$$

In the last step, we have used the analogue of Eq. (3.69) for E.

Equation (3.72) allows us to switch back and forth between f and E. It is an equation we will use a lot. It doesn't work in the special case

$$E = f = \pi = 180^\circ. \qquad (3.73)$$

However, in this case, we don't really need an equation. The conversion is obvious.

Equations (3.68) and (3.71) give r and f in terms of E. This is the justification for the earlier statement that the eccentric anomaly E is a useful quantity.

At this point, we introduce another so-called anomaly, in other words another angle which evolves in time,

$$M = \frac{2\pi t}{T}, \qquad (3.74)$$

Where t = time elapsed and T is the time to complete one orbit. This anomaly M is called the "mean anomaly". It is considered to be an angle about the focus containing the Sun (or other overwhelmingly massive body as appropriate). It is shown in Fig. 5.16.

The area of our ellipse is

$$A_{\text{ellipse}} = \pi ab, \tag{3.75}$$

which is straightforward to prove by imagining that every infinitesimal "cell" of the area is squashed by the ratio b/a compared to the analogous infinitesimal "cell" of the area of a circle, whose area is well known to be πa^2.

The area PF_2S enclosed by our ellipse is

$$
\begin{aligned}
\text{Area } PF_2S &= \left(\frac{t}{T}\right)(\text{Area enclosed by orbit}) \\
&= \frac{\pi ab \; t}{T} \\
&= \frac{\pi a^2 \sqrt{1-e^2} \; t}{T} \\
&= \frac{1}{2} Ma^2 \sqrt{1-e^2}.
\end{aligned}
\tag{3.76}
$$

It is also true that

$$
\begin{aligned}
\text{Area } PF_2S &= \text{Area } PF_2T + \text{Area } PTS \\
&= \frac{1}{2}a(\cos E - e)(r \sin f) + \frac{b}{a}\text{Area } QTS \\
&= \frac{1}{2}ab(\cos E - e)\sin E + \frac{b}{a}\left[\left(\frac{\pi a^2}{2\pi}\right)E - \left(\frac{1}{2}a^2 \sin E \cos E\right)\right] \\
&= \frac{1}{2}a^2 \sqrt{1-e^2}\,(E - e \sin E).
\end{aligned}
\tag{3.77}
$$

In the third line of Eq. (3.77), we make use of Eq. (3.66) in the first term on the right-hand side. The second term in the third line is the area of circle sector OSQ. The final term is the area of the triangle FQT, which is subtracted from sector OSQ. The factor b/a allows for the fact that the areas we really want are on the ellipse, not the circumscribing circle.

Collecting terms from Eqs. (3.74), (3.76) and (3.77),

$$M = E - e \sin E. \tag{3.78}$$

Just to emphasize that M and E evolve over time, we rewrite Eq. (3.78) as

$$\boxed{M(t) = E(t) - e \sin E(t).} \tag{3.79}$$

This is the equation derived by Kepler to represent progress over time in an elliptical orbit. It bears his name: Kepler's equation. It was mentioned in the section on Kepler above. Now at least you can see what it looks like. I have in my possession a textbook by Peter Colwell devoted to the question of solving Kepler's equation [85]. While Kepler himself had a rough and ready iterative method of solving his equation, this was by no means the last word on the subject. Papers are still being published on ways to solve it.

If we only want to solve a two-body problem for orbits of low-to-moderate eccentricities, and are not attempting to investigate perturbations such as planet-to-planet interactions, there is no need to delve into all these. Indeed Professor Colwell was well aware that, after a while, the astronomers stopped taking much notice of the mathematicians' games with Kepler's equation. They had what they needed.

We can obtain a perfectly good solution to Eq. (3.79) using the well-known, and very simple, Newton-Raphson technique. The limitations of this technique simply do not come out and bite us. The first limitation is that it is not frightfully robust about converging. The second is that it is not terrifically efficient. So what? I found by trial and error that it converges up to eccentricities of at least 0.9. We are not terribly constrained by the power of domestic computers nowadays. And we don't have to use quill pens, unlike the early scientists. It is not as if we need to do thousands of these calculations on the on-board computer of a spacecraft, where the computer needs to be as small and light as possible. Microsoft Excel™ on my one-year-old PC zips through these calculations in the blink of an eye.

The Newton-Raphson technique can be found in virtually any introductory undergraduate textbook on mathematical methods for scientists and/or engineers, or indeed on Wikipedia [86].

The idea is to start from an initial guess. My initial guess was to assume that instead of Eq. (3.79) I was solving

$$M(t) \approx E(t). \tag{3.80}$$

This worked just fine. The principle of the Newton-Raphson method is to rearrange the equation so that one side is zero and solve the resulting equation by iteration, i.e. using one trial solution as a guess for the next one. Let's rearrange Eq. (3.79) to the form

$$0 = E(t) - e\sin E(t) - M(t). \tag{3.81}$$

Now let's define a function of E

$$\lambda(E) = E(t) - e\sin E(t) - M(t), \tag{3.82}$$

And give λ and E subscripts 0 for the initial guess, 1 for the next guess, 2 for the next, and so on.

Then,

$$
\begin{aligned}
\lambda_0(E) &\approx M(t) \\
\lambda_1(E) &\approx \lambda_0(E) - \frac{\lambda(E)}{d\lambda(e)/dE} \\
&\approx \lambda_0(E) - \frac{E_0 - e\sin(E_0) - M}{1 - e\cos(E_0)} \\
\lambda_2(E) &\approx \lambda_1(E) - \frac{E_1 - e\sin(E_1) - M}{1 - e\cos(E_1)} \\
\lambda_3(E) &\approx \lambda_2(E) - \frac{E_2 - e\sin(E_2) - M}{1 - e\cos(E_2)}
\end{aligned}
\tag{3.83}
$$

etc.

Repeat until sufficient accuracy is obtained. You, dear reader, are the judge of what constitutes sufficient accuracy. I never found a case where more than six iterations were needed to achieve repeatability to seven significant figures (Fig. 3.9).

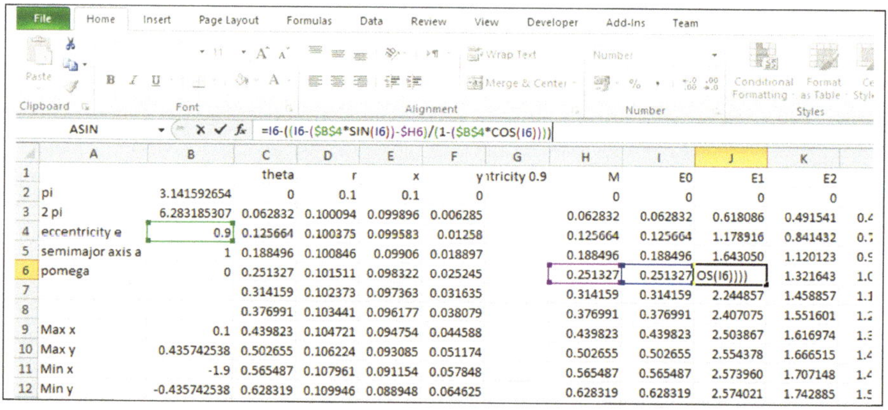

Fig. 3.9 Screen shot showing how the Newton-Raphson method is programmed in Excel. The formula is shown for cell J6. You could be clever and tell me that I don't need so many brackets. I am unrepentant. Brackets remove ambiguity, and, in my humble opinion, remove worry. They are color-coded to help prevent errors

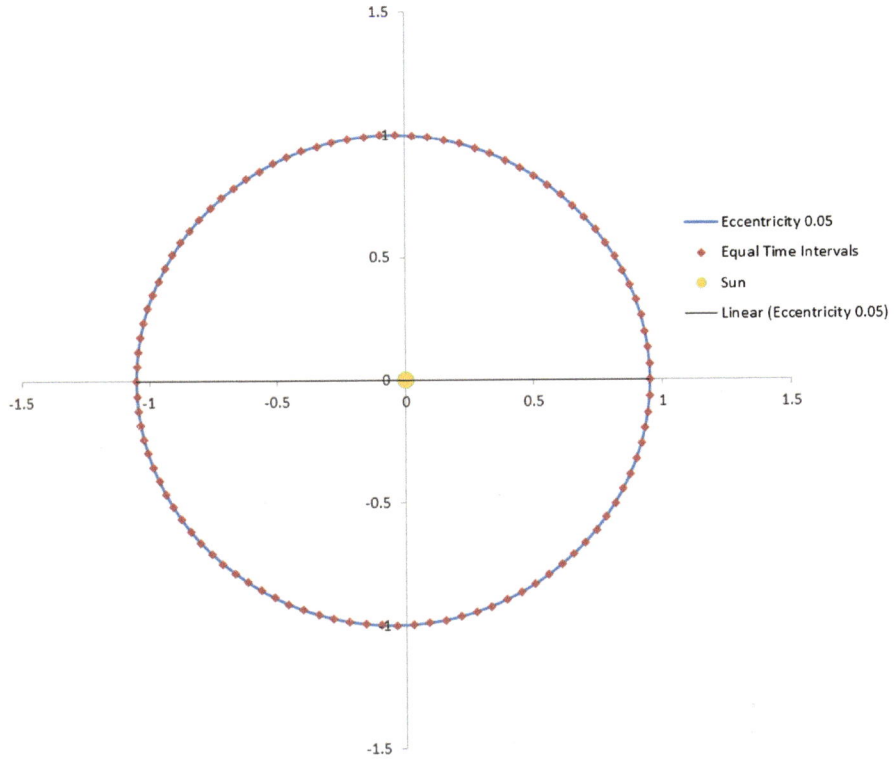

Fig. 3.10 Showing an elliptical orbit with eccentricity 0.05, with arbitrary coordinates, whose origin is at the centre of the Sun. The red marks are spaced such that the time interval between neighbouring tick marks is always the same. The Sun is noticeably off-centre. At the perihelion on the right, it is just about possible to see that the tick marks are spaced further apart than at the aphelion on the left

In the figures that follow I show how a planet would move in orbits with eccentricities of 0.05, 0.09, 0.3, 0.6 and 0.967. The last was chosen just for fun, because 0.967 is the eccentricity of the orbit of Halley's Comet (Figs. 3.10, 3.11, 3.12, 3.13, and 3.14).

We have all the ingredients we need to derive Kepler's third law of planetary motion. Consider the area swept out by a planet during one complete orbit, taking time T. From Eq. (3.38), the magnitude of this area is then

$$S = \frac{1}{2} hT = \pi ab = \pi a^2 \sqrt{1 - e^2}, \tag{3.84}$$

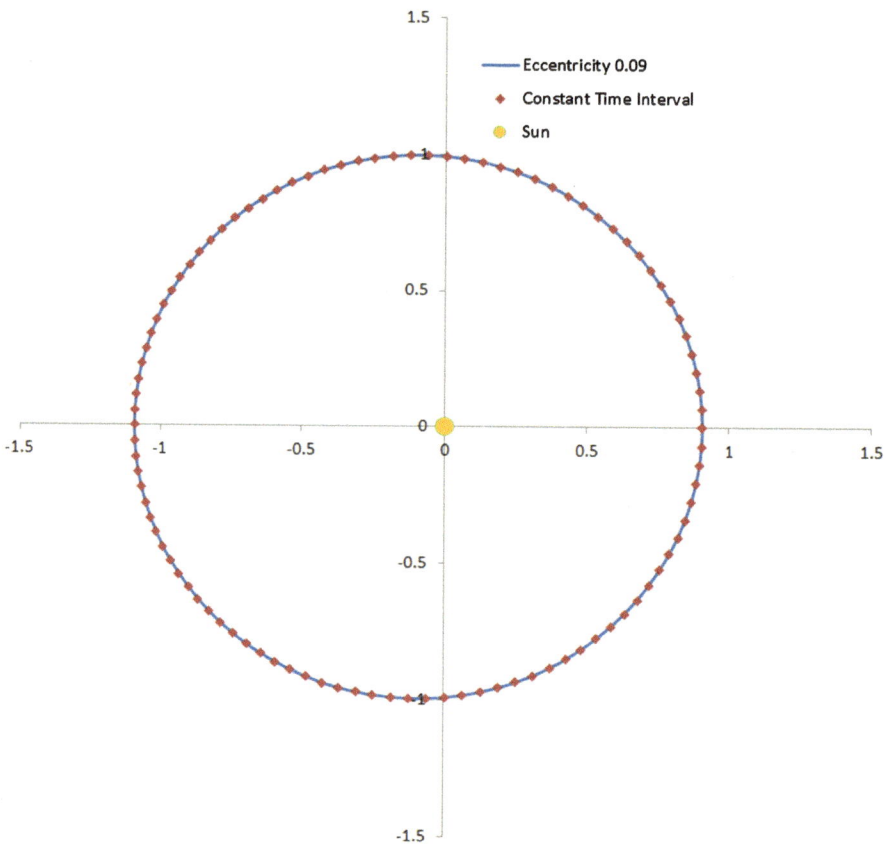

Fig. 3.11 Showing an elliptical orbit with eccentricity 0.09, with arbitrary coordinates, whose origin is at the centre of the Sun. The red marks are spaced such that the time interval between neighbouring tick marks is always the same. The Sun is noticeably off-centre. At the perihelion on the right, it is unmistakable that the tick marks are spaced further apart than at the aphelion on the left

where we have used Equation (3.75) for the area of an ellipse and Eq. (2.6) to substitute for b.

We know from Fig. 2.8 that when $\theta - \varpi = 0$, Eq. (3.60) becomes

$$r = \frac{h^2}{\mu(1+e)} = a(1-e).$$ (3.85)

We can manipulate Eq. (3.85) to give

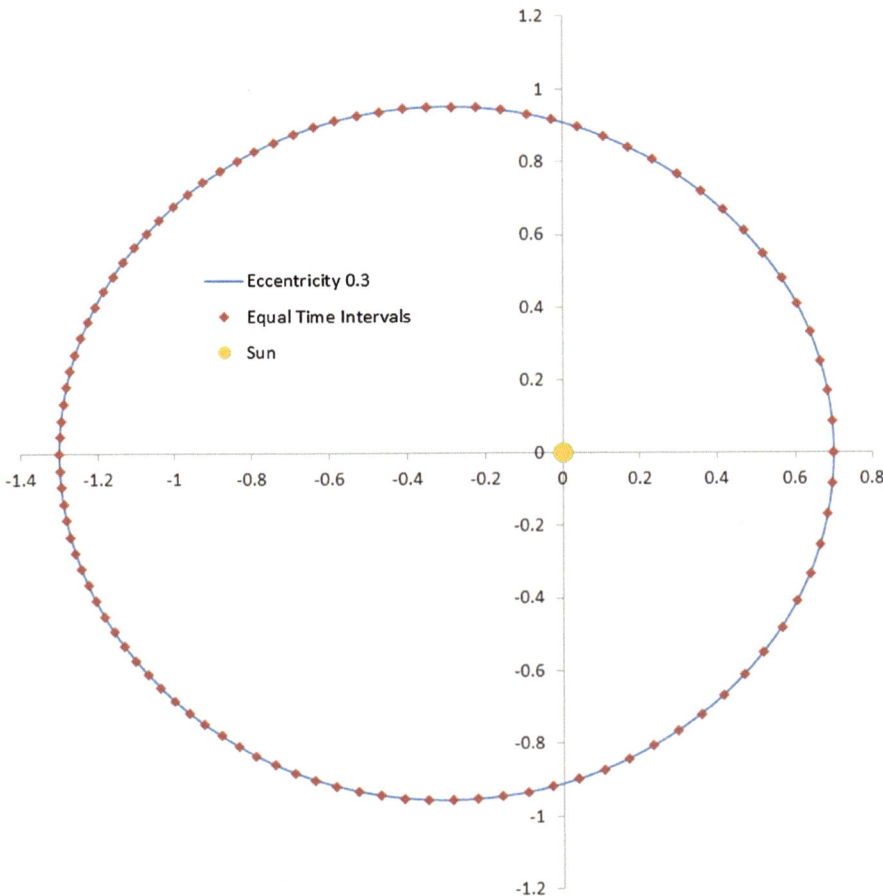

Fig. 3.12 Showing an elliptical orbit with eccentricity 0.3, with arbitrary coordinates, whose origin is at the centre of the Sun. The red marks are spaced such that the time interval between neighbouring tick marks is always the same. The Sun is very clearly off-centre. At the perihelion on the right, it is easy to see that the tick marks are spaced further apart than at the aphelion on the left

$$\frac{h^2}{\mu} = a(1-e)(1+e)$$

$$= a(1-e^2).$$

$$h = \sqrt{\mu a(1-e^2)}.$$

(3.86)

Combining Eqs. (3.84) and (3.86) gives

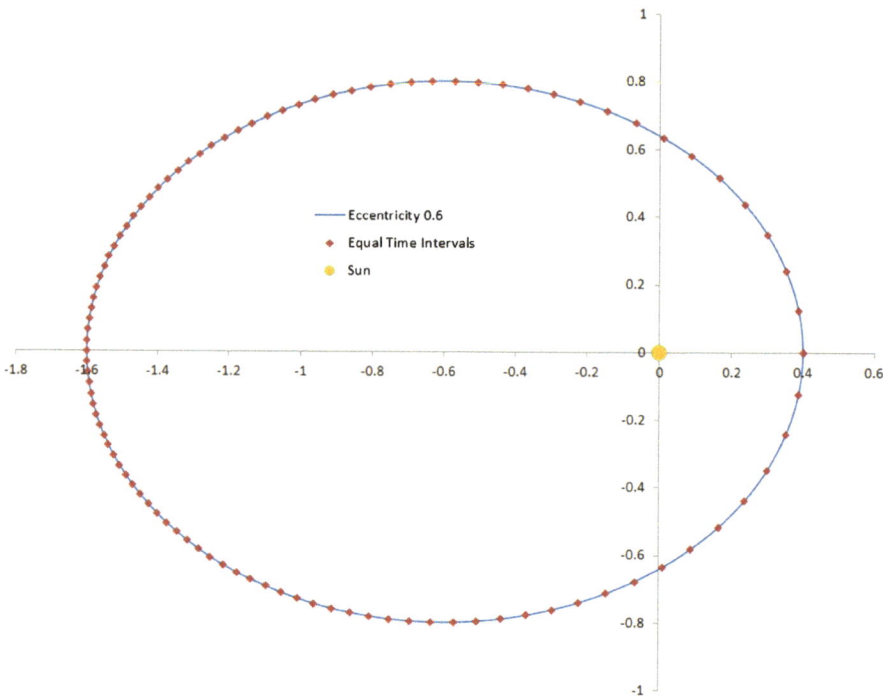

Fig. 3.13 Sowing an elliptical orbit with eccentricity 0.6, with arbitrary coordinates, whose origin is at the centre of the Sun. The red marks are spaced such that the time interval between neighbouring tick marks is always the same. The Sun is far away from the centre. At the perihelion on the right, the tick marks are spaced more than twice as far apart as those at the aphelion on the left. The planet is going much faster at periphelion than at aphelion

$$\frac{hT}{2} = \pi a^2 \sqrt{(1-e^2)} = \frac{T}{2} \sqrt{\mu a (1-e^2)}. \tag{3.87}$$

Squaring

$$
\begin{aligned}
\pi^2 a^4 (1-e^2) &= \frac{T^2}{4} \mu a (1-e^2) \\
\frac{a^3}{T^2} &= \frac{\mu}{4\pi^2} \\
&\approx \frac{Gm_1}{4\pi^2} \\
&= \text{constant}
\end{aligned}
\tag{3.88}
$$

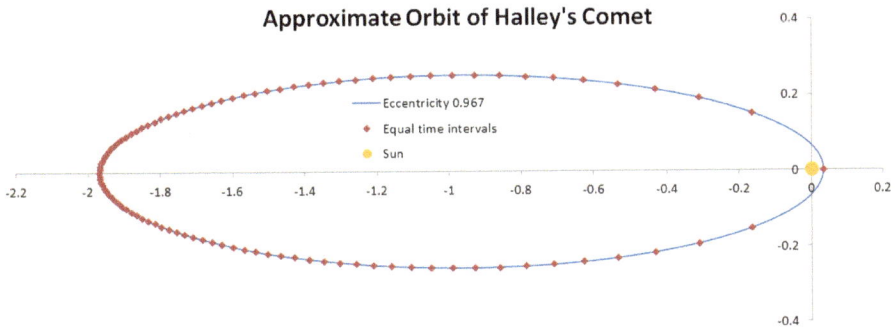

Fig. 3.14 Sowing an elliptical orbit with eccentricity 0.967, with arbitrary coordinates, whose origin is at the centre of the Sun. This ellipse is as eccentric as that describing the orbit of Halley's Comet around the Sun. The red marks are spaced such that the time interval between neighbouring tick marks is always the same. The Sun is now much closer to the aphelion than to the perihelion. At the perihelion on the right, it is very obvious that the tick marks are spaced further apart than at the aphelion on the left. Indeed, by counting tick-marks, you can see that the comet spends almost 90% of its time further to the left of the point -1 on the major axis, i.e. further from the Sun than -1

for a planet orbiting m_1. We have used Eq. (3.22). Since Eq. (3.88) is Kepler's third law, we have now derived all of Kepler's laws of planetary motion.

We have also demonstrated that the orbital period does not depend on eccentricity: it only depends on the semi-major axis a. This striking fact is illustrated in Fig. 3.15.

We have now seen that Kepler's laws of planetary motion follow from Newton's laws of motion and his law of universal gravitation. Kepler's hunch that his laws had a physical cause turned out to be justified.

While we are here, we are in a position quickly to derive one result that we will need in Chapter 6, viz. the time derivative of the mean anomaly. In radians, the mean anomaly M at time T, i.e. after one orbit, is 2π. Since the mean anomaly advances at a constant rate, we can write that

$$M = M_0 + \frac{2\pi}{T} t \qquad (3.89)$$

Where M_0 is a constant. Therefore, using Eq. (3.88),

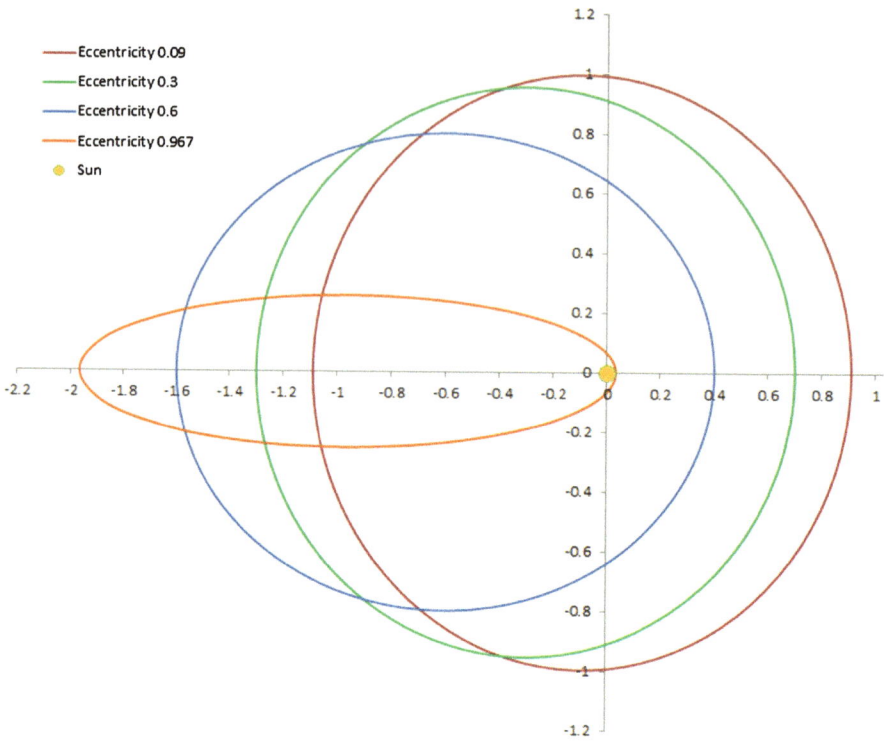

Fig. 3.15 Bodies orbiting in each of the four ellipses shown, with $a = 1$ in the units shown, would all have the same orbital period. (This assumes that none of the masses is significant compared to that of the Sun)

$$
\begin{aligned}
\frac{dM}{dt} &= \frac{2\pi}{T} \\
&= \frac{2\pi}{2\pi}\sqrt{\mu}\; a^{-3/2} \\
&= \sqrt{\mu}\; a^{-3/2}.
\end{aligned}
\tag{3.90}
$$

The Moon: A Challenge for Newtonian Physics

The difficulty is not hard to see: the Earth's Moon is significantly acted upon by two bodies: the Earth and the Sun. Compared to its mother planet, it is big enough that, unlike, say, Jupiter's moons, it does not orbit in a simple ellipse, with a single overwhelmingly dominant attracting body. This makes the Earth's Moon a three-body problem. Newton was not able to solve this – indeed it took a further 200 years to obtain a satisfactory model of the Moon's orbit [87] – but he made enough progress to show that this apparent counter-example did not disprove his theory.

Indeed Newton was wont to tell Halley, whenever the latter chased his book-writing progress, to say that the Moon "made his head ache, and kept him awake so often, that he would think of it no more" [88].

The story of the Moon's orbit is a long one, which is beyond the scope of this book. Interested readers are referred to Danby [87] or, for an account of Newton's treatment, Chandrasekhar [82] and the *Principia* itself [79].

Canonical Units

This is a very fancy name for a simple concept, designed to make life a bit simpler as we consider orbits other than that of the Earth.

In Eq. (3.88) we set

$$Gm_1 \approx \mu = 1 \tag{3.91}$$

so that for the Earth, symbol $\oplus$, where

$$a = 1 \text{ AU }, \tag{3.92}$$

$$\frac{a_\oplus^3}{T_\oplus^2} \approx \frac{Gm_\odot}{4\pi^2}$$

$$= \frac{\mu}{4\pi^2} \tag{3.93}$$

$$\frac{1}{4\pi^2} = \frac{(1\,\text{AU})^3}{T_\oplus^2}$$

$$T_\oplus = 2\pi.$$

In Eq. (3.93), the symbol $\odot$ stands for the Sun, as is usual.

If we were to imagine a circular orbit with radius 1 AU, a planet in this orbit would travel a distance $2\pi a$ in a time 2π, so that its orbital speed would be

$$|\mathbf{v}| = \frac{2\pi a}{2\pi} = a = 1. \tag{3.94}$$

Thus, the advantage of this set of units is that distance, time and speed are all of the same order of magnitude. In digital numerical computation, this would reduce round-off errors compared to using other systems of units such as SI units.

The unit of angle in a system that makes an Earth year equal to 2π time units can only be the radian.

The angular momentum per unit mass of the Earth in these units

$$h_\oplus = \left| \mathbf{r}_\oplus \times \mathbf{v}_\oplus \right| = \left| 1 \times 1 \right| = 1. \tag{3.95}$$

It should be pointed out that the length of a year for the Earth varies measurably on a timescale of centuries, by about one part in 10^7 per century. The value of the astronomical unit is customarily adjusted slightly to allow for this and keep the length of an Earth year equal to 2π in canonical units [89]. Such a level of precision is far beyond that necessary for this book. Indeed, the age of the Solar system is often reckoned to be of the order of 4.6×10^9 years $= 4.6 \times 10^7$ centuries [89], so that the change of year length is only large on a timescale of the order of the age of the Solar System.

Energy

We do need to think about the energy of an orbit in order to obtain some useful formulae.

First, let us recall Eq. (3.14)

$$\frac{d^2\mathbf{r}}{dt^2} + \frac{\mu}{r^3}\mathbf{r} = 0 \tag{3.14}$$

We take the scalar product of this with $d\mathbf{r}/dt$ and use the fact, noted earlier, that

$$\frac{d\mathbf{r}}{dt} = \frac{dr}{dt}\hat{\mathbf{r}} + r\frac{d\theta}{dt}\hat{\boldsymbol{\theta}}. \tag{3.12}$$

Then, since anything multiplied by zero is zero

$$
\begin{aligned}
\frac{d\mathbf{r}}{dt} \bullet \frac{d^2\mathbf{r}}{dt^2} + \mu \frac{d\mathbf{r}}{dt} \bullet \frac{\mathbf{r}}{r^3} &= \frac{d\mathbf{r}}{dt} \bullet \frac{d^2\mathbf{r}}{dt^2} + \mu \left(\frac{dr}{dt}\hat{\mathbf{r}} + \frac{d\theta}{dt}\hat{\boldsymbol{\theta}} \right) \bullet \frac{\hat{\mathbf{r}}}{r^2}. \\
&= \frac{d\mathbf{r}}{dt} \bullet \frac{d^2\mathbf{r}}{dt^2} + \frac{\mu}{r^2}\frac{dr}{dt} \\
&= 0.
\end{aligned}
\tag{3.96}
$$

Since the scalar product of anything with zero is zero, Eq. (3.96) is equal to zero. It can quickly be integrated, more by using slight cunning than by applying brute force. First, notice that

$$\frac{d}{dt}\left(\frac{d\mathbf{r}}{dt}\bullet\frac{d\mathbf{r}}{dt}\right) = \frac{d\mathbf{r}}{dt}\bullet\frac{d^2\mathbf{r}}{dt^2}+\frac{d^2\mathbf{r}}{dt^2}\bullet\frac{d\mathbf{r}}{dt}$$

$$= 2\frac{d\mathbf{r}}{dt}\bullet\frac{d^2\mathbf{r}}{dt^2}. \tag{3.97}$$

Second, notice that

$$\frac{d}{dt}\left(\frac{\mu}{r}\right) = \frac{d}{dr}\left(\frac{\mu}{r}\right)\frac{dr}{dt}$$

$$= -\frac{\mu}{r^2}\frac{dr}{dt}. \tag{3.98}$$

With these little facts, we can see that if we integrate Eq. (3.96) with respect to time, we get

$$\int\left(\frac{d\mathbf{r}}{dt}\bullet\frac{d^2\mathbf{r}}{dt^2}+\frac{\mu}{r^2}\frac{dr}{dt}\right)dt = \frac{1}{2}\left(\frac{dr}{dt}\right)^2-\frac{\mu}{r}-C=0, \tag{3.99}$$

i.e.

$$\frac{1}{2}\left(\frac{dr}{dt}\right)^2-\frac{\mu}{r}=C, \tag{3.100}$$

where C is a constant of integration. The textbooks delphically tell us that Eq. (3.100) is known by its Latin name the "vis viva" equation, where "vis viva" is pronounced "weess wee-wa". It is informative to multiply Eq. (3.100) through by m_2 to give

$$\frac{m_2}{2}\left(\frac{dr}{dt}\right)^2-\frac{\mu m_2}{r} = m_2 C$$

$$= \frac{m_2}{2}\left(\frac{dr}{dt}\right)^2-\frac{G(m_1+m_2)m_2}{r} \tag{3.101}$$

$$\approx \frac{m_2}{2}\left(\frac{dr}{dt}\right)^2-\frac{Gm_1m_2}{r}.$$

We have used the fact that for the Sun and a rocky planet, $m_1\gg m_2$. The first term of Eq. (3.101) is the kinetic energy of the orbiting body, and the second term can be interpreted as the potential energy of this body in the Sun's gravitational field.

We deduce the potential energy as follows. The work done to move the planet away from the Sun to infinity is obtained by integrating infinitesimal elements of

$$\text{Work done} = \text{Force applied} \times \text{Distance over which it is applied.}$$

$$
\begin{aligned}
V &= \int_r^\infty -\frac{Gm_1 m_2}{\xi^2} \hat{\mathbf{r}} \cdot \hat{\mathbf{r}} d\xi \\
&= \left[\frac{Gm_1 m_2}{\xi} \right]_r^\infty \\
&= \frac{-Gm_1 m_2}{r} \\
&\approx -\frac{\mu}{r} m_2,
\end{aligned}
\tag{3.102}
$$

in which ξ is a dummy variable of integration. The minus sign in the first line appears because the work is being done in the opposite direction to the gravitational pull.

Collecting terms in Eq. (3.100), C is the total energy of per unit mass of the orbiting planet.

We can go a step further and deduce the value of C. At the perihelion of the orbit, we know that

$$r = a(1-e). \tag{3.103}$$

We can see by inspection that this is true in Fig. 3.6, for example. Furthermore, the apses, perihelion and aphelion, if the Sun is at the attracting focus, are places where the radius and velocity are perpendicular to one another.

The angular momentum per unit mass

$$
\begin{aligned}
|\mathbf{h}| &= \left| \mathbf{r} \times \frac{d\mathbf{r}}{dt} \right| \\
&= r \frac{dr}{dt} \sin \frac{\pi}{2} \\
&= a(1-e) \frac{dr}{dt} \sin \frac{\pi}{2} \\
&= a(1-e) \frac{dr}{dt} \\
&= \sqrt{\mu a(1-e^2)}
\end{aligned}
\tag{3.104}
$$

The last line of Eq. (3.104) is true by virtue of Equation (3.62). Squaring,

$$\mu a\left(1-e^2\right) \quad = \quad a^2\left(1-e\right)^2\left(\frac{dr}{dt}\right)^2$$

$$\mu\left(1+e\right)\left(1-e\right) = a\left(1-e\right)\left(1-e\right)\left(\frac{dr}{dt}\right)^2 \qquad (3.105)$$

$$\frac{\mu\left(1+e\right)}{a\left(1-e\right)} \quad = \quad \left(\frac{dr}{dt}\right)^2.$$

Substituting this value into Eq. (3.100) at the perihelion gives

$$\frac{1}{2}\left(\frac{\mu\left(1+e\right)}{a\left(1-e\right)}\right) - \frac{\mu}{r} \quad = \quad C$$

$$C \quad = \quad \frac{1}{2}\left(\frac{\mu\left(1+e\right)}{a\left(1-e\right)}\right) - \frac{\mu}{a\left(1-e\right)}$$

$$= \quad \frac{1}{2}\left(\frac{\mu\left(1+e\right)-2\mu}{a\left(1-e\right)}\right)$$

$$= \quad \frac{1}{2}\left(\frac{\mu\left(1+e\right)-2\mu}{a\left(1-e\right)}\right) \qquad (3.106)$$

$$= \quad \frac{1}{2}\left(\frac{\mu\left(e-1\right)}{a\left(1-e\right)}\right)$$

$$= \quad \frac{-\mu}{2a}.$$

Thus, our vis viva equation becomes

$$\frac{1}{2}\left(\frac{dr}{dt}\right)^2 - \frac{\mu}{r} \quad = \quad -\frac{\mu}{2a}$$

$$\left(\frac{dr}{dt}\right)^2 \quad = \quad \frac{2\mu}{r} - \frac{\mu}{a} \qquad (3.107)$$

$$\left(\frac{dr}{dt}\right)^2 \quad = \quad \mu\left(\frac{2}{r} - \frac{1}{a}\right).$$

It will turn out that Eq. (3.107) is a very convenient way to evaluate the semimajor axis a once both the radius and the velocity are known at any point along the orbit.

So where did the term "vis viva" come from? It comes from early in the history of scientists' attempts to wrap their heads around the concept of energy. It was actually coined by Leibniz, the co-inventor of calculus with Newton, who, it seems, first derived Eq. (3.100), which was used by Johann Bernoulli, a member of a famous family of mathematical physicists, who championed the ideas of Leibniz. He stumbled upon the idea that there was something important about mass x velocity squared and noticed that it was conserved in certain mechanical systems. The English scientists unwisely pooh-poohed this idea in favour of the law of conservation of momentum, which they had discovered. The English and Leibniz were both right, and everyone clearly would have done better not to play the "not invented here" game.

Anyway, Leibniz called the quantity mass x velocity squared the "vis viva" or "life force" of the body.

Leibniz himself contributed to an astonishingly wide range of subjects including philosophy, metaphysics, mathematics, physics, computation, law, psychology, and worked as a diplomat.

From there, the history of the development of energy goes onto the European continent for a while. Willem's Gravesande of Leiden did a series of experiments in which he dropped brass balls into soft clay and measured the depth of the dent produced. He found that this depth was proportional to the vis viva. Similar observations were independently published by Giovanni Poleni, an Italian aristocrat. 's Gravesande also worked as a diplomat, which is how he came into contact with the English scientific establishment.

From here, the story takes another turn. Émilie du Châtelet, whose translation of Newton's Principia into French is still the standard, was the first to postulate that the total energy of systems was constant over time. She was a French aristocrat who took the lax view of marriage vows common to the European aristocracy, had three children by her husband, and became the mistress of Voltaire, who strongly encouraged her scientific work. Unfortunately, she did not survive a late pregnancy.

It took a further 100 years for the concept of potential energy to be expressed, by one of the early thermodynamicists based at the University of Glasgow, which also housed William Thomson, later ennobled as Lord Kelvin (Fig. 3.16).

The Orbit Tells Us What the Force Law Is

What happens if we pose the question the other way round? In the above account, I told you what Newton's law of universal gravitation was, and then derived the orbit as an ellipse. If instead, I start with an elliptical orbit,

Gottfried Wilhelm (von) Leibniz 1646-1716	Willem Jacob 's Gravesande 1688-1742	Johann Bernoulli 1667-1748	Gabrielle Émilie Le Tonnelier de Breteuil, Marquise du Châtelet, 1706-49	William John Macquorn Rankine 1820 – 1872

Fig. 3.16 A "rogue's gallery" of the people who groped towards the concept of energy as a transferable, but conserved, quantity. (All images in public domain)

or more generally a conic-sectional orbit – circle, ellipse, parabola or hyperbola – what can I deduce about the law of gravity? This was not a historically trivial question, because some of Newton's contemporaries, notably Huyghens, did not immediately buy the argument about the law of universal gravitation [77].

Lest anyone should think that this was silly of Huyghens, I might diplomatically remind them that we have had similar debates in our own time. One response to the challenge of missing mass in galaxies and galaxy clusters was to question whether Newtonian gravity applies on a distance scale of galaxies. Newtonian gravity explains very well the orbital behaviour of the Solar System, and of binary stars, as long as neither of the pair is compact enough to be relativistic – i.e. a white dwarf, a neutron star or a black hole – but what about on the much bigger scale of a whole galaxy or a cluster of galaxies? The Israeli astrophysicist Mordechai Milgrom developed a theory of modified Newtonian dynamics (MOND) to try to avoid invoking invisible dark matter [91]. In fact, dark matter seems to be winning the argument [92], but Milgrom asked a very reasonable question.

Incidentally, my comment about relativistic gravitational fields only applies close to the surface of a white dwarf in a binary pair. For example, the mutual orbit of the brightest star in our sky Sirius A and its white dwarf companion Sirius B is well known to be accurately described by Newtonian dynamics [93]. The mutual orbit of the pair is elliptical, but Eq. (3.22) does not apply, as the masses of the two components are of similar masses. They orbit about a common centre of mass.

I have followed the treatment of Danby [87], except that I have removed one major irritant. Danby spreads this treatment throughout his book, which means that it takes you ages to find your way through the argument, and you keep having to stop and go back 20 pages. The difficulty of doing this reminds me of the story that stars form in clusters, which eventually break up and scatter through the galaxy. Our own star presumably so formed [90]. People have tried to trace back and find other stars which are chemically so similar that they would almost certainly have condensed from the same nebula. Using some of the best kit we have, to date exactly one such star has so far been identified as a probable co-evolver with the Sun [94]. Finding my way through Danby similarly felt like looking for needles in a haystack. I have therefore collected the entire argument together here. I figured that repeating a couple of equations was a lesser sin than scattering the argument to the four winds.

Let me assume that the potential energy V due to my force law only depends on the line between the planet and the Sun

$$\frac{\partial V}{\partial r} = q(r)$$

$$\frac{1}{r}\frac{\partial V}{\partial \phi} = 0.$$

(3.108)

$q(r)$ is a function of r but not the angle ϕ, as shown in Fig. 3.17.

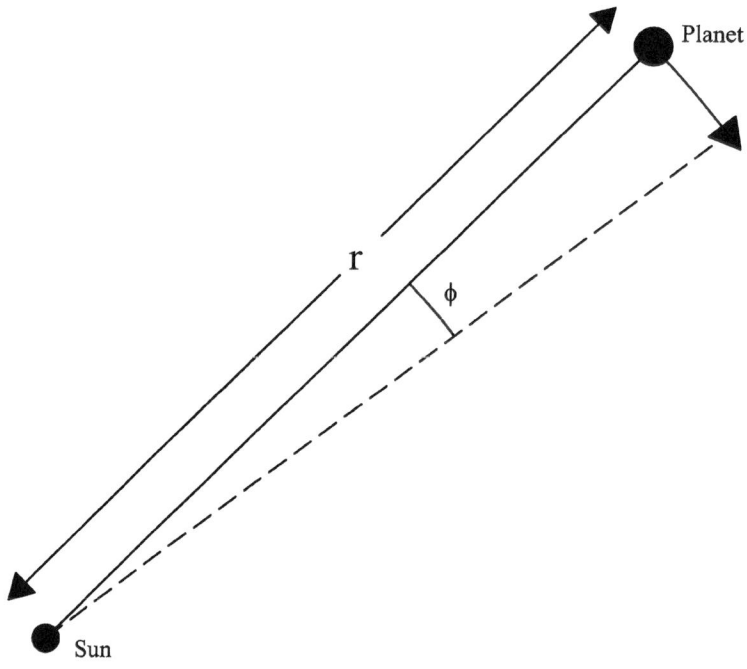

Fig. 3.17 A planet orbiting in a force field whose potential energy per unit mass does not depend on ϕ but does depend on r

Now let me poke around a little and draw out some properties of my potential energy.

By integrating Eq. (3.108), know in principle that

$$V(r) = \int q(r) \; dr \qquad (3.109)$$

where for now I have not given limits and made Eq. (3.109) a definite integral.

The point about setting up a potential energy with no ϕ-dependence is that the planet does not pick up energy during every orbit and hence (in classical mechanics at least) building up tangential velocity indefinitely. Thus, after the next orbit, the planet will have the same kinetic energy as it does now. Such a potential field is called *conservative* in the jargon, because it is not feeding kinetic energy to a planet in a stable orbit. This term has nothing to do with politics. We can use it quite liberally in this sense without fear of causing offence. Furthermore, it is a pretty safe bet that our law of gravity is not going to be non-conservative. We have not seen the planets being sped up in several millennia of observation since the Babylonians (for example) began recording their movements. Our predecessors more than 100 years ago were able to detect a 43 arcsec/century precession of the axes of the orbit of Mercury, which eventually turned out to be due to the correction for general relativity. They'd have picked up a non-conservative gravitational potential.

The equation of motion in my potential would be

$$\frac{\partial^2 \mathbf{r}}{\partial t^2} = -\frac{\partial V}{\partial r} \hat{\mathbf{r}} = -q(r) \; \hat{\mathbf{r}}. \qquad (3.110)$$

In other words

$$\frac{\partial^2 \mathbf{r}}{\partial t^2} + \frac{\partial V}{\partial r} \; \hat{\mathbf{r}} = \mathbf{0} \; . \qquad (3.111)$$

I can take the dot product of both sides of Eq. (3.111) with anything I like, since it is zero. I then obtain

$$\frac{d\mathbf{r}}{dt} \cdot \frac{\partial^2 \mathbf{r}}{\partial t^2} + \frac{d\mathbf{r}}{dt} \cdot \frac{\partial V}{\partial r} \; \hat{\mathbf{r}} = \mathbf{0} \; . \qquad (3.112)$$

I told you above, but will repeat to save you leafing through pages, that

$$\frac{d}{dt}\left(\frac{d\mathbf{r}}{dt} \bullet \frac{d\mathbf{r}}{dt}\right) = \frac{d\mathbf{r}}{dt} \bullet \frac{d^2\mathbf{r}}{dt^2} + \frac{d^2\mathbf{r}}{dt^2} \bullet \frac{d\mathbf{r}}{dt}$$

$$= 2\frac{d\mathbf{r}}{dt} \bullet \frac{d^2\mathbf{r}}{dt^2}. \tag{3.97}$$

Furthermore,

$$\frac{d\mathbf{r}}{dt} \bullet \frac{\partial V}{\partial r}\hat{\mathbf{r}} = \frac{d\mathbf{r}}{dt}\frac{\partial V}{\partial r}. \tag{3.113}$$

This enables me to do an integration by substitution of the term containing $V(r)$.

Collecting terms from Eqs. (3.97) and (3.113) into Eq. (3.112) and integrating gives

$$\frac{1}{2}\left(\frac{d\mathbf{r}}{dt} \bullet \frac{d\mathbf{r}}{dt}\right) + V(r) = C, \tag{3.114}$$

where C is a constant of integration. I can make some immediate statements because the leftmost term of Eq. (3.114) cannot be negative. I can choose C to be zero when $r \to \infty$, and then call C the energy of the orbit. If $C > 0$, and $V(r)$ never changes sign, then the velocity will tend to a finite limit when $r \to \infty$, and the planet can escape into the Galaxy. The minimum value of C occurs when

$$\left(\frac{d\mathbf{r}}{dt} \bullet \frac{d\mathbf{r}}{dt}\right) = 0, \tag{3.115}$$

in other words when the magnitude of the velocity is constant. This state of affairs occurs when the orbit is circular. If there is a very low level of damping external to the equations discussed here, orbits will circularize and then spiral inwards. We see this with many of the moons around gas giants. (Other mechanisms involving tides tend to push the moons out, so that they don't spiral inwards. These are beyond the scope of this book.) I suspect that ultimately this is one why planetary orbits in the Solar System don't have high eccentricities. Another reason is discussed in Chap. 6.

Anyway, I digress. Above, I proved that

$$\frac{d^2\mathbf{r}}{dt^2} = \frac{d}{dt}\left(\frac{dr}{dt}\hat{r}\right) + \frac{d}{dt}\left(r\hat{\theta}\frac{d\theta}{dt}\right)$$

$$= \frac{d^2r}{dt^2}\hat{r} + \frac{dr}{dt}\frac{d\hat{r}}{dt} + r\hat{\theta}\frac{d^2\theta}{dt^2} + r\frac{d\hat{\theta}}{dt}\frac{d\theta}{dt} + \frac{dr}{dt}\hat{\theta}\frac{d\theta}{dt}$$

$$= \frac{d^2r}{dt^2}\hat{r} + \frac{dr}{dt}\hat{\theta}\frac{d\theta}{dt} + r\hat{\theta}\frac{d^2\theta}{dt^2} - r\hat{r}\frac{d\theta}{dt}\frac{d\theta}{dt} + \frac{dr}{dt}\hat{\theta}\frac{d\theta}{dt}$$

$$= \frac{d^2r}{dt^2}\hat{r} - r\hat{r}\frac{d\theta}{dt}\frac{d\theta}{dt} + \frac{dr}{dt}\hat{\theta}\frac{d\theta}{dt} + r\hat{\theta}\frac{d^2\theta}{dt^2} + \frac{dr}{dt}\hat{\theta}\frac{d\theta}{dt}$$

$$= \left(\frac{d^2r}{dt^2} - r\frac{d\theta}{dt}\frac{d\theta}{dt}\right)\hat{r} + \left(\frac{dr}{dt}\frac{d\theta}{dt} + r\frac{d^2\theta}{dt^2} + \frac{dr}{dt}\frac{d\theta}{dt}\right)\hat{\theta}. \qquad (3.13)$$

I can use this result to resolve Eq. (3.110) into components. In the radial direction

$$\ddot{r} - r\dot{\theta}^2 = -q(r) \ . \qquad (3.116)$$

In the transverse direction

$$r\ddot{\theta} + 2\dot{r}\dot{\theta} = 0. \qquad (3.117)$$

If I integrate Equation (3.117) by noticing that the left-hand side looks like the derivative of $r^2\dot{\theta}$, then I get

$$r^2\dot{\theta} = h_a \ , \qquad (3.118)$$

where the symbol h_a is given the subscript a to indicate that I am still dealing with an arbitrary potential.

Above, I also proved that

$$\frac{dr}{dt} = \frac{dr}{dt}\hat{r} + r\hat{\theta}\frac{d\theta}{dt}. \qquad (3.12)$$

If I take the dot product of Eq. (3.12) with itself, I obtain from Eq. (3.114) that

$$\frac{1}{2}\left(\dot{r}^2 + r^2\dot{\theta}^2\right) + V(r) = C \ . \qquad (3.119)$$

This is the energy equation for our arbitrary potential. I need this in a slightly different form. Let me write

$$\frac{dr}{d\theta} = \frac{dr/dt}{d\theta/dt} = \frac{\dot{r}}{\dot{\theta}} \ . \qquad (3.120)$$

I can use Eq. (3.120) to eliminate time derivatives from Eqs. (3.118) and (3.119).

$$\frac{1}{2}\left(\dot{r}^2 + r^2\dot{\theta}^2\right) + V(r) \quad = \quad C$$

$$\frac{1}{2}\left(\dot{\theta}^2\left(\frac{dr}{d\theta}\right)^2 + r^2\dot{\theta}^2\right) \quad = \quad C - V(r)$$

$$\frac{1}{2}\left(\frac{h_a^2}{r^4}\left(\frac{dr}{d\theta}\right)^2 + r^2\frac{h_a^2}{r^4}\right) \quad = \quad C - V(r) \qquad (3.121)$$

$$\frac{h_a^2}{2r^4}\left(\frac{dr}{d\theta}\right)^2 \quad = \quad C - V(r) - \frac{h_a^2}{2r^2}$$

$$\left(\frac{dr}{d\theta}\right)^2 \quad = \quad \frac{2r^4}{h_a^2}\left[C - V(r) - \frac{h_a^2}{2r^2}\right].$$

Now, Let us suppose that we have an orbit which we know to be elliptical, with the Sun at one focus. That's where Kepler got to. We already know what such an equation looks like in polar coordinates: We showed in Chapter 2 that if θ is the true anomaly of the ellipse, then

$$r = \frac{a\left(1 - e^2\right)}{\left(1 + e\cos\theta\right)} \qquad (3.122)$$

From Eq. (3.122), it follows that

$$\frac{a\left(1 - e^2\right)}{r} = 1 + e\cos\theta \ . \qquad (3.123)$$

For convenience, let

$$p = a\left(1 - e^2\right). \qquad (3.124)$$

Now, in Eq. (3.123)

$$p\frac{d\left(1/r\right)}{d\theta} = -e\sin\theta$$

$$p\frac{d\left(1/r\right)}{dr}\frac{dr}{d\theta} = -e\sin\theta \qquad (3.125)$$

$$p\frac{-1}{r^2}\frac{dr}{d\theta} = -e\sin\theta \ .$$

Therefore,

$$\left(\frac{dr}{d\theta}\right)^2 = \frac{r^4 e^2}{p^2}\sin^2\theta$$

$$= \frac{r^4 e^2}{p^2}\left(1 - \frac{1}{e^2}\left[\frac{p}{r} - 1\right]^2\right) \tag{3.126}$$

$$= \frac{r^4}{p^2}\left(e^2 - \left[\frac{p}{r} - 1\right]^2\right).$$

I can plug this into Eq. (3.121) and get

$$\frac{r^4}{p^2}\left(e^2 - \left[\frac{p}{r} - 1\right]^2\right) = \frac{2r^4}{h_a^2}\left[C - V(r) - \frac{h_a^2}{2r^2}\right]$$

$$\frac{h_a^2}{2p^2}\left(e^2 - \frac{p^2}{r^2} - 1 + \frac{2p}{r}\right) + \frac{h_a^2}{2r^2} = C - V(r)$$

$$\frac{h_a^2(e^2 - 1)}{2p^2} - \frac{h_a^2}{2r^2} + \frac{h_a^2}{pr} + \frac{h_a^2}{2r^2} = C - V(r) \tag{3.127}$$

$$\frac{h_a^2(e^2 - 1)}{2p^2} + \frac{h_a^2}{pr} = C - V(r).$$

The first term on each side of Eq. (3.127) is constant. If we specify that as $r \to \infty$, the potential tends to be zero, then

$$C = \frac{h_a^2(e^2 - 1)}{2p^2} \tag{3.128}$$

and

$$\frac{h_a^2}{p}\left(\frac{1}{r}\right) = V(r). \tag{3.129}$$

That is, by imposing an ellipse as the orbit, we end up with a potential proportional to $(1/r)$. This is where we were with Eq. (3.102), except that in that case V was potential energy, i.e. potential × mass of the orbiting body.

Newton proved this by a different method, and also proved some other theorems, viz. that a potential proportional to r gives an ellipse orbiting about the centre, not the focus, and that a potential proportional to $(1/r^2)$ gives a spiral orbit. Danby gives details [87].

Chapter 4

Least Squares Fit to Sets of Equations

This is a method that we will need repeatedly. It is therefore put into a stand-alone chapter to make it easier to see that you don't need to read whole other chapters to pick up this information.

The least squares idea is originally due to Gauss. He applies it to astronomical observations in his write-up of the determination of the orbit of Ceres [95]. This technique is so important that it has been developed considerably since Gauss' time.

Consider a set of equations

$$
\begin{aligned}
t_i &\approx A_{i1}s_1 + \cdots + A_{im}s_m \quad \text{for } i = 1 \ldots n \\
&= \sum_{j=1}^{m} A_{ij}s_j \quad \text{for } i = 1 \ldots n,
\end{aligned}
\tag{4.1}
$$

Because Eq. (4.1) is approximate, taking the t_i over to the right-hand side will leave nonzero residuals:

$$
r_i = \sum_{j=1}^{m} A_{ij}s_j - t_i \quad \text{for } i = 1 \ldots n.
\tag{4.2}
$$

A best fit is reached when some measure of the r_i is minimized. The least squares minimum would be

$$
S = \sum_{i=1}^{n} r_i^2.
\tag{4.3}
$$

J. Clark, *Calculate the Orbit of Mars!*,
https://doi.org/10.1007/978-3-030-78267-2_4,
© Springer Nature Switzerland AG 2021

I don't know about you, but when I think of minimizing quantities, I think of differentiating and setting something equal to zero. In this instance, I'm going to show you a simple, and very cunning, method from the early days of digital computing, due to Wallace Givens [96], which avoids the need to differentiate.

What Givens did was to transform the set of Eq. (4.2) into a set where the terms towards the lower left are populated with zeros if the equations are written in the form

$$
\begin{aligned}
r_1' &= A_{11}'s_1 + A_{12}'s_2 + A_{13}'s_3 + \cdots + A_{1j}'s_j + \cdots + A_{1m}'s_m - t_1' \\
r_2' &= 0 + A_{22}'s_2 + A_{23}'s_3 + \cdots + A_{2j}'s_j + \cdots + A_{1m}'s_m - t_2' \\
r_3' &= 0 + 0 + A_{33}'s_3 + \cdots + A_{3j}'s_j + \cdots + A_{3m}'s_m - t_3' \\
&\cdots \\
r_j' &= 0 + 0 + 0 + \cdots + A_{jj}'s_j + \cdots + A_{jm}'s_m - t_j' \\
&\cdots \\
r_j' &= 0 + 0 + 0 + \cdots + 0 + \cdots + A_{mm}'s_m - t_j' \\
r_{j+1}' &= 0 + 0 + 0 + \cdots + 0 + \cdots + 0 - t_j' \\
&\cdots \\
r_n' &= 0 + 0 + 0 + \cdots + 0 + \cdots + 0 + 0 - t_n'
\end{aligned}
\tag{4.4}
$$

Let p and q be two real numbers such that

$$
p^2 + q^2 = 1.
\tag{4.5}
$$

Now, let

$$
\begin{aligned}
r_1' &= pr_1 - qr_2, \\
r_2' &= qr_1 + pr_2.
\end{aligned}
\tag{4.6}
$$

Then,

$$
\begin{aligned}
\left(r_1'\right)^2 + \left(r_2'\right)^2 &= p^2r_1^2 - 2pqr_1r_2 + q^2r_2^2 + q^2r_1^2 + 2pqr_1r_2 + p^2r_2^2 \\
&= p^2r_1^2 + q^2r_2^2 + q^2r_1^2 + p^2r_2^2 \\
&= \left(p^2 + q^2\right)\left(r_1^2 + r_2^2\right) \\
&= r_1^2 + r_2^2.
\end{aligned}
\tag{4.7}
$$

Equation (4.7) gives us a transformation such that the sum of the squares of the residuals, Eq. (4.3), does not change.

Now, let

$$h = \sqrt{A_{11}^2 + A_{21}^2},\tag{4.8}$$

and let

$$p = +\frac{A_{11}}{h},$$
$$q = -\frac{A_{21}}{h}.\tag{4.9}$$

The choice in Equations (4.8) and (4.9) satisfies Eq. (4.5). Furthermore, from Eq. (4.6)

$$\begin{aligned}
A_{21}' &= qA_{11} + pA_{21} \\
&= \frac{-A_{21}A_{11} + A_{11}A_{21}}{h} \\
&= 0.
\end{aligned}\tag{4.10}$$

We can exploit the fact that for $j > m$,

$$r_j' = t_j'\tag{4.11}$$

so that

$$S = \sum_{i=1}^{m} r_i'^2 + \sum_{i=m+1}^{n} t_i^2.\tag{4.12}$$

Only the first term on the right-hand side of Equation (4.12) is dependent on the unknown s_k. Since both terms on the right-hand side of Eq. (4.12) are positive, if the first sum is zero, then S is at a minimum and the least squares fit is found.

This happens when

$$s_m = \frac{t_m'}{A_{mm}'} \quad \text{and}$$

$$s_k = \frac{t_k - \sum_{l=k+1}^{m} A_{xl}' s_l}{A_{kk}'}.\tag{4.13}$$

The numerical algorithm works its way through and ensures that the conditions (4.13) are satisfied.

The numerical algorithm was coded by myself and Jeannette Fine into Fortran. The code is given below.

Least Squares Fitting Subroutine

Listed by permission of Dr. Jeannette M. Fine

Line numbers come from one particular instance where the subroutine was used.

```
336 SUBROUTINE LSQFIT (A,N,M,S)
337 !---------------------------------------------
338 ! LSQFIT: Per Equations 3.1 through 3.13
above
339 ! solution of an overdetermined system of
linear equations
340 ! A(i,1)*s(1)+...A(i,m)*s(m) - A(i,m+1) =    0
(i=1,..,n)
341 ! according to the method of least squares
using Givens rotations
342 ! A: matrix of coefficients
343 ! N: number of equations (rows of A)
344 ! M: number of unknowns (M+1=columns of A,
M=elements of S)
345 ! S: solution vector
346 !---------------------------------------------
347 INTEGER I,J,K,M,N
348 double precision A(1000,8),P,Q,H,H2,H3,
S(M),EPS
349
350 EPS = 1.0D-10
351 DO J=1,M ! loop over columns 1...M of A
352 ! eliminate matrix elements A(i,j) with i>j
from column j
353 DO I=J+1,N
354 IF (A(I,J).NE.0.0D0)THEN
355 ! calculate p, q and new A(j,j); set A(i,j)=0
356 IF (DABS(A(J,J)). LT.EPS*DABS(A(I,J))) THEN
357 P=0.0D0
358 Q=1.0D0
```

```
359 A(J,J)=-A(I,J)
360 A(I,J)=0.0D0
361 ELSE
362 H=DSQRT(A(J,J)*A(J,J)+A(I,J )*A(I,J))
363 IF (A(J,J).LT.0.0D0) H=-H
364 P=A(J,J)/H
365 Q=-A(I,J)/H
366 A(J,J)=H
367 A(I,J)=0.0D0
368 END IF
369 ! calculate rest of the line
370 DO K=J+1,M+1
371 H2 = P*A(J,K) - Q*A(I,K)
372 A(I,K) = Q*A(J,K) + P*A(I,K)
373 A(J,K) = H2
374 END DO
375 END IF
376 END DO
377 END DO
378 ! backsubstitution
379
380 DO I = M,1,-1
381 H3=A(I,M+1)
382 DO K=I+1,M
383 H3=H3-A(I,K)*S(K)
384 END DO
385 S(I) = H3/A(I,I)
386 END DO
387 !
388 RETURN
389 END SUBROUTINE LSQFIT
```

Chapter 5

The Orbit of the Earth

We need to know where the Earth is at any point in its journey around the Sun. Otherwise we don't know where the heck we are or where we are going. Were someone to claim that I conduct my life as if I neither know where I am nor where I'm going, I would have some difficulty defending myself against this wicked allegation. However, I have worked out where the Earth is and where it's going.

Believe it or not, with a little bit of cunning, this can be done from your sunrise and sunset times. I am about to tell the story of how this is done.

I have lived in places, like King's Lynn, Norfolk, England, where it is so flat that they have to look hard for somewhere to include a hill start in the driving test. There you can see sunrise and sunsets without much difficulty, apart from the dreaded clouds (Fig. 5.1). Where I live now, in Risca, South Wales, there is not much flat land. The locals will happily call anything more than a 1000 feet high a mountain. The highest nearby mountain, Pen-y-Fan, is a smidgin under 3000 feet high and is 30 miles away. This low altitude may cause Americans from mountainous states to scoff that these are not proper mountains. Nevertheless, they, and we, have the same problem: we can't directly witness either sunrises or sunsets, because of the hills and mountains (Fig. 5.2).

J. Clark, *Calculate the Orbit of Mars!*,
https://doi.org/10.1007/978-3-030-78267-2_5,
© Springer Nature Switzerland AG 2021

Fig. 5.1 Checking sunset times in a flat area is not without its pleasures. King's Lynn, 14 January 2009. (Image by the author. © Jane Clark)

Fig. 5.2 In hilly areas, you do not directly witness the sunset. Risca, Wales, 7 October 2018. (Image by the author © Jane Clark)

Indirect methods are available to check local sunset times. One simple method is that many GPS-based satellite navigation devices change the colour of the map background from white to black at the moment of local sunset and do the opposite at the moment of sunrise. I have checked this phenomenon against witnessing the sunrise many times, and my Garmin device knows if it has travelled a little to the east or west and still gets the rise and set times of the sun correct. There is a three-minute difference between the times at my home and my office, which is just across the border in Bristol, England. In late winter, I witness some glorious sunrises as I cross the Severn estuary, passing over a three-mile-long bridge from where you get a good view of the flood plain. The Garmin device gets the sunrise times right wherever on the journey I am.

Less precisely, you can see to within a few minutes when sunset is by watching the cloud patterns.

I claim no artistic merit for Figs. 5.3, 5.4, 5.5, 5.6, 5.7, 5.8, 5.9, 5.10 and 5.11. Their purpose is to show what happens to cloud patterns around sunset, or conversely, around sunrise. The glorious reds that you sometimes see only really occur right around the moment of sunset. If you look towards the sunset, an afterglow will remain for a few minutes after sunset, but in other directions, the clouds rapidly darken and turn blue just after sunset. That's how you can detect the time of sunrise or sunset with only your eyes.

Fig. 5.3 Looking north-west towards sunset. The Sun has disappeared behind the mountains, but we are still 6 minutes away from sunset. 24 May 2019 at 21:04. (Image by the author)

Fig. 5.4 Looking north-west towards sunset. This is about a minute after sunset where I was standing. 24 May 2019 at 21:11. You can see "sunbeams", which are actually shadows cast by the sun shining onto the undersides of the clouds and casting shadows. When looking towards the Sun, seeing sunbeams is a good indicator of the point of sunset. I have darkened the image to bring the sunbeams out. (Image by the author)

Fig. 5.5 Looking north-west towards sunset. Nine minutes after sunset. 24 May 2019 at 21:19. The sunset has "moved" on towards the west. (Image by the author)

Fig. 5.6 Looking south-east away from sunset, three minutes before local sunset. Looking this way the clouds are not illuminated by the sun to anything like the same extent as looking towards sunset. You can see that only the top sides of the bigger clouds are sunlit. The lower sides are in shadow. 24 May 2019 at 21:07. (Image by the author)

Fig. 5.7 Looking south-east away from sunset, at the local sunset time. The clouds are noticeably less sunlit than a few minutes before. 24 May 2019 at 21:10. (Image by the author)

Fig. 5.8 Looking south-east away from the Sun, 7 minutes after sunset. The clouds now show no reddening by sunlight. 24 May 2019 at 21:17. (Image by the author)

Fig. 5.9 Looking north, not quite at right angles to the sunset. The angle was constrained by the presence of trees at my vantage point, about 600 feet up. I actually live a couple of hundred feet further down the hill at whose summit I was. About 3 minutes before sunset, you can see that the bottoms of the clouds are in shadow, but the top halves are not. The reddening of the clouds is only just beginning. 24 May 2019 at 21:07. (Image by the author)

Fig. 5.10 Looking north, not quite at right angles to the sunset. Actually at sunset, the clouds are reddening noticeably. 24 May 2019 at 21:10. (Image by the author)

Not every sunset is as clear-cut as the one shown. This is not an exact science, and it obviously won't work if there are no clouds. But the rules of thumb are reddening of the clouds occurs right around sunrise/sunset. Looking towards the Sun, the clue is sunbeams if you can see them. Looking away from the Sun, the clue is that the redness very quickly disappears after sunset.

Figures 5.3, 5.4, 5.5, 5.6, 5.7, 5.8, 5.9, 5.10 and 5.11 also show quite a long dusk. The pictures were taken half a mile from my house, which is at about 51½°N, further north than the "lower 48" United States, which end at the 49th parallel. Even in Cleveland, at 41°N, where I was at Case Western Reserve University, I noticed that the dawns and dusks were much more rapid than I was used to in the United Kingdom. You certainly would not see such a long dusk in Texas or Florida. There, the times of sunrise and sunset would be even more obvious.

I therefore rest my case that even if you do not verify sunrise and sunset times every day, you will be broadly aware of when they occur, especially if you are an active astronomical observer.

Fig. 5.11 Looking north, not quite at right angles to the sunset. Seven minutes after sunset, the redness has gone. 24 May 2019 at 21:17. (Image by the author)

From Solar Midday to Orbit

I define the solar midday to be the exact midpoint of the time between sunrise and sunset. As we will see, this is not *quite* the same as defining the solar midday as that moment at which the sun is due south for you, but the two definitions are close. The latter is more accurate, but presents a measurement difficulty for amateur astronomers. Most of us have to be at work at midday, even if the sky is then clear. On the other hand, by means described above, we can follow sunrise and sunset times.

There are three causes of deviation of the midpoint between sunrise and sunset and 12:00 local time, which in the United Kingdom is equal to universal time for practical purposes. The first cause is easy to fix. Every degree of longitude away from the meridian defining your time zone makes your local time 4 minutes earlier if you are to the east of this meridian or makes it 4 minutes later if you are to the west of this meridian. I'm ignoring daylight savings time, summer time or any similar evils, and just working in winter time. Much though I would like to blame changing the clocks twice a year on Satan, its origin is alas all too human. We inflicted this inconvenience on ourselves. For the life of me I could never see why.

To take a simple example, suppose you live in Evanston, Illinois, at 81°47' W. The clock time is Central Standard Time, which is correct for longitude 90°W. That's approximately where Evanston's latitude crosses the Mississippi River. You are therefore

$$90^{\circ} - 81^{\circ}47' = 8^{\circ}13' = 8.2167^{\circ} \qquad (5.1)$$

east of the time zone meridian. This means that your midday is 32.8667 minutes earlier than your time zone would lead you to believe. Subtract this time from your local midday and you will get the figures for your time zone. If you live west of your time zone meridian, say in Des Moines, Iowa, then you add the time corresponding to the longitude difference.

The second cause of deviation is the tilt of the earth's axis relative to the ecliptic. We know that zero Right Ascension is defined by an equinox, so we also know that the tilt points at right angles to this direction. The northern hemisphere points towards Right Ascension 6 h 0 m 0 s and the southern hemisphere is tilted towards Right Ascension 18 h 0 m 0 s.

The third cause of the deviation is that, as we now know, the Earth's orbit is elliptical to an excellent approximation.

Therefore, if we are smart enough, we ought to be able to back-calculate the tilt, eccentricity and time of perihelion from the deviation of solar midday from mean or standard time midday.

Sorting Out Some Definitions

A year is the time the Sun takes to return to the same position in the sky. The system of having a leap year every 4 years keeps excellent track of the seasons. It requires minor corrections, with no leap year at the turn of three out of four centuries.

Nevertheless, we can define a year as 365.25 days with an error only in the fifth significant figure. Over 4 years, the four quarters of a day are balanced out by the addition of an extra day, February 29th. So

$$
\begin{aligned}
1 \text{ Solar year} \ &= \ 365.25 \text{ mean days} \\
&= \ 8766 \text{ hours} \\
&= \ 525{,}960 \text{ minutes} \qquad (5.2) \\
&= \ 31{,}557{,}600 \text{ seconds} \\
&= \ T_{\text{Earth}}.
\end{aligned}
$$

A sidereal day is defined by the time the Earth takes to rotate. Because it is moving round the Sun, this is not quite the same as a solar day. In fact, because the rotation direction is prograde, the Earth has to make exactly one extra rotation over and above the 365.25 to allow for the fact that it has been around the Sun over the course of 1 year. So

$$
\begin{aligned}
1 \text{ Mean solar day} &= 24 \text{ hours} \\
&= 1440 \text{ minutes} \\
&= t_{\text{day mean solar}}.
\end{aligned}
\tag{5.3}
$$

$$
\begin{aligned}
1 \text{ Sidereal day} &= \frac{365.25 \times 24}{366.25} \text{ hours} \\
&= 23.934471 \text{ hours} \\
&= 23 \text{ h } 56 \text{ m } 4.1 \text{ s} \\
&= 1436.068 \text{ minutes} \\
&= t_{\text{day sidereal}}.
\end{aligned}
\tag{5.4}
$$

Figure 5.12 illustrates why a solar day is longer than a sidereal day. The vast distance to the stars relative to the Solar System is made clear by the fact that the nearest known star to us, Proxima Centauri, is about a quarter of a million astronomical units away.

As stated above the length of a solar day is not quite constant. Solar midday can differ from mean midday by up to a quarter of an hour.

The formula used to calculate the solar midday from the data used is computed as follows.

The source of the sunrise and sunset data is the US Naval Observatory's website [99]. You type in the year, the longitude, latitude and name, if required, of your site to the nearest minute of longitude or latitude. To get data spanning across two calendar years, simply run the calculation for both years and copy the data into Microsoft Excel, or your favorite spreadsheet, and format it up.

The time before Greenwich Mean Time is

$$
t_{\text{zone}} = -4 \times (26/60) \text{ minutes} = -1.7333 \text{ minutes},
\tag{5.5}
$$

since the observation site, King's Lynn, Norfolk, England, is at 52°46'N and 0°26'E. The minus sign is because the solar events take place later at the Greenwich Meridian, longitude 0°0'.

The time of solar midday is

$$
t = 12 - \frac{t_{\text{sunrise}} + t_{\text{sunset}}}{2} + t_{\text{zone}}.
\tag{5.6}
$$

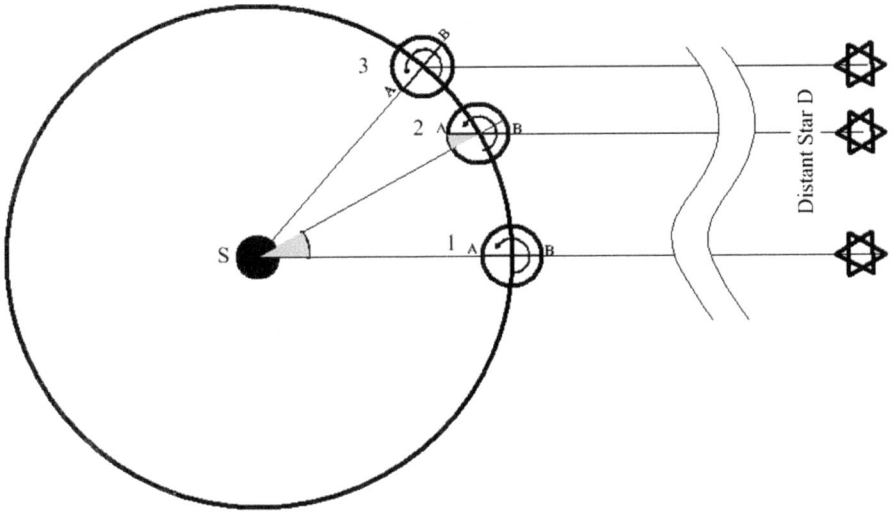

Fig. 5.12 Showing the difference (exaggerated) between a sidereal day and a solar day. We imagine the Earth orbiting the Sun S. At time zero Earth is at position 1. Observer A records that it is midday. Observer B at the antipodes of A sees a distant star D (which is millions of AU away) overhead. By the time position 2 is reached, the Earth has performed exactly one rotation relative to the distant stars, as is confirmed by Observer B. Notice also that the two angles shaded light grey, one at the Sun and one at the Earth, are identical. Observer A, however, does not observe that it is midday. One sidereal day has elapsed. Observer A still has to wait for solar midday. The Earth has to rotate some more. At Position 3, Observer A now observes and records midday. Thus, a solar day has now passed. Observer B no longer sees Star D to be overhead. Therefore, a solar day is longer than a sidereal day. (Image by the author)

The value of Δt given by Eq. (5.6) is listed in Tables 5.1, 5.2 and 5.3.

These data are plotted in Figs. 5.13, 5.14 and 5.15. To me it is quite astonishing that you can deduce the orbit of the Earth from these data, but you can.

Let me show you how.

The noise in the data in Fig. 5.13 bothers me: it may be a source of error. I tried a do-it-yourself smoothing operation in which I made the value of each point the average of itself, the four points to its left and the four points to its right. In other words, each point on the smoothed line is a nine-point moving average. The resultant curve is shown in Fig. 5.14.

Table 5.1 Sunrise and sunset times for King's Lynn during the year 1 July 2009–30 June 2010, the period over which Mars was observed. The value of Δt is given by Eq. (5.6). Part 1: July–October

	July 2009					Aug.					Sept.					Oct.				
	Rise		Set			Rise		Set			Rise		Set			Rise		Set		
Day	h	m	h	m	Δt	h	m	h	m	Δt	h	m	h	m	Δt	h	m	h	m	Δt
1	3	38	20	26	-3.7333	4	17	19	51	-5.7333	5	9	18	47	0.2667	6	0	17	35	10.7667
2	3	39	20	25	-3.7333	4	19	19	49	-5.7333	5	10	18	44	1.2667	6	1	17	33	11.2667
3	3	40	20	25	-4.2333	4	20	19	48	-5.7333	5	12	18	42	1.2667	6	3	17	30	11.7667
4	3	41	20	24	-4.2333	4	22	19	46	-5.7333	5	14	18	40	1.2667	6	5	17	28	11.7667
5	3	42	20	24	-4.7333	4	23	19	44	-5.2333	5	16	18	37	1.7667	6	7	17	26	11.7667
6	3	42	20	23	-4.2333	4	25	19	42	-5.2333	5	17	18	35	2.2667	6	8	17	23	12.7667
7	3	43	20	23	-4.7333	4	27	19	40	-5.2333	5	19	18	32	2.7667	6	10	17	21	12.7667
8	3	44	20	22	-4.7333	4	28	19	38	-4.7333	5	21	18	30	2.7667	6	12	17	19	12.7667
9	3	45	20	21	-4.7333	4	30	19	36	-4.7333	5	22	18	28	3.2667	6	14	17	16	13.2667
10	3	46	20	20	-4.7333	4	32	19	34	-4.7333	5	24	18	25	3.7667	6	15	17	14	13.7667
11	3	48	20	19	-5.2333	4	33	19	32	-4.2333	5	26	18	23	3.7667	6	17	17	12	13.7667
12	3	49	20	18	-5.2333	4	35	19	30	-4.2333	5	27	18	21	4.2667	6	19	17	10	13.7667
13	3	50	20	17	-5.2333	4	37	19	28	-4.2333	5	29	18	18	4.7667	6	21	17	7	14.2667
14	3	51	20	16	-5.2333	4	38	19	26	-3.7333	5	31	18	16	4.7667	6	22	17	5	14.7667
15	3	52	20	15	-5.2333	4	40	19	24	-3.7333	5	32	18	13	5.7667	6	24	17	3	14.7667
16	3	54	20	14	-5.7333	4	42	19	22	-3.7333	5	34	18	11	5.7667	6	26	17	1	14.7667
17	3	55	20	13	-5.7333	4	43	19	20	-3.2333	5	36	18	9	5.7667	6	28	16	59	14.7667
18	3	56	20	12	-5.7333	4	45	19	18	-3.2333	5	37	18	6	6.7667	6	30	16	56	15.2667
19	3	58	20	11	-6.2333	4	47	19	16	-3.2333	5	39	18	4	6.7667	6	31	16	54	15.7667
20	3	59	20	9	-5.7333	4	49	19	14	-3.2333	5	41	18	1	7.2667	6	33	16	52	15.7667
21	4	0	20	8	-5.7333	4	50	19	11	-2.2333	5	43	17	59	7.2667	6	35	16	50	15.7667
22	4	2	20	7	-6.2333	4	52	19	9	-2.2333	5	44	17	57	7.7667	6	37	16	48	15.7667
23	4	3	20	5	-5.7333	4	54	19	7	-2.2333	5	46	17	54	8.2667	6	39	16	46	15.7667
24	4	5	20	4	-6.2333	4	55	19	5	-1.7333	5	48	17	52	8.2667	6	41	16	44	15.7667
25	4	6	20	2	-5.7333	4	57	19	3	-1.7333	5	49	17	49	9.2667	6	42	16	41	16.7667
26	4	8	20	1	-6.2333	4	59	19	0	-1.2333	5	51	17	47	9.2667	6	44	16	39	16.7667
27	4	9	19	59	-5.7333	5	0	18	58	-0.7333	5	53	17	45	9.2667	6	46	16	37	16.7667
28	4	11	19	58	-6.2333	5	2	18	56	-0.7333	5	54	17	42	10.2667	6	48	16	35	16.7667
29	4	12	19	56	-5.7333	5	4	18	53	-0.2333	5	56	17	40	10.2667	6	50	16	33	16.7667
30	4	14	19	55	-6.2333	5	5	18	51	0.2667	5	58	17	38	10.2667	6	52	16	31	16.7667
31	4	15	19	53	-5.7333	5	7	18	49	0.2667						6	54	16	30	16.2667

To work out which physical parameters influence the curve shown in Figs. 5.13, 5.14 and 5.15 is a three-stage task.

- First I will make relevant deductions about the geometry of an elliptical orbit.
- Then I will show you what the obliquity, or tilt, of the earth's axis relative to the ecliptic, does.
- Third I will put all this information together.

Table 5.2 Sunrise and sunset times for King's Lynn during the year 1 July 2009–30 June 2010, the period over which Mars was observed. The value of Δt is given by Eq. (5.6). Part 2: November–February

Nov.					Dec.					Jan. 2010					Feb.				
Day	Rise		Set			Rise		Set			Rise		Set			Rise		Set		
	h	m	h	m	Δt	h	m	h	m	Δt	h	m	h	m	Δt	h	m	h	m	Δt
1	6	55	16	28	16.7667	7	48	15	47	10.7667	8	10	15	53	-3.2333	7	41	16	43	-13.7333
2	6	57	16	26	16.7667	7	49	15	46	10.7667	8	10	15	55	-4.2333	7	39	16	45	-13.7333
3	6	59	16	24	16.7667	7	51	15	45	10.2667	8	10	15	56	-4.7333	7	38	16	47	-14.2333
4	7	1	16	22	16.7667	7	52	15	45	9.7667	8	10	15	57	-5.2333	7	36	16	49	-14.2333
5	7	3	16	20	16.7667	7	53	15	44	9.7667	8	9	15	58	-5.2333	7	34	16	51	-14.2333
6	7	5	16	18	16.7667	7	55	15	44	8.7667	8	9	15	59	-5.7333	7	33	16	53	-14.7333
7	7	7	16	17	16.2667	7	56	15	44	8.2667	8	9	16	1	-6.7333	7	31	16	55	-14.7333
8	7	8	16	15	16.7667	7	57	15	43	8.2667	8	8	16	2	-6.7333	7	29	16	57	-14.7333
9	7	10	16	13	16.7667	7	58	15	43	7.7667	8	8	16	4	-7.7333	7	27	16	59	-14.7333
10	7	12	16	12	16.2667	7	59	15	43	7.2667	8	7	16	5	-7.7333	7	25	17	1	-14.7333
11	7	14	16	10	16.2667	8	0	15	43	6.7667	8	6	16	6	-7.7333	7	23	17	2	-14.2333
12	7	16	16	9	15.7667	8	1	15	43	6.2667	8	5	16	8	-8.2333	7	21	17	4	-14.2333
13	7	18	16	7	15.7667	8	2	15	43	5.7667	8	5	16	10	-9.2333	7	19	17	6	-14.2333
14	7	19	16	5	16.2667	8	3	15	43	5.2667	8	4	16	11	-9.2333	7	17	17	8	-14.2333
15	7	21	16	4	15.7667	8	4	15	43	4.7667	8	3	16	13	-9.7333	7	15	17	10	-14.2333
16	7	23	16	3	15.2667	8	5	15	43	4.2667	8	2	16	14	-9.7333	7	13	17	12	-14.2333
17	7	25	16	1	15.2667	8	6	15	43	3.7667	8	1	16	16	-10.2333	7	11	17	14	-14.2333
18	7	27	16	0	14.7667	8	6	15	43	3.7667	8	0	16	18	-10.7333	7	9	17	16	-14.2333
19	7	28	15	59	14.7667	8	7	15	44	2.7667	7	59	16	19	-10.7333	7	7	17	18	-14.2333
20	7	30	15	57	14.7667	8	8	15	44	2.2667	7	58	16	21	-11.2333	7	5	17	20	-14.2333
21	7	32	15	56	14.2667	8	8	15	45	1.7667	7	57	16	23	-11.7333	7	3	17	22	-14.2333
22	7	33	15	55	14.2667	8	9	15	45	1.2667	7	55	16	25	-11.7333	7	1	17	24	-14.2333
23	7	35	15	54	13.7667	8	9	15	46	0.7667	7	54	16	27	-12.2333	6	59	17	25	-13.7333
24	7	37	15	53	13.2667	8	10	15	46	0.2667	7	53	16	28	-12.2333	6	57	17	27	-13.7333
25	7	38	15	52	13.2667	8	10	15	47	-0.2333	7	52	16	30	-12.7333	6	54	17	29	-13.2333
26	7	40	15	51	12.7667	8	10	15	48	-0.7333	7	50	16	32	-12.7333	6	52	17	31	-13.2333
27	7	42	15	50	12.2667	8	10	15	49	-1.2333	7	49	16	34	-13.2333	6	50	17	33	-13.2333
28	7	43	15	49	12.2667	8	10	15	49	-1.2333	7	47	16	36	-13.2333	6	48	17	35	-13.2333
29	7	45	15	48	11.7667	8	11	15	50	-2.2333	7	46	16	38	-13.7333					
30	7	46	15	47	11.7667	8	11	15	51	-2.7333	7	44	16	40	-13.7333					
31						8	11	15	52	-3.2333	7	43	16	41	-13.7333					

I invented none of this. I read several accounts and concluded that while they all looked superficially straightforward, the more I looked into them, the more confused I became.

I have mixed and matched the best bits from where I could and present below what I believe to be the least laborious way to solve for the orbit of the Earth. Since it all caused me so much confusion, I will show all my workings.

Table 5.3 Sunrise and sunset times for King's Lynn during the year 1 July 2009–30 June 2010, the period over which Mars was observed. The value of Δt is given by Eq. (5.6). Part 3: March–June

	Mar. Rise		Set			Apr. Rise		Set			May Rise		Set			June Rise		Set		
Day	h	m	h	m	Δt	h	m	h	m	Δt	h	m	h	m	Δt	h	m	h	m	Δt
1	6	46	17	37	-13.2333	5	33	18	33	-4.7333	4	27	19	25	2.2667	3	40	20	13	1.7667
2	6	43	17	39	-12.7333	5	31	18	34	-4.2333	4	25	19	27	2.2667	3	39	20	14	1.7667
3	6	41	17	40	-12.2333	5	28	18	36	-3.7333	4	23	19	29	2.2667	3	39	20	15	1.2667
4	6	39	17	42	-12.2333	5	26	18	38	-3.7333	4	21	19	31	2.2667	3	38	20	16	1.2667
5	6	36	17	44	-11.7333	5	23	18	40	-3.2333	4	19	19	32	2.7667	3	37	20	17	1.2667
6	6	34	17	46	-11.7333	5	21	18	41	-2.7333	4	17	19	34	2.7667	3	37	20	18	0.7667
7	6	32	17	48	-11.7333	5	19	18	43	-2.7333	4	15	19	36	2.7667	3	36	20	19	0.7667
8	6	30	17	50	-11.7333	5	17	18	45	-2.7333	4	13	19	37	3.2667	3	35	20	20	0.7667
9	6	27	17	51	-10.7333	5	14	18	47	-2.2333	4	11	19	39	3.2667	3	35	20	20	0.7667
10	6	25	17	53	-10.7333	5	12	18	48	-1.7333	4	10	19	41	2.7667	3	35	20	21	0.2667
11	6	23	17	55	-10.7333	5	10	18	50	-1.7333	4	8	19	42	3.2667	3	34	20	22	0.2667
12	6	20	17	57	-10.2333	5	7	18	52	-1.2333	4	6	19	44	3.2667	3	34	20	23	-0.2333
13	6	18	17	59	-10.2333	5	5	18	54	-1.2333	4	5	19	46	2.7667	3	34	20	23	-0.2333
14	6	16	18	0	-9.7333	5	3	18	56	-1.2333	4	3	19	47	3.2667	3	33	20	24	-0.2333
15	6	13	18	2	-9.2333	5	1	18	57	-0.7333	4	1	19	49	3.2667	3	33	20	24	-0.2333
16	6	11	18	4	-9.2333	4	58	18	59	-0.2333	4	0	19	50	3.2667	3	33	20	25	-0.7333
17	6	8	18	6	-8.7333	4	56	19	1	-0.2333	3	58	19	52	3.2667	3	33	20	25	-0.7333
18	6	6	18	8	-8.7333	4	54	19	3	-0.2333	3	57	19	54	2.7667	3	33	20	26	-1.2333
19	6	4	18	9	-8.2333	4	52	19	4	0.2667	3	55	19	55	3.2667	3	33	20	26	-1.2333
20	6	1	18	11	-7.7333	4	49	19	6	0.7667	3	54	19	57	2.7667	3	33	20	26	-1.2333
21	5	59	18	13	-7.7333	4	47	19	8	0.7667	3	53	19	58	2.7667	3	34	20	27	-2.2333
22	5	57	18	15	-7.7333	4	45	19	10	0.7667	3	51	20	0	2.7667	3	34	20	27	-2.2333
23	5	54	18	17	-7.2333	4	43	19	11	1.2667	3	50	20	1	2.7667	3	34	20	27	-2.2333
24	5	52	18	18	-6.7333	4	41	19	13	1.2667	3	49	20	2	2.7667	3	34	20	27	-2.2333
25	5	50	18	20	-6.7333	4	39	19	15	1.2667	3	48	20	4	2.2667	3	35	20	27	-2.7333
26	5	47	18	22	-6.2333	4	37	19	17	1.2667	3	46	20	5	2.7667	3	35	20	27	-2.7333
27	5	45	18	24	-6.2333	4	35	19	18	1.7667	3	45	20	6	2.7667	3	36	20	27	-3.2333
28	5	42	18	25	-5.2333	4	33	19	20	1.7667	3	44	20	8	2.2667	3	36	20	27	-3.2333
29	5	40	18	27	-5.2333	4	31	19	22	1.7667	3	43	20	9	2.2667	3	37	20	26	-3.2333
30	5	38	18	29	-5.2333	4	29	19	24	1.7667	3	42	20	10	2.2667	3	37	20	26	-3.2333
31	5	35	18	31	-4.7333						3	41	20	11	2.2667					

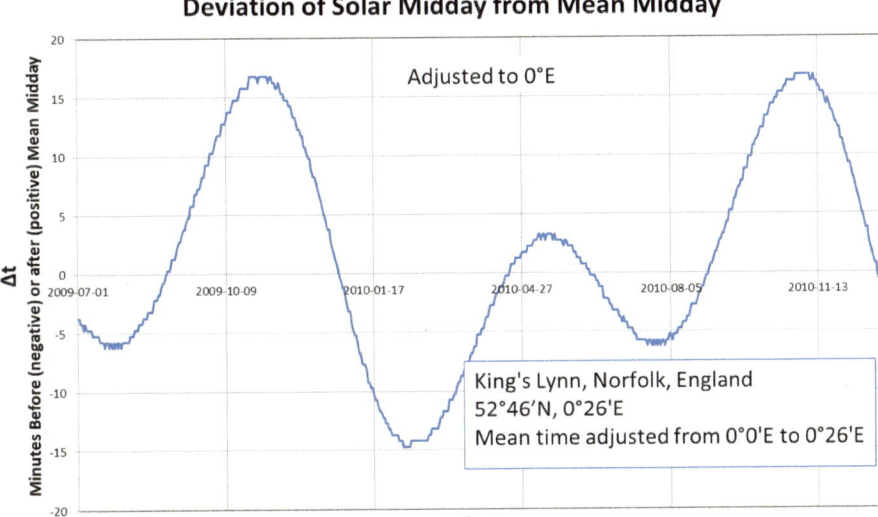

Fig. 5.13 Deviation of Solar midday from mean midday at my observing site. These are the data I have used to compute the orbit of the Earth. (Image by the author)

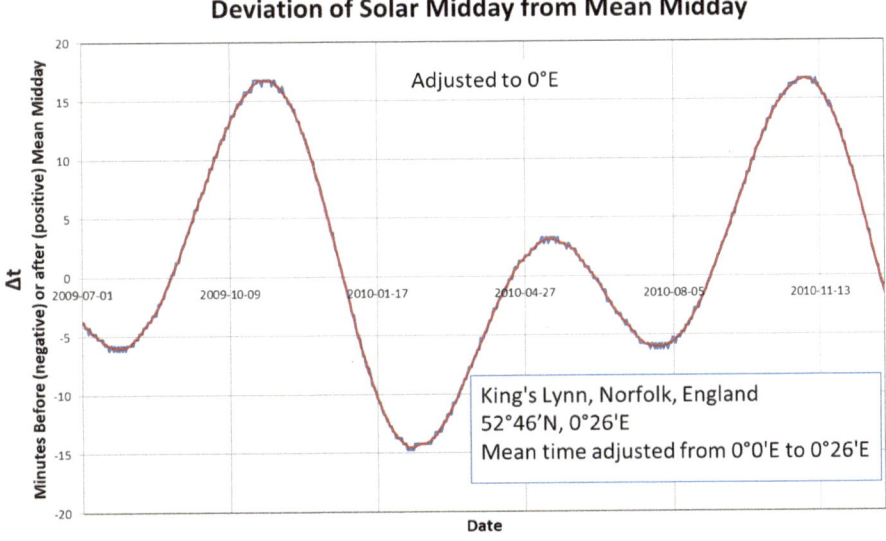

Fig. 5.14 Deviation of Solar midday from mean midday at my observing site. These are the data I have used to compute the orbit of the Earth. Blue line: USNO data as in Fig. 5.13. Red line: 9-point moving average of data. (Image by the author)

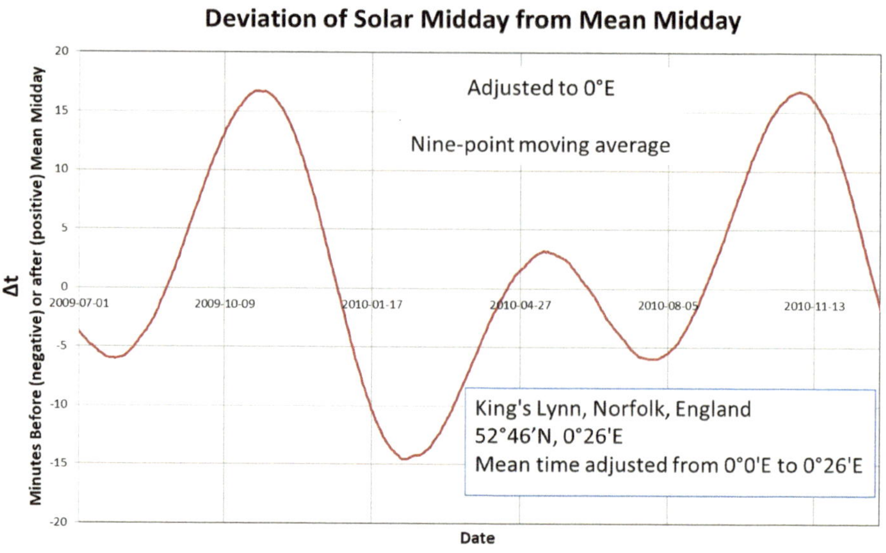

Fig. 5.15 Deviation of Solar midday from mean midday at my observing site. These are the data I have used to compute the orbit of the Earth. Red line: 9-point moving average of data taken from Fig. 5.14. (Image by the author)

Further Geometric Gymnastics with Ellipses

The true anomaly, f, could be calculated from the eccentric anomaly E using Eq. (3.71):

$$\tan\frac{f}{2} = \sqrt{\frac{1+e}{1-e}} \, \tan\frac{E}{2}, \tag{3.71}$$

where e is the eccentricity of the elliptical orbit as before. In principle, I could disentangle this equation and derive an expression for f in terms of E. In practice, my impression is that, at least for small eccentricities, this is a dumb way to proceed. There are easier alternatives.

How do I know the eccentricity of the earth's orbit is low? Two clues are that the four seasons each last approximately 3 months and that the size of the Sun in the sky does not change much. If we were to live on a planet with a highly eccentric orbit, neither of these would be true.

The recipe for computing Δt is that we use Kepler's Eq. (3.79) to calculate E for a given M and then calculate f. We then substitute M and f into Eq. (5.48).

Kepler's Equation is

$$M = E - e\sin E. \tag{3.79}$$

Kepler's own approach to solving this was iterative: start with a guess and hope that the next iteration gives a better solution. This method actually predates Kepler, albeit using the same equation for other purposes. Colwell [85] states that it was known to ninth-century mathematician Habash Al-Hasib al-Marwazi [100], who seems to have been the earliest known user of sines, cosines and tangents [98] and that the method may even have been Hindu.

Anyway, first, please let me rearrange Eq. (3.79) into a more convenient form,

$$E = M + e\sin E. \tag{5.7}$$

The idea of iteration is to consider an ith guess at the value of E, E_i. Then you would iterate Eq. (5.7) like this:

$$E_{i+1} \approx M + e\sin E_i. \tag{5.8}$$

My first guess is going to be

$$E_1 \oplus M. \tag{5.9}$$

Then from Eq. (5.8),

$$E_2 \approx M + e\sin M. \tag{5.10}$$

Substituting E_2 into Eq. (5.8) gives

$$
\begin{aligned}
E_3 &\approx & M + e\sin\left(M + e\sin M\right) \\
&= & M + e\left[\sin M \cos\left(e\sin M\right) + \cos M \sin\left(e\sin M\right)\right].
\end{aligned} \tag{5.11}
$$

For small eccentricities e, we could say that each successive approximation is accurate to order $1, e, e^2, e^3, \ldots$ or generally E_i is precise to order e^{i-1}. As will eventually become clear, there is no point in keeping terms more precise than order e^2 in Eq. (5.11). Multiplying out the square bracket in Eq. (5.11) gives

$$\sin(e\sin M) = e\sin M - \frac{e^3\sin^3 M}{3!} + \frac{e^5\sin^5 M}{5!} - \cdots$$

$$\approx e\sin M$$

$$\cos(e\sin M) = 1 - \frac{e^2\sin^2 M}{2!} + \frac{e^4\sin^4 M}{4!} - \cdots \tag{5.12}$$

$$\approx 1 - \frac{e^2\sin^2 M}{2!}$$

to order e^2, using the usual series expansions for sines and cosines. Substituting Eq. (5.12) into Eq. (5.11) gives

$$E_3 \approx M + e\sin(M + e\sin M)$$

$$\approx M + e\left[\sin M\left(1 - \frac{1}{2}e^2\sin^2 M\right) + \cos M e\sin M\right]$$

$$= M + e\sin M - \frac{1}{2}e^3\sin^3 M + e^2\sin M\cos M \tag{5.13}$$

$$\approx M + e\sin M + e^2\sin M\cos M.$$

We could continue to refine our value of the eccentric anomaly E, but for now we won't need to.

We are going to embark on a slightly long journey to find f in terms of M, a variable whose value we know.

Using Eqs. (3.64) and (3.68)

$$a(1 - e\cos E) = \frac{a(1 - e^2)}{1 + e\cos f}$$

$$1 + e\cos f = \frac{1 - e^2}{(1 - e\cos E)}$$

$$e\cos f = \frac{1 - e^2}{(1 - e\cos E)} - 1$$

$$= \frac{1 - e^2 - 1 + e\cos E}{(1 - e\cos E)} \tag{5.14}$$

$$= \frac{-e^2 + e\cos E}{(1 - e\cos E)}$$

$$\cos f = \frac{\cos E - e}{(1 - e\cos E)}$$

We will also need sin f in terms of E.

$$1-\cos^2 f = \sin^2 f$$

$$= 1 - \frac{(\cos E - e)^2}{(1 - e\cos E)^2}$$

$$= \frac{(1-e\cos E)^2 - (\cos E - e)^2}{(1-e\cos E)^2}$$

$$= \frac{(1+e^2\cos^2 E - 2e\cos E) - (\cos^2 E + e^2 - 2e\cos E)}{(1-e\cos E)^2}$$

$$= \frac{1+e^2\cos^2 E - 2e\cos E - \cos^2 E - e^2 + 2e\cos E}{(1-e\cos E)^2} \tag{5.15}$$

$$= \frac{1+e^2\cos^2 E - \cos^2 E - e^2}{(1-e\cos E)^2}$$

$$= \frac{(1-e^2)(1-\cos^2 E)}{(1-e\cos E)^2}$$

$$= \frac{(1-e^2)\sin^2 E}{(1-e\cos E)^2}$$

Thus $\sin f = \dfrac{\sqrt{1-e^2}\,\sin E}{(1-e\cos E)}.$

Let me now take you along a slightly non-obvious alleyway. This idea is sketched out by Danby [101], but he does not give the details. This one took me a few days to sort out. My hope is that by spelling it all out, you can get to the same point in a couple of hours.

I'm going to differentiate $\cos f$ with respect to M.

First, from Kepler's Eq. (3.79)

$$M = E - e\sin E. \tag{3.79}$$

Hence

$$\frac{dM}{dE} = 1 - e\cos E$$

$$\text{So } \frac{dE}{dM} = \frac{1}{1-e\cos E}. \tag{5.16}$$

Now

$$
\begin{aligned}
\frac{d\cos f}{dM} &= \frac{d\cos f}{dE}\frac{dE}{dM} \\[2mm]
&= \frac{d}{dE}\left(\frac{\cos E - e}{\left(1 - e\cos E\right)}\right)\frac{dE}{dM} \\[2mm]
&= \left(\frac{\left(1 - e\cos E\right)\dfrac{d}{dE}\left(\cos E - e\right) - \left(\cos E - e\right)\dfrac{d}{dE}\left(1 - e\cos E\right)}{\left(1 - e\cos E\right)^2}\right)\frac{dE}{dM} \\[2mm]
&= \left(\frac{\left(1 - e\cos E\right)\left(-\sin E\right) - \left(\cos E - e\right)e\sin E}{\left(1 - e\cos E\right)^2}\right)\frac{dE}{dM} \\[2mm]
&= \left(\frac{\left(e\cos E\sin E - \sin E\right) - \left(e\sin E\cos E - e^2\sin E\right)}{\left(1 - e\cos E\right)^2}\right)\frac{dE}{dM} \\[2mm]
&= \left(\frac{-\left(1 - e^2\right)\sin E}{\left(1 - e\cos E\right)^2}\right)\frac{dE}{dM} \\[2mm]
&= \left(\frac{-\sqrt{\left(1 - e^2\right)}\sqrt{\left(1 - e^2\right)}\sin E}{\left(1 - e\cos E\right)^2}\right)\frac{dE}{dM} \\[2mm]
&= \left(\frac{-\sqrt{\left(1 - e^2\right)}\sin f}{\left(1 - e\cos E\right)}\right)\frac{dE}{dM} \\[2mm]
&= -\sqrt{\left(1 - e^2\right)}\sin f\left(\frac{dE}{dM}\right)^2 \\[2mm]
&= \frac{d\cos f}{dM} \\[2mm]
&= \frac{d\cos f}{df}\frac{df}{dM} \\[2mm]
&= -\sin f\frac{df}{dM}
\end{aligned}
$$

(5.17)

Thus $\dfrac{df}{dM} = \sqrt{\left(1 - e^2\right)}\left(\dfrac{dE}{dM}\right)^2.$

Now I substitute Eqs. (5.16) and (5.11) to get

$$
\begin{aligned}
\frac{df}{dM} &= \sqrt{\left(1-e^2\right)}\left(\frac{d}{dM}\left(M+e\sin M+e^2\sin M\cos M+\ldots\right)\right)^2 \\
&= \sqrt{\left(1-e^2\right)}\left(\frac{d}{dM}\left(M+e\sin M+\frac{1}{2}e^2\sin 2M+\ldots\right)\right)^2 \\
&= \sqrt{\left(1-e^2\right)}\left(1+e\cos M+\frac{2}{2}e^2\cos 2M+\ldots\right)^2 \\
&= \sqrt{\left(1-e^2\right)}\left(\begin{array}{l}1+e^2\cos^2 M+e^4\cos^2 2M+2e\cos M+\\2e^2\cos 2M+2e^3\cos M\cos 2M+\ldots\end{array}\right)
\end{aligned}
$$

(5.18)

When we're trying to expand in powers of e, the square root term is just a perishing nuisance. Let's get rid of it. The easiest way to do this is to expand it as a series in powers of e about the point $e=0$. First, let

$$u=e^2. \tag{5.19}$$

Then a series in a function of u about $u=0$ would look like this.

$$g(u)=g(0)+u\frac{dg(0)}{du}+\frac{1}{2!}u^2\frac{d^2g(0)}{du^2}+\ldots. \tag{5.20}$$

Substituting $g(u)=\sqrt{1-u}$ into Eq. (5.20) gives

$$\sqrt{1-u}=1+u\left(\frac{d\sqrt{1-u}}{du}\right)_{u=0}+\frac{1}{2!}u^2\left(\frac{d^2\sqrt{1-u}}{du^2}\right)_{u=0}+\ldots \tag{5.21}$$

First I evaluate the first derivative.

$$
\begin{aligned}
\frac{d}{du}\sqrt{1-u} &= \frac{d\sqrt{w}}{dw}\frac{dw}{du}\quad(w=1-u) \\
&= \frac{1}{2\sqrt{w}}\frac{d(1-u)}{du} \\
&= \frac{1}{2\sqrt{1-u}}(-1) \\
&= -\frac{1}{2}\quad\text{when }u=0.
\end{aligned}
$$

(5.22)

Now let me do the second derivative.

$$\frac{d^2}{du^2}\sqrt{1-u} = \frac{d}{du}\left(\frac{-1}{2\sqrt{1-u}}\right)$$

$$= -\frac{1}{2}\frac{d}{du}\left(\frac{1}{\sqrt{1-u}}\right)$$

$$= -\frac{1}{2}\frac{d}{dw}\left(\frac{1}{\sqrt{w}}\right)\frac{dw}{du} \quad \text{where} \quad w = 1-u$$

$$= -\frac{1}{2}\left(\frac{-1}{2w^{3/2}}\right)(-1)$$

$$= -\frac{1}{4}\left(\frac{1}{(1-u)^{3/2}}\right)$$

$$= -\frac{1}{4} \quad \text{when} \quad u = 0.$$

(5.23)

Substituting Eqs. (5.22) and (5.23) into Eq. (5.21) gives

$$\sqrt{1-u} = 1 - \frac{u}{2} - \frac{u^2}{8} - \dots \ .$$

$$\sqrt{1-e^2} = 1 - \frac{e^2}{2} - \frac{e^4}{8} - \dots \ .$$

(5.24)

Equation (5.19) has been used in the second line.
I can now rewrite Eq. (5.18) as

$$\frac{df}{dM} = \sqrt{(1-e^2)}\left(\begin{array}{l}1 + e^2\cos^2 M + e^4\cos^2 2M + 2e\cos M + \\ 2e^2\cos 2M + 2e^3\cos M\cos 2M + \dots\end{array}\right)$$

$$= \left(1 - \frac{e^2}{2} - \frac{e^4}{8}\right)\left(\begin{array}{l}1 + e(2\cos M) + e^2\left(2\cos 2M + \cos^2 M\right) + \\ 2e^3\cos M\cos 2M + e^4\cos^2 2M + \dots\end{array}\right)$$

$$= \left(1 + e(2\cos M) + e^2\left(2\cos 2M + \frac{\cos 2M + 1}{2} - \frac{1}{2}\right) + O(e^3)\right)$$

$$= \left(1 + e(2\cos M) + e^2\left(\frac{5}{2}\cos 2M\right) + O(e^3)\right),$$

(5.25)

Where the last step in Eq. (5.25) can be proved by using the formula

$$
\begin{aligned}
\cos 2M &= \cos^2 M - \sin^2 M \\
&= \cos^2 M - \left(1 - \cos^2 M\right) \\
&= 2\cos^2 M - 1; \\
\text{so } -2\cos^2 M &= -1 - \cos 2M \\
\text{or } \cos^2 M &= \frac{1 + \cos 2M}{2}.
\end{aligned}
\tag{5.26}
$$

I have manipulated df/dM into a form that I can conveniently integrate.

$$
\begin{aligned}
f &= \int \left(1 + e(2\cos M) + e^2\left(\frac{5}{2}\cos 2M\right) + O\left(e^3\right)\right) dM \\
&= \int \left(1 + 2e\cos M + \frac{5}{2}e^2\cos 2M + O\left(e^3\right)\right) dM \\
&= \left[M + 2e\sin M + \frac{5}{4}e^2\left(\sin 2M\right) + O\left(e^3\right)\right] + \text{constant.}
\end{aligned}
\tag{5.27}
$$

Since $M = 0$ at perihelion when $f = 0$, the constant vanishes. What remains is

$$
f \approx M + 2e\sin M + \frac{5}{4}e^2\left(\sin 2M\right)
\tag{5.28}
$$

You may see Eq. (5.28) referred to as the "Equation of the Centre". It seems to be obligatory to tell you that. Every author I have read mentions it.

Instead of just enacting the ritual of naming the equation, I'm going to do something radical and tell you where the name comes from. Imagine yourself in a frame of reference centred at the focus of your ellipse. This frame co-rotates with the mean anomaly M. Then as the orbit progresses, sometimes the true anomaly f will be ahead of the mean anomaly M, and sometimes it will be behind it. The formula governing this is

$$
f - M \approx 2e\sin M + \frac{5}{4}e^2\left(\sin 2M\right)
\tag{5.29}
$$

Changing tack a little, comparing Figs. 5.16 and 5.20 tells us that the right ascension of the Sun is given by Eq. (5.30).

$$
\lambda = f + \varpi.
\tag{5.30}
$$

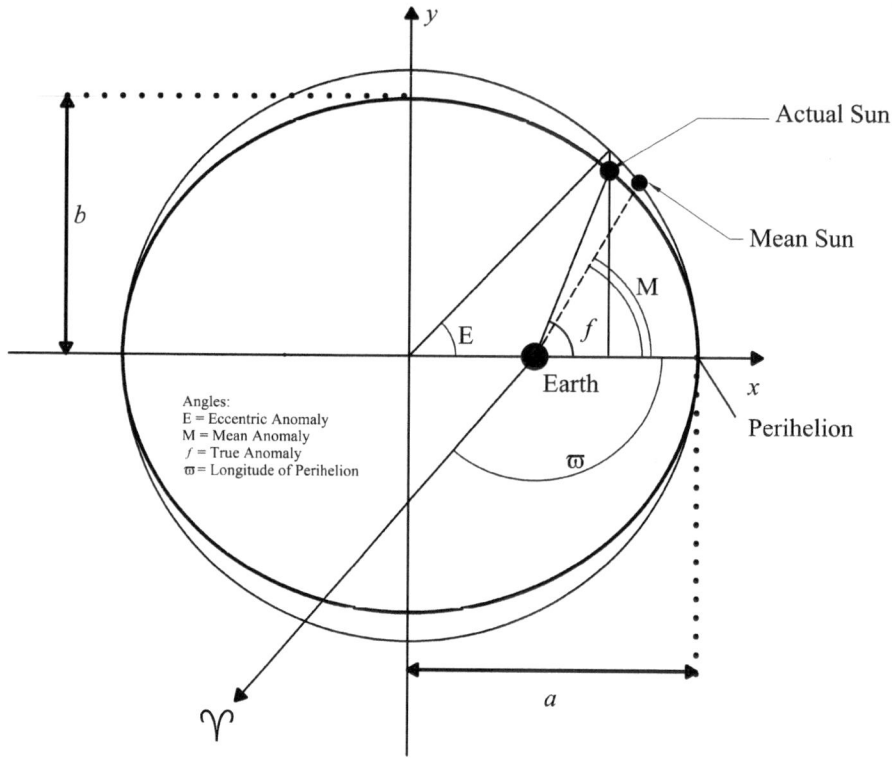

Fig. 5.16 A second visit to Fig. 3.7, now showing the mean anomaly M as well as the eccentric and true anomalies, E and f, respectively. An unphysical "Mean Sun" is shown with anomaly M. The drift of the Sun relative to the stars is eastward. This figure therefore looks down from north of the ecliptic. Notice that the eccentric anomaly E is the angle to the line going to the circumscribing circle, not to the Sun

Using Eq. (5.28),

$$\lambda \approx M + \varpi + 2e\sin M + \frac{5}{4}e^2\left(\sin 2M\right). \tag{5.31}$$

If the Earth's axis were normal to the ecliptic, we would only need to analyse the difference between the true and mean anomalies and our job would be done. But it is at a jaunty angle to the ecliptic, and we have to account for that.

Tilting My Lance at Obliquity

Why does obliquity matter? At the equinoxes it has no effect. Let us again imagine an observing site in the American Midwest, at about 42°N 90°W, which would be at roughly where the Mississippi divides Illinois from Iowa. In the figures that follow I have painted a red dot at the location of this observer and used black tape to make the 90°W line very obvious (Fig. 5.17).

In Fig. 5.13, on the right-hand side there is no difference between solar noon and mean noon. They coincide at the equinoxes. As shown in Fig. 5.18, the same is true at the solstices.

Away from equinoxes and solstices it is a different story. Then there is a slight but noticeable difference between the mean and solar noon. This is illustrated in Fig. 5.19.

In the not-so-good old days before vector methods became popular, people used to solve three-dimensional problems using a largely lost art called spherical trigonometry.

Modern vector notation, incidentally, was the brainchild of the thermodynamicist J. W. Gibbs. He was too busy on other matters that earned him fame to publish his notation, but his student E. B Wilson wrote up Gibbs' lecture notes as a book in 1901. This book is out of copyright and is freely available to download [102]. It has a charming preface written by Gibbs himself.

Anyway, we need to prove one theorem from spherical trigonometry. Fortunately, we don't need to master the whole subject. Spherical trigonometry

Fig. 5.17 A globe showing my observer in the American Midwest. Black tape has been used to show the 90°W line which goes through the observing site. The camera is in the direction of the Sun, and the globe has been set so that we are at an equinox. The red dot shows the location of the observer. Left: Well before noon, the 90°W meridian appears to be arc-shaped. Right: At noon, the 90°W meridian now looks like a straight line

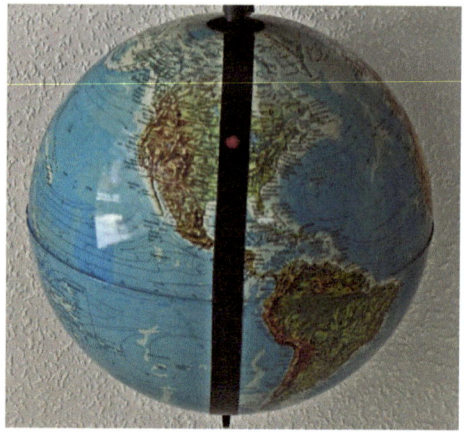

Fig. 5.18 A globe showing my observer in the American Midwest. Black tape has been used to show the 90°W line which goes through the observing site. The camera is in the direction of the Sun, and the globe has been set so that we are at The December solstice (left) and the June solstice (right). The red dot shows the location of the observer, who is actually barely visible in December, because I could not move my camera far enough away to see half of the globe

Fig. 5.19 A globe showing my observer in the American Midwest. Black tape has been used to show the 90°W line which goes through the observing site. The red dot shows the location of the observer. The camera is in the direction of the Sun, and the globe has been set so that we midway between a solstice and an equinox. It is no longer quite true that the solar day, when the red dot is midway across the globe at it transits through its day (left). It is at mean noon. A yellow line has been added to that you can see the slight apparent arc shape of the black tape. In the right-hand image, slightly later in the day, we are now at the solar noon. In other words, the solar and mean noon are no longer the same

is largely concerned with the geometry of the surfaces of spheres. The classic account of it is another electronically downloadable book by the Victorian mathematician Isaac Todhunter [103]. (Todhunter also produced a very readable translation of Euclid's Elements, the classic geometry text [104].)

Consider the situation in Fig. 5.20. This is from a frame of reference centred on the Earth, which does not rotate relative to the International Celestial Reference Frame [125]. This diagram shows the Earth, the Sun, the Celestial North Pole and the Celestial Equator. It also shows a meridian going through the Sun. The direction of the ascending node of the ecliptic, which is the Υ direction, is shown. These are all in black. In grey, we can see the plane of the ecliptic, which passes through the Sun. The normal direction to the ecliptic is also shown.

Three angles are shown:

- The obliquity, that is, the tilt of the Earth's axis relative to the ecliptic φ
- The sum of the true anomaly of the Sun and the longitude of the ascending node ϖ
- The angle from Υ to the Sun along the ecliptic, λ

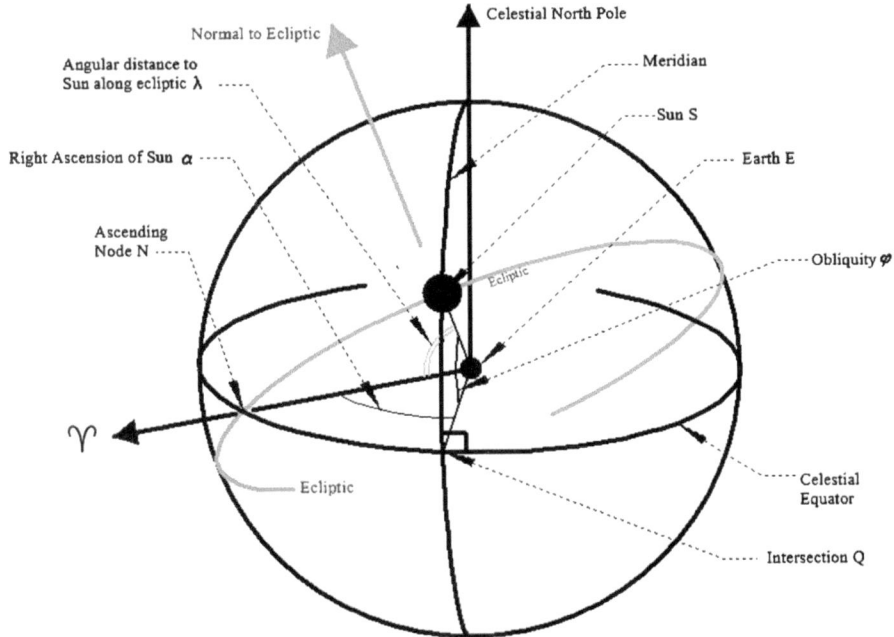

Fig. 5.20 The apparent motion of the Sun around the Earth is along the ecliptic plane. The grey arc and normal arrow show the plane of the ecliptic. It is tilted by an angle of obliquity φ. The anomaly of the Sun relative to its ascending node, which is along the line to Υ, is represented by the angle $\lambda = \varpi + f$. The Celestial Equator is represented by a black arc. The meridian passing through the Sun is also shown. (Image by the author)

In Fig. 5.21, another view of the same situation, we introduce a spherical triangle SNQ, shown grey. Spherical triangles are triangles on the surface of a sphere, each of whose sides form part of a great circle, a circle on the surface of the sphere whose centre coincides with the centre of the sphere. Notice that the angle at Q is a right angle.

The sides of spherical triangles can be given lengths, just like those of planar triangles. Unlike with planer triangles, the sides of spherical triangles can also be expressed as angles subtended at the centre.

Figure 5.22 is another, closer view of the spherical triangle SNQ. We have introduced a planar triangle UPR, so defined that ∠PRE and ∠PUE are right angles.

We follow §62 of Todhunter (work cited). Since the angle at Q is a right angle, it follows that ∠PRU is also a right angle. Now, we make repeated use of Pythagoras' theorem to see that

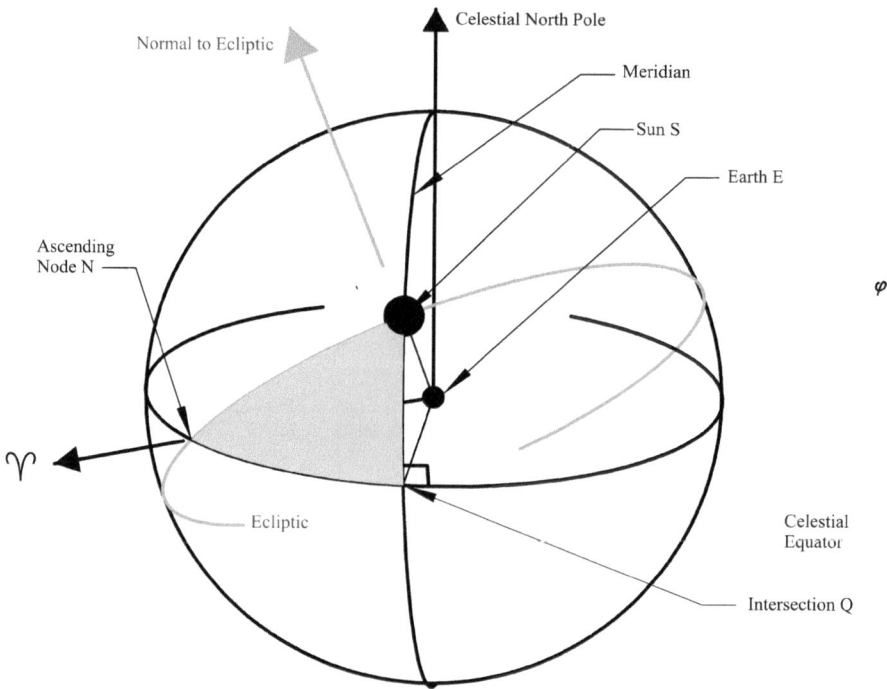

Fig. 5.21 There is useful information to be gleaned from analysing the spherical triangle SNQ shown in light grey. It has a right angle at Q. Spherical triangles are triangles on the surface of a sphere, each of whose sides form part of a great circle, a circle on the surface of the sphere whose centre coincides with the centre of the sphere. (Image by the author)

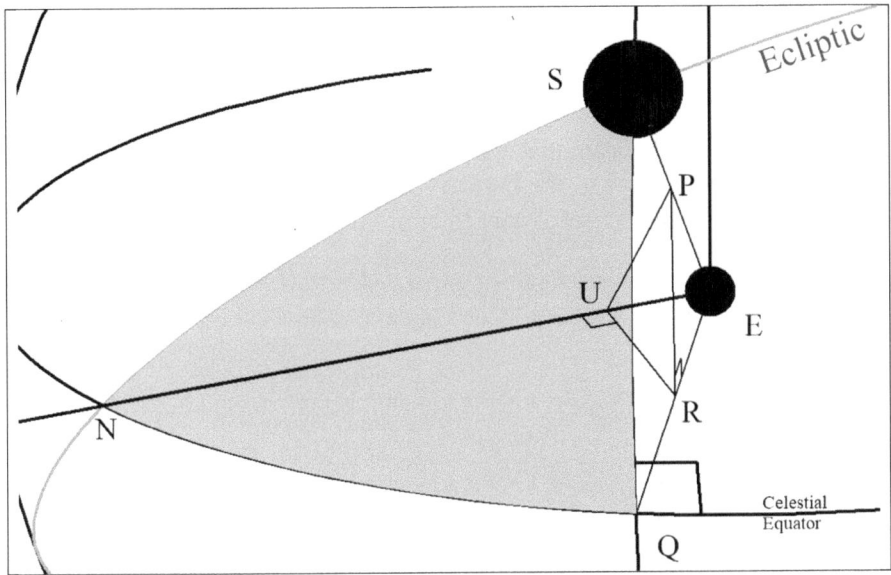

Fig. 5.22 Another view of spherical triangle SNQ, with some (straight) construction lines added for theorem proving, viz. the lines PU, UR and RP. Note that ∠PRE and ∠PUE are right angles. The line EN and the point U, behind the spherical triangle, are also shown. (Image by the author)

$$
\begin{aligned}
PU^2 \;&=\; PR^2 + RU^2 \\
&=\; \left(EP^2 - ER^2\right) + \left(ER^2 - EU^2\right) \\
&=\; EP^2 + EU^2.
\end{aligned}
\tag{5.32}
$$

Therefore, ∠PUE is a right angle.

$$
\frac{UR}{EU} = \frac{UR}{PU} \times \frac{PU}{EU}.
\tag{5.33}
$$

But

$$
\begin{aligned}
\frac{UR}{EU} &= \tan \alpha, \\
\frac{UR}{PU} &= \cos \varphi, \\
\frac{PU}{EU} &= \tan \lambda.
\end{aligned}
\tag{5.34}
$$

Substituting into Eq. (5.33) gives

$$\tan \alpha = \cos \varphi \tan \lambda. \tag{5.35}$$

This is the little theorem that we need to prove. Although it is hardly the greatest equation known to the human race, it is needed for what follows.
Milne [105] shows a useful way to approximate Eq. (5.35). Since

$$\cos \varphi = \frac{\tan \alpha}{\tan \lambda}, \tag{5.36}$$

$$
\begin{aligned}
\tan^2 \frac{\varphi}{2} &= \frac{2 \sin^2 \frac{\varphi}{2}}{2 \cos^2 \frac{\varphi}{2}} \\
&= \frac{1 + \sin^2 \frac{\varphi}{2} - \cos^2 \frac{\varphi}{2}}{1 - \sin^2 \frac{\varphi}{2} + \cos^2 \frac{\varphi}{2}} \\
&= \frac{1 - \cos\left(\frac{2\varphi}{2}\right)}{1 + \cos\left(\frac{2\varphi}{2}\right)} \\
&= \frac{1 - \cos \varphi}{1 + \cos \varphi} \\
&= \frac{1 - \left(\dfrac{\tan \alpha}{\tan \lambda}\right)}{1 + \left(\dfrac{\tan \alpha}{\tan \lambda}\right)} \\
&= \frac{\tan \lambda - \tan \alpha}{\tan \lambda + \tan \alpha} \\
&= \frac{(\tan \lambda - \tan \alpha) \cos \alpha \cos \lambda}{(\tan \lambda + \tan \alpha) \cos \alpha \cos \lambda} \\
&= \frac{(\tan \lambda - \tan \alpha) \cos \alpha \cos \lambda}{(\tan \lambda + \tan \alpha) \cos \alpha \cos \lambda} \\
&= \frac{\sin \lambda \cos \alpha - \sin \alpha \cos \lambda}{\sin \lambda \cos \alpha + \sin \alpha \cos \lambda} \\
&= \frac{\sin(\lambda - \alpha)}{\sin(\lambda + \alpha)}.
\end{aligned}
\tag{5.37}
$$

Therefore

$$\sin(\lambda - \alpha) = \tan^2 \frac{\varphi}{2} \sin(\lambda + \alpha) \qquad (5.38)$$

This expression will simplify to the level of approximation we need. First, let's see what we can do with the tangent squared term. The series expansion of $\tan x$ is

$$\tan(x) = \tan(0) + x \frac{d\tan(0)}{dx} + \frac{1}{2!} x^2 \frac{d^2\tan(0)}{dx^2} + \frac{1}{3!} x^3 \frac{d^3\tan(0)}{dx^3} + \dots \ (5.39)$$

We now work out the derivatives term by term.

$$
\begin{aligned}
\frac{d\tan x}{dx} &= \frac{d}{dx}\left(\frac{\sin x}{\cos x}\right) \\[2mm]
&= \frac{\cos x \dfrac{d\sin x}{dx} - \sin x \dfrac{d\cos x}{dx}}{\cos^2 x} \qquad (5.40)\\[2mm]
&= \frac{\cos^2 x + \sin^2 x}{\cos^2 x} \\[2mm]
&= \sec^2 x.
\end{aligned}
$$

$$
\begin{aligned}
\frac{d^2\tan x}{dx^2} &= \frac{d\sec^2 x}{dx} \\[2mm]
&= \frac{d}{dx}\left(\frac{1}{\cos^2 x}\right) \\[2mm]
&= \frac{\cos^2 x \dfrac{d(1)}{dx} - (1)\dfrac{d\cos^2 x}{dx}}{\cos^4 x} \qquad (5.41)\\[2mm]
&= \frac{0 - 2\cos x \dfrac{d\cos x}{dx}}{\cos^4 x} \\[2mm]
&= \frac{0 - 2\cos x(-\sin x)}{\cos^4 x} \\[2mm]
&= 2\tan x \sec^2 x.
\end{aligned}
$$

$$\frac{d^3 \tan x}{dx^3} = \frac{d}{dx}\left(2 \tan x \sec^2 x\right)$$

$$= 2 \tan x \frac{d \sec^2 x}{dx} + 2 \sec^2 x \frac{d \tan x}{dx}$$

$$= 2 \tan x \left(2 \tan x \sec^2 x\right) + 2 \sec^2 x \left(\sec^2 x\right)$$

$$= 2 \sec^2 x \left(2 \tan^2 x + \sec^2 x\right). \tag{5.42}$$

Fortunately, the first two derivatives can be used to speed up working out the third. Substituting Eqs. (5.40), (5.41) and (5.42) into Eq. (5.39) gives

$$\tan x = x + \frac{1}{3}x^3 + \dots . \tag{5.43}$$

Squaring Eq. (5.43) gives

$$\tan^2 x = \left(x + \frac{1}{3}x^3 + \dots\right)^2$$

$$= x^2 + \frac{x^6}{9} + \frac{2}{3}x^4 + \dots$$

$$= x^2 + \frac{2}{3}x^4 + \dots \tag{5.44}$$

$$\text{So} \quad \tan^2\left(\frac{\varphi}{2}\right) = \frac{\varphi^2}{4} + \frac{2}{3 \times 16}\varphi^4 + \dots$$

$$= \frac{\varphi^2}{4} + \dots .$$

Equation (5.38) can be further simplified and approximated to

$$\sin(\lambda - \alpha) = \tan^2\frac{\varphi}{2}\sin(\lambda + \alpha)$$

$$\lambda - \alpha \approx \frac{\varphi^2}{4}\sin(\lambda + \lambda). \tag{5.45}$$

Therefore

$$\alpha \approx \lambda - \frac{\varphi^2}{4}\sin(2\lambda). \tag{5.46}$$

We now have what we will need: α in terms of λ.

The reason we need to know α in terms of λ is that the time of day is governed by the rotation of the Earth around its axis, not around the ecliptic.

Getting Around to Earth's Orbit

Let us now return to the difference Δt between mean and solar midday. We wrote this difference earlier, in Eq. (5.6) as

$$t = 12 - \frac{t_{sunrise} + t_{sunset}}{2} + t_{zone}. \tag{5.6}$$

Let me show you how to quantify the physical cause of this time difference. The Mean anomaly measured from the aphelion of the Earth (or equivalently in the Earth's frame of reference the apogee of the Sun) is M. Measured from the vernal Equinox, the mean anomaly of the Earth would be

$$\lambda = M + \varpi. \tag{5.47}$$

To compare the time of the solar midday with midday in mean time, we need to compare right ascensions, not positions along the ecliptic.

Using orbital mechanics, we measure the time difference from Υ and do not worry too closely about the time of day where we are on the surface of the Earth. The actual time the Sun takes to get from the Vernal Equinox depends on

$$t_{solar} = t_{sidereal-\Upsilon} - \alpha, \tag{5.48}$$

because the Earth rotates with axes normal to the place of the celestial equator, not the ecliptic. Let

$$\omega_{Earth\ Orbit} = \frac{2\pi}{T_{Earth}}, \tag{5.49}$$

where T_{Earth} was defined as one solar year in Eq. (5.2). We can rewrite Eq. (5.48) as

$$t = \frac{M + \varpi - \alpha}{\omega_{Earth\ Orbit}}. \tag{5.50}$$

We have terms for all these values now. Substituting first for $M + \varpi$ from Eq. (5.31)

$$M + \varpi \approx \lambda - 2e \sin M, \tag{5.51)}$$

where we have neglected higher terms of $O(e^2)$. Taking our value for α from Eq. (5.46), we have

$$
\begin{aligned}
t &\approx \frac{\lambda - 2e \sin M - \left(\lambda - \dfrac{\varphi^2}{4} \sin(2\lambda) \right)}{\omega_{\text{Earth-Orbit}}} \\[2ex]
&= \frac{-2e \sin M + \dfrac{\varphi^2}{4} \sin(2\lambda)}{\omega_{\text{Earth-Orbit}}} \\[2ex]
&= \frac{-2e \sin M + \dfrac{\varphi^2}{4} \sin(2M + 2\varpi)}{\omega_{\text{Earth-Orbit}}} .
\end{aligned}
\tag{5.52}
$$

Here we have an equation which relates three parameters governing the orbit of the Earth to Δt, viz. the eccentricity e, the obliquity φ and the longitude ϖ of the periapsis from Υ. Equation (5.52) is known, rather grandly, as the Equation of Time.

At this level of approximation, there's a term proportional to $\sin(M)$ due to ellipticity, which has a repeating period of a year, and a term proportional to $\sin(2(M + \varpi))$, due to obliquity, which has a repeating period of half a year and is offset in phase relative to the term proportional to $\sin(M)$. An offset in phase is to be expected because the passage of perihelion is not in phase with the seasons.

At this point I am going to use trial values of these parameters in my data for Δt, as shown in Fig. 5.15. I substitute my trial values into Eq. (5.52) and calculate the difference of squares

$$
S_{SQ} = \sum_{n}^{i=1} \left(t_{\text{sunrise\_set}} - t_{\text{Eq\_of\_Time}} \right)^2
\tag{5.53}
$$

I might have to do a Monte-Carlo simulation where I repeat this exercise thousands of times with trial values to find the set that gives the lowest value of S_{SQ}. In practice, this proved not to be necessary. I set up an Excel spreadsheet and tried a handful of cases and soon found the lowest value of S_{SQ} I could get (Fig. 5.23).

How did I get the Vernal Equinox? The zeroth order answer is to look for 12 hours between sunrise and sunset in March 2009. However, a closer look at the definitions of sunrise and sunset from the US Naval Observatory web site [106]

Sunrise and sunset conventionally refer to the times when the upper edge of the disk of the Sun is on the horizon. Atmospheric conditions are assumed to be average, and the location is in a level region on the Earth's surface.

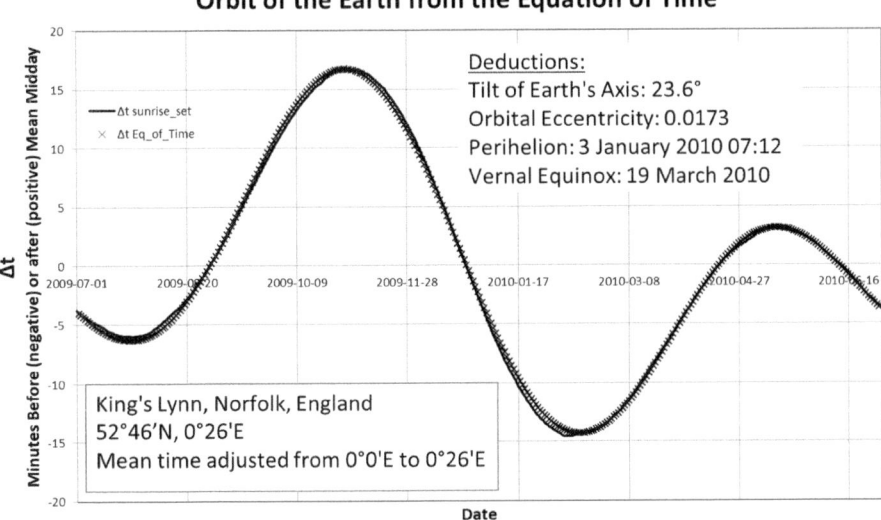

Fig. 5.23 My deduction of the orbit of the Earth from the Equation of Time

Table 5.4 Data used to select the March Equinox date	Date in March 2009	(Sunrise-Sunset) time (hours)
	17	11.9667
	18	12.0333
	19	12.1167
	20	12.1833
	21	12.25

For computational purposes, sunrise or sunset is defined to occur when the geometric zenith distance of centre of the Sun is 90.8333 degrees. That is, the centre of the Sun is geometrically 50 arcminutes below a horizontal plane. For an observer at sea level with a level, unobstructed horizon, under average atmospheric conditions, the upper limb of the Sun will then appear to be tangent to the horizon. The 50-arcminute geometric depression of the Sun's centre used for the computations is obtained by adding the average apparent radius of the Sun (16 arcminutes) to the average amount of atmospheric refraction at the horizon (34 arcminutes).

So I looked for a little more than 12 hours' daylight in Tables 5.2, 5.3 and 5.4 to allow for these effects. The data in Table 5.4 suggest March 19th as the best fit.

Otherwise my orbital parameters are:

$$
\begin{array}{l}
\varpi = 285.25^\circ = 4.978 \ \text{radians} \\
e = 0.0173 \\
\varphi = 23.6^\circ = 0.412 \ \text{radians}
\end{array}
\tag{5.54}
$$

Table 5.5 List of Earth Positions at Observing Times, 2009–10 Apparition of Mars

Year	Month	Date	Hour (UT)	Minute (UT)	Components of the Position of the Earth		
					X (AU)	Y (AU)	Z (AU)
2009	10	22	03	58	−0.87227	−0.47869	0.00
2009	10	23	01	52	−0.86434	−0.49234	0.00
2009	10	25	02	51	−0.84583	−0.5224	0.00
2009	10	29	01	52	−0.80692	−0.57881	0.00
2009	11	2	02	29	−0.76344	−0.63336	0.00
2009	11	7	02	29	−0.70423	−0.69675	0.00
2009	11	8	02	21	−0.69181	−0.70874	0.00
2009	11	13	03	51	−0.62539	−0.7664	0.00
2009	11	15	02	36	−0.5985	−0.78701	0.00
2009	11	17	02	17	−0.57036	−0.80709	0.00
2009	11	22	02	27	−0.49646	−0.85331	0.00
2009	12	11	01	3	−0.18732	−0.96614	0.00
2009	12	16	01	25	−0.10053	−0.97843	0.00
2009	12	23	01	10	0.021578	−0.9828	0.00
2009	12	29	06	13	0.129857	−0.97415	0.00
2010	1	1	02	50	0.179259	−0.96623	0.00
2010	1	8	02	17	0.297648	−0.9366	0.00
2010	1	18	01	51	0.458747	−0.86969	0.00
2010	1	26	22	22	0.589927	−0.78773	0.00
2010	2	1	0	15	0.658885	−0.73194	0.00
2010	2	14	21	21	0.819074	−0.55117	0.00
2010	2	17	19	13	0.846802	−0.50872	0.00
2010	2	21	02	45	0.875682	−0.4588	0.00
2010	2	21	19	43	0.881464	−0.44794	0.00
2010	2	26	22	31	0.919528	−0.36677	0.00
2010	3	2	01	09	0.938646	−0.31703	0.00
2010	3	2	19	59	0.943106	−0.30416	0.00
2010	3	4	21	06	0.953905	−0.27033	0.00
2010	3	6	19	35	0.962453	−0.24009	0.00
2010	3	7	21	01	0.967525	−0.22018	0.00
2010	3	13	20	03	0.986772	−0.1188	0.00
2010	3	21	20	15	0.996006	0.019225	0.00
2010	4	2	20	24	0.974171	0.224658	0.00
2010	4	4	20	07	0.966493	0.257983	0.00
2010	4	5	21	12	0.962003	0.275383	0.00
2010	4	8	21	17	0.947379	0.324873	0.00
2010	5	6	21	27	0.698257	0.728789	0.00
2010	5	15	22	14	0.580927	0.827957	0.00

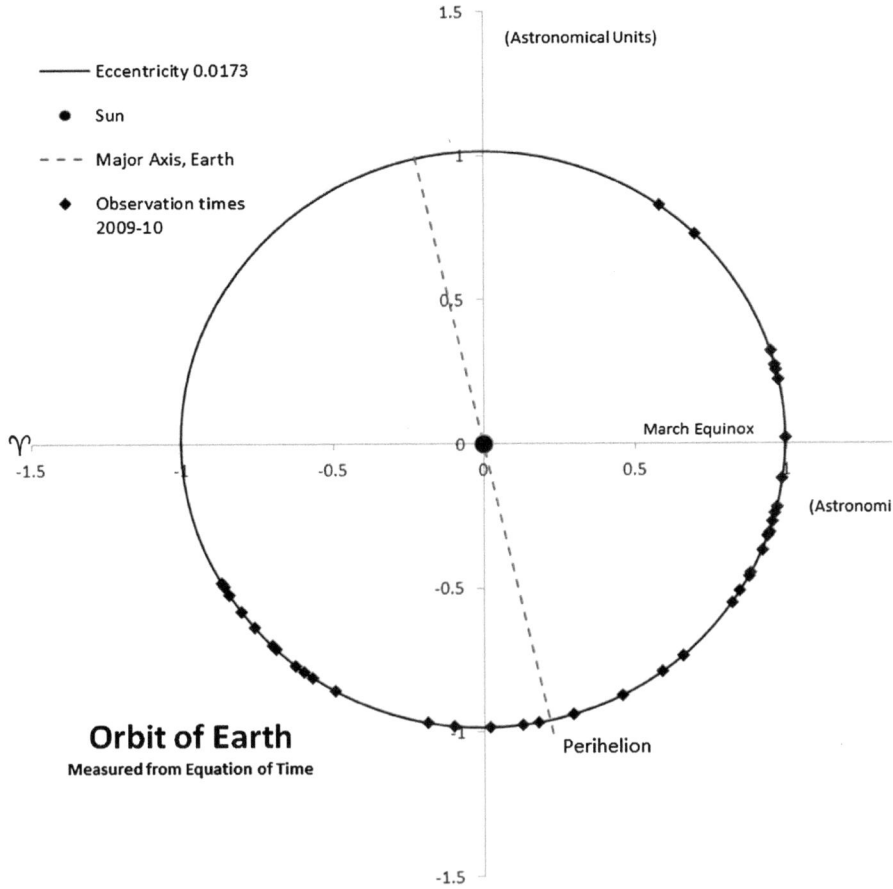

Fig. 5.24 The orbit of the Earth as calculated from sunrise and sunset times. The diamonds show the times observations of the positions of Mars were made during the 2009–10 apparition

The value of ϖ is deduced from the dates and shown in Table 5.4.

Figure 5.24 shows the orbit of the Earth. You have to look hard to see that it is not quite circular. The most obvious way to see this is to note that it does not quite cross the vertical axis of the diagram at distances of ±1 Astronomical Unit from the Sun.

Later on, we will need to know the distances from the Sun at the observation points. This is because, as Fig. 5.24 makes very plain, Mars was observed from a platform which was anything but stationary. We will need to be able to subtract the movement of the Earth away from that of Mars.

Table 5.5 lists the coordinates of the Earth in a coordinate system where the ecliptic plane is the $Z = 0$ plane, and the negative X axis points towards ♈. That is, we are looking "down" onto the ecliptic from the northward normal to the ecliptic.

Chapter 6

Here's Looking at You, Mars!

My technique for observing Mars turned out to be a little unusual. It never struck me as such, but when other people tried it, they did not seem to be able to copy me. People have had it dinned into them that they need to stack lots of images. Bad, bad idea. Mars moves too much with respect to the starry background. You lose vital information if you stack.

Equipment

The images I used were taken with a common or garden Canon digital single-lens reflex (DSLR) camera, to be precise, a Canon EOS 1000D™. By today's standards, this is a dated beast, but I don't care. It works great, and it takes much better landscape pictures than my smart phone. You will gather that it has not been modified for astronomy, e.g. by removing the IR filter. Its resolution is 3888 × 2592 pixels on a 22.2 × 14.8 mm CMOS sensor.

The main thing you need is to have a camera which can provide live view shooting. This means that it bypasses the mirror and pentaprism and sends the image straight from the chip to a screen on the back of the camera. My experience has been that without this feature, you can't focus at night.

The lens I used is a 500 mm catadioptric lens. It doesn't even have a diaphragm. It is f/8 or f/8. If you don't like that you can have f/8. The optical quality isn't great, but it is good enough. The optical arrangement does not appear to be published anywhere, but it looks like a Schmidt-Cassegrain

J. Clark, *Calculate the Orbit of Mars!*,
https://doi.org/10.1007/978-3-030-78267-2_6,
© Springer Nature Switzerland AG 2021

layout with an extra lens behind the primary: this would be the last refraction the light gets before it reaches the sensor chip.

The thing about these lenses is that people who have never had to focus a camera buy them, and then discover the joys of losing your target bird or whatever because they couldn't focus in time. So they sell them on eBay for next to nothing.

In fact I eventually learned that I could exploit the crummy optics to focus. Once I learned that this lens has a bit of off-axis astigmatism, I would put the planet just slightly to one side of the centre, with the live view magnified to 10×, and as I moved the lens in and out of focus, the image of the planet would switch from being oval with the major axis vertical to being oval with the major axis horizontal. Just rocking it through focus until the planet stopped being oval meant I could focus in 10 seconds. Before I learned that trick, focus was a real problem.

The camera is shown in Fig. 6.1.

Eventually I discovered that you can almost eliminate the orange glow from streetlights with a so-called "Redhancer" filter (Fig. 6.2). These transmit most visible light just fine, but absorb orange. They are sold to make pictures of autumn leaves look less brown. But they stop sodium light brilliantly.

Because there is an irritant in the sky called the Moon, I soon realized that I needed a good lens hood. The ones you buy are useless for astronomy, so I made one to fit over the one I bought. My first effort was a large yoghurt pot. That was a disaster, not least because it was white. So I tried black electrical insulation tape. I was a bit concerned it might be too shiny, but in practice that is not much of a problem. Regular use soon removes the shine.

Fig. 6.1 The Canon camera with 500 mm catadioptric lens. (Image by the author)

Fig. 6.2 The "Redhancer" filter, a didymium filter. (Image by the author)

Fig. 6.3 Making a lens hood from a cola bottle with black electrical insulting tape. Images by the author

Figure 6.3 shows how I made it. It is pretty well impossible to apply the tape circumferentially to the interior. You have to do it axially. Then you can do the exterior circumferentially. It has lasted very well.

Guiding

I piggy-backed the camera onto a telescope. I used a jubilee clip to hold a home-made holder my dad and I made onto an altitude-azimuth of the type that goes onto camera tripods. You can buy them separately. The arrangement is shown in Fig. 6.4.

The telescope itself I would use for autoguiding. As the years went by, I have bought and sold different telescopes. I had an 8" Newtonian when I took these pictures, on a Sky Watcher HEQ5 mount with the Rajiva autoguiding system, a modification to allow autoguiding. The Rajiva website seems to have disappeared, presumably because computer-controlled mounts are now mainsteam, not just for electronics hobbyists. However, it was a good system while it lasted.

For guiding I used a Phillips SPC900NC webcam, with no modifications, fastened to a Plössl eyepiece (Fig. 6.5). With this arrangement, the planet being tracked came out on the guiding image as more or less a point. This enabled me to guide on the planet itself.

The guiding software I used was Garzarolli's Guidemaster. This software probably predated the now-fashionable PHD2. While it is not as reliable as PHD2 – it does have an infuriating tendency to freeze up your laptop from time to time – it certainly did the job as far as I was concerned. A guide to using it appears in my book *Viewing and Imaging the Solar System: A Guide for Amateur Astronomers* [107].

Gradually I learned that I got better guiding in a broadly southerly direction than in a broadly easterly or westerly direction. I never found out whether this was a limitation of this particular equipment. The guiding to the east or west wasn't bad, but nor was it spot-on.

Fig. 6.4 My home-made mount for piggy-backing the DLSR onto a telescope. Images by the author

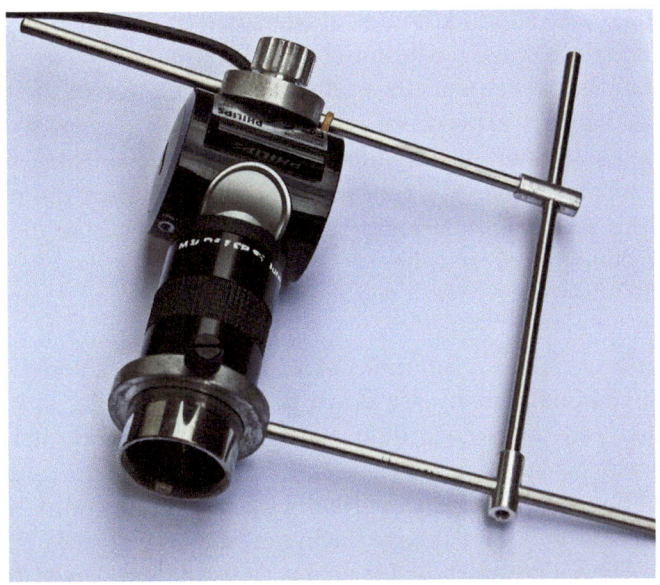

Fig. 6.5 My first guiding camera was a webcam in eyepiece projection mode, with its own lens not removed. The eyepiece is a 28 mm Plössl lens. The metal framework is to hold the webcam onto the eyepiece. I plugged this into the focuser of the telescope from which I was piggybacking. Image by the author

With Mars, an additional difficulty was that it is a fast moving body well away from opposition. This meant that except close to opposition, my stars came out as short lines if I guided on the planet.

Photographic Technique

By trial and error, I developed a process whereby I shot two exposures: one for one second to capture the planet and another for four minutes to capture the starry background. It was not necessary to take "dark" frames because I also discovered by messing about with raw files generated by the camera (CR2 files in Canons) that the camera does quite a bit of processing to remove noise before it makes a JPEG file. There would still be a very few hot pixels, but nowhere near as many as in the raw file. Since they were always in red green or blue primary colours, I soon learned not to mistake them for stars. Eventually I modified this technique. First I did shoot a dark frame and used Deep Sky Stacker to subtract the dark signal. Second, I learned to take *very* short exposures of Mars before and after the main shot and deemed their average position to be the location of Mars.

I needed about twenty stars to get a good triangulation of the position of Mars. Occasionally, I exposed an intermediate image for 20 seconds. I'm not sure I had any grand plan to doing this, but whenever I did, bright stars were better resolved than in a four-minute exposure. They did not spill all over the place into neighbouring pixels. If I could wind the clock back, I would have done this every time.

Gradually I discovered that it was only worth trying such photography on really clear nights. Bad seeing due to wind was not an issue: it would average out over four minutes. But milky skies meant that you could not clearly see dim stars. I could easily see down to 14th magnitude: at one point I photographed 14th-magnitude Pluto with no difficulty. I hesitate to claim 16th magnitude, but I'm sure I managed that on the clearest nights, especially after I bought a "Redhancer" filter. I did not get this until 2011, so the pictures I present will have orange backgrounds.

The Moon was also a formidable enemy. It would back-light the slightest cloud if it was close, especially if it was more than half full. I stopped bothering if it was within about 45 degrees of Mars, unless the weather meant it was the only night I was going to get anything.

I certainly did not need hours and hours of clear sky. A few minutes was enough to set up and grab my pictures. I could put my stuff away in any weather, and occasionally it was raining when I did so.

Normally I left my mount outside, because that way I didn't have to keep polar-aligning it. I would tie a large black refuse bag over it and use rubber bands to shape the bag to the mount so that it did not get damaged in the wind. A bag would last about a month except in summer, when it only lasted a couple of weeks, before sunlight rotted it.

I would need to polar align about once a month. My back yards in King's Lynn and Bristol were well secluded. I did worry about theft, but gambled on nobody being able to sell a heavy mount in a seedy bar somewhere nearby. I never suffered a theft. I moved 35 miles to Wales in 2016, mostly because housing here is a quarter the price of houses in Bristol. Now I have a very exposed back yard on a hillside. This more or less forced me to build a permanent observatory, which is something I wanted to do anyway.

Output from a typical night's observing is shown in Figs. 6.6 and 6.7. I would usually enhance the mid-tone contrast to make more stars show up. You can do this in Photoshop Elements or Microsoft Picture Manager.

Fig. 6.6 A typical one-second exposure to locate Mars on the photograph. 29 December 2009, 06:13. Nowadays I would only expose for maybe 1/5 s. (Image by the author)

How I Made Measurements from Pictures Like Fig. 6.7

I imported the figures into a computer-aided drafting (CAD) software package. Although you can download free ones like Alibre™, I chose to pay for one, partly because it is good and partly because I am very familiar with it. These packages are sophisticated. The one I have is TurboCAD™ 2018. I have in fact used it to draw many of the diagrams in this book. It is not so hot for 3-D CAD work, for which I use an old version of SolidWorks. Nowadays I have to use that on an old laptop because my version won't run on Windows 10, and I am darned if I am going to pay over $5000 for a newer version. You can always get free CAD systems. Google will find plenty of them. You need one with "layers" capability and the ability to import jpeg images.

Anyway, back to TurboCAD, I imported the JPEG image from my DSLR and used the "snap" tools to make the imported image exactly 10 units wide. The "snap" feature in CAD systems enables you to make the system pick up the nearest position, such as a point on a grid, a circle centre, or a vertex

Fig. 6.7 A typical four-minute exposure, showing the starry background around Mars. I can see many more stars than I need in this photo. Reproduction in a book may reduce the numbers that are visible. 29 December 2009, Exposure begun at 06:13. (Image by the author)

of an imported image to machine-level accuracy, which of course is absurdly accurate on a 64-bit computer. This is where CAD systems really come into their own and can do things Photoshop™ can't. The snap feature is the most important feature. The other advantage is that you can zoom in almost indefinitely.

Talking of Photoshop, Fig. 6.7, I drew a point wherever I could see a star in the image of Fig. 6.7 after using every trick I knew in Photoshop to make the stars visible. There are about 120 stars. That's about 100 more than I will need. Most of them are too faint to be able to guess their positions accurately anyway.

I have used the snap tools in TurboCAD to mark the centre of the image by drawing cross-hairs from the corners. I have also drawn the largest complete circle I can, centred on the image centre. Experience has taught me that this circle is pretty much the limit of good image quality from my 500 mm catadioptric telephoto lens. Measuring star positions outside this circle is asking for errors. The circle I drew in the middle is for convenience to enable me to snap onto the image centre quickly and easily (Fig. 6.8).

Figure 6.9 shows how I next divide the circle of interest into four quarters. This is purely a practical aide memoire, as it reminds me that in two of the quarters, x-coordinates are positive, while in the other two, x-coordinates

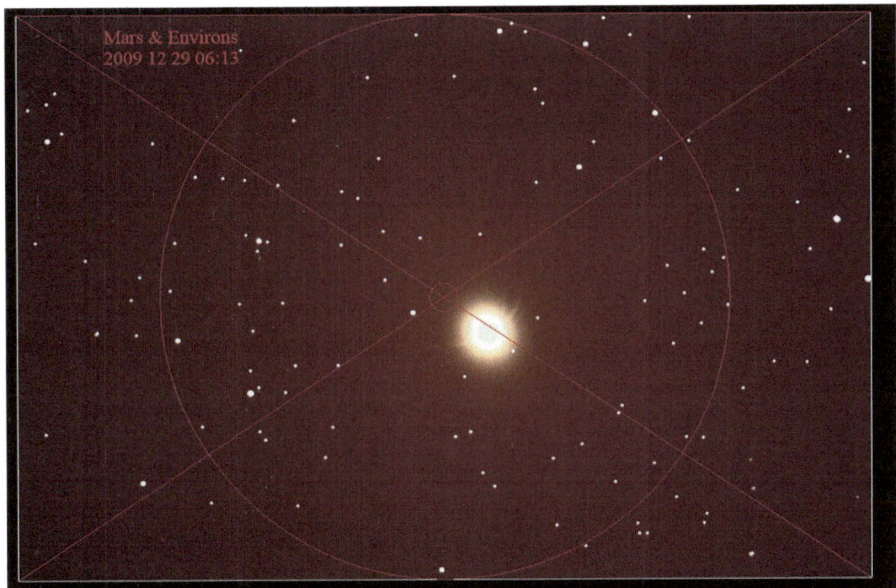

Fig. 6.8 I used Photoshop to pencil in dots where I could unequivocally see a star in the image in Fig. 6.7. There were a few doubtful cases. One was blurred and extended so it may have been a galaxy. I have not attempted to put the dots in very exact positions. The objective was to make the star positions visible after the photograph has been put into a book. I have also used the CAD system to mark out positions on the photograph. (Image by the author)

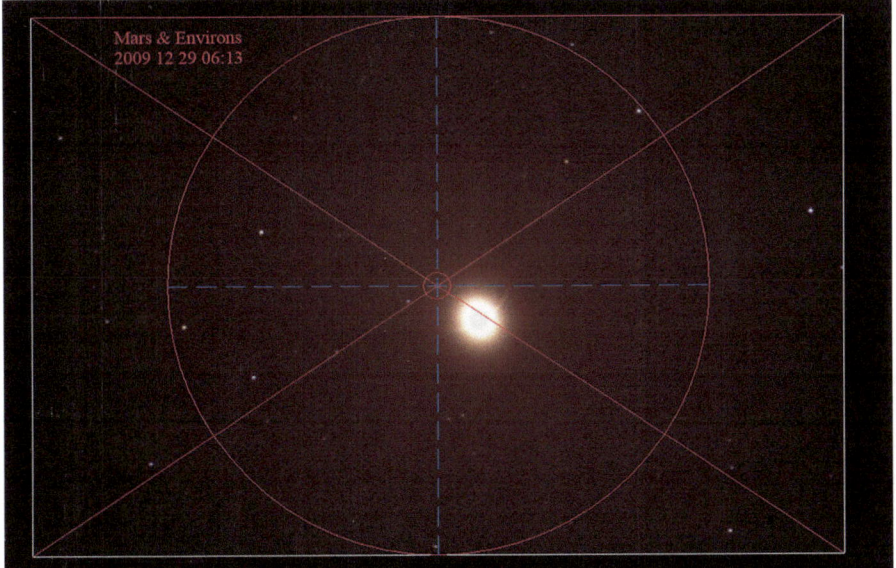

Fig. 6.9 I have divided the circle in which I am interested into quarters, for convenience. (Image by the author)

are negative. Similarly y is positive in the upper two quarters and negative in the lower two.

The next step requires a little bit of skill and practice, but it isn't that difficult. You zoom in to each star and draw a circle sing the centre-and-radius option. The point on which you click is the centre, then you draw the circle out to a radius of your choice. It costs nothing but time to hit the undo button and repeat until you are happy. Fig. 6.10 shows the result.

Next I need to identify these stars. You can do this using any one of the many planetarium software packages that are now available. The one I have on my PC (as opposed to my iPhone™ and iPad™) is Cartes du Ciel [108]. It is free, and after a dozen years, I have become accustomed to its foibles. While in principle I could import the photo in Fig. 6.7 into Cartes du Ciel and overlay the catalogue stars onto it, in practice that's time-consuming. Instead, I just set the observatory and time and date to correspond to the photo and rotate the image in the Cartes du Ciel screen in 30° jumps until it has roughly the same orientation as the photo.

By the way, you need to download the Tycho2 and UCAC4 star catalogs [109] and make sure that you switch them on to get enough stars (Fig. 6.11)

Then, click on the stars you think you recognize from the photo, until you have recorded all their names in the CAD system. In practice, you can copy

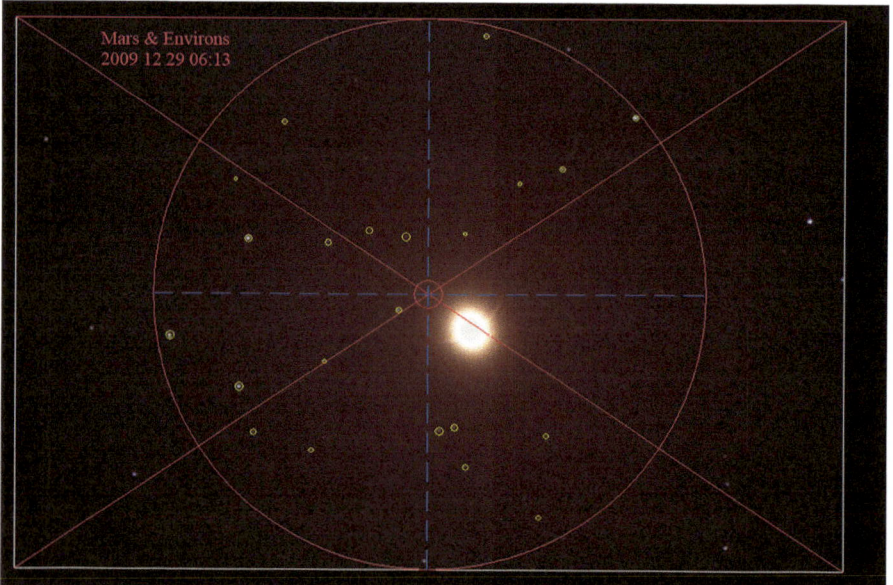

Fig. 6.10 Now I have hand-drawn circles centred on twenty stars of interest. (Image by the author)

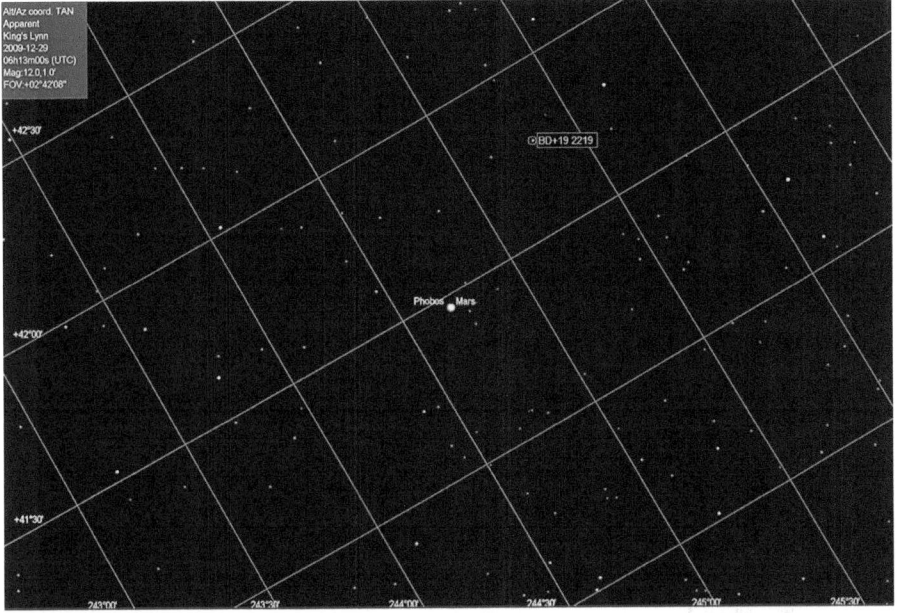

Fig. 6.11 The same star field shown in Cartes du Ciel. By clicking on stars, you can identify them. (Image by the author)

and paste your star lists from one photo to the next, which makes for a considerable labour saving. If you know what you are doing, you can copy, move and rotate the list of names.

Inside the CAD systems, you have what they call "layers". You are meant to imagine that each layer corresponds to sheet of tracing paper from the "good" old days of doing design on drawing boards. You can make the layers visible or invisible, and you can assign a default colour to each layer. That way, you can keep the circles centred on the stars on one layer and colour it, say, yellow, and put the names on another layer, which you might colour cyan. You can set the copying only to copy one layer so that you don't copy all sorts of stuff you don't want to. In TurboCAD, you do this by first selecting objects, then copying the selected objects. Each system has its own little ways (Fig. 6.12).

One very labour-saving feature of Cartes du Ciel is that it will produce a savable list of all the stars on the screen (Fig. 6.13). You can choose to save it as a tab-separated or comma-separated file. You then import this into your favourite spreadsheet program, such as Excel, and use the "text-to-columns" feature to arrange the data conveniently. You are going to throw a lot of it away, and you might as well make this process as easy as possible.

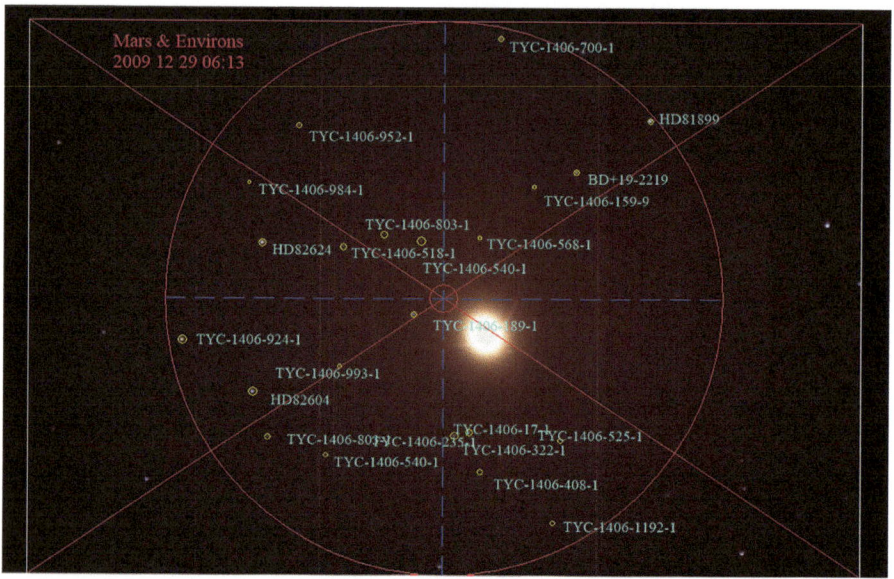

Fig. 6.12 By repeatedly clicking on stars, you can identify the stars in the photograph from Fig. 6.7. (Image by the author)

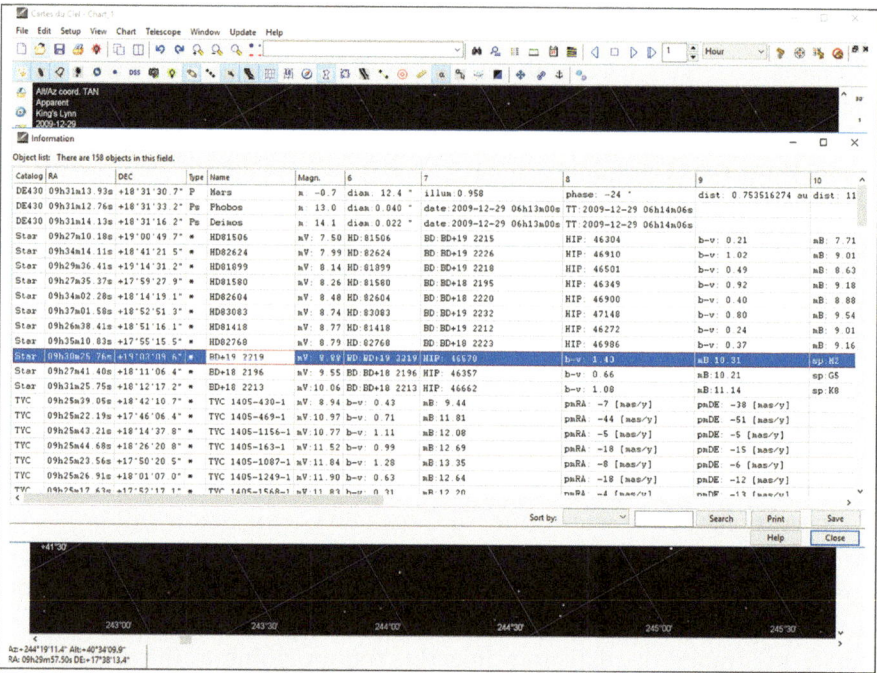

Fig. 6.13 Cartes du Ciel will produce a table of all the stars, and other objects if you request them, visible on the screen. (Image by the author)

Figure 6.14 shows how these data might look once imported into a spreadsheet and tidied up.

Figure 6.15 shows how the coordinates of a star are measured using the CAD system. By instructing the measurements to "snap"; to circle centres, I can quickly measure the dimension from the center to the circle I hand drew to within machine accuracy. Three significant figures are enough, and with skill, achievable. In practice I measured the positions of all the stars in

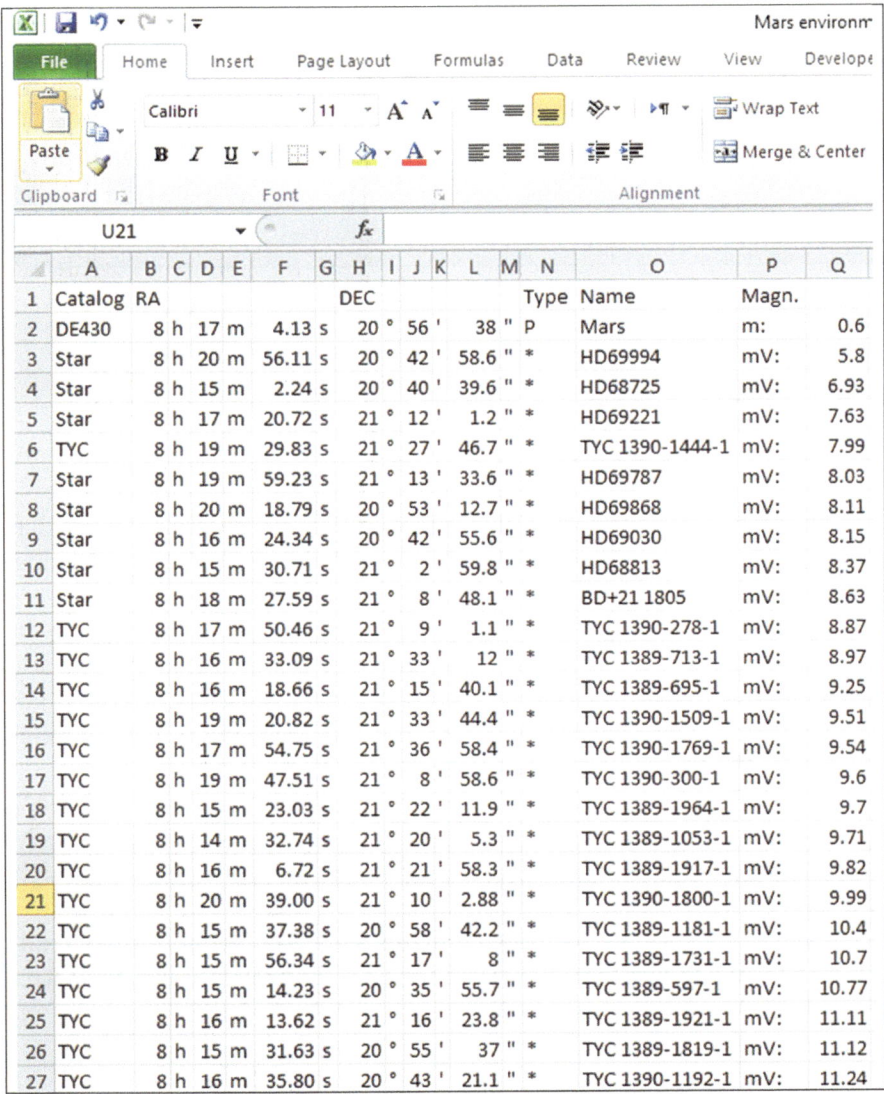

	A	B	C	D	E	F	G	H	I	J	K	L	M	N	O	P	Q
1	Catalog	RA						DEC						Type	Name	Magn.	
2	DE430	8 h	17 m	4.13 s		20 °	56 '	38 "	P						Mars	m:	0.6
3	Star	8 h	20 m	56.11 s		20 °	42 '	58.6 "	*						HD69994	mV:	5.8
4	Star	8 h	15 m	2.24 s		20 °	40 '	39.6 "	*						HD68725	mV:	6.93
5	Star	8 h	17 m	20.72 s		21 °	12 '	1.2 "	*						HD69221	mV:	7.63
6	TYC	8 h	19 m	29.83 s		21 °	27 '	46.7 "	*						TYC 1390-1444-1	mV:	7.99
7	Star	8 h	19 m	59.23 s		21 °	13 '	33.6 "	*						HD69787	mV:	8.03
8	Star	8 h	20 m	18.79 s		20 °	53 '	12.7 "	*						HD69868	mV:	8.11
9	Star	8 h	16 m	24.34 s		20 °	42 '	55.6 "	*						HD69030	mV:	8.15
10	Star	8 h	15 m	30.71 s		21 °	2 '	59.8 "	*						HD68813	mV:	8.37
11	Star	8 h	18 m	27.59 s		21 °	8 '	48.1 "	*						BD+21 1805	mV:	8.63
12	TYC	8 h	17 m	50.46 s		21 °	9 '	1.1 "	*						TYC 1390-278-1	mV:	8.87
13	TYC	8 h	16 m	33.09 s		21 °	33 '	12 "	*						TYC 1389-713-1	mV:	8.97
14	TYC	8 h	16 m	18.66 s		21 °	15 '	40.1 "	*						TYC 1389-695-1	mV:	9.25
15	TYC	8 h	19 m	20.82 s		21 °	33 '	44.4 "	*						TYC 1390-1509-1	mV:	9.51
16	TYC	8 h	17 m	54.75 s		21 °	36 '	58.4 "	*						TYC 1390-1769-1	mV:	9.54
17	TYC	8 h	19 m	47.51 s		21 °	8 '	58.6 "	*						TYC 1390-300-1	mV:	9.6
18	TYC	8 h	15 m	23.03 s		21 °	22 '	11.9 "	*						TYC 1389-1964-1	mV:	9.7
19	TYC	8 h	14 m	32.74 s		21 °	20 '	5.3 "	*						TYC 1389-1053-1	mV:	9.71
20	TYC	8 h	16 m	6.72 s		21 °	21 '	58.3 "	*						TYC 1389-1917-1	mV:	9.82
21	TYC	8 h	20 m	39.00 s		21 °	10 '	2.88 "	*						TYC 1390-1800-1	mV:	9.99
22	TYC	8 h	15 m	37.38 s		20 °	58 '	42.2 "	*						TYC 1389-1181-1	mV:	10.4
23	TYC	8 h	15 m	56.34 s		21 °	17 '	8 "	*						TYC 1389-1731-1	mV:	10.7
24	TYC	8 h	15 m	14.23 s		20 °	35 '	55.7 "	*						TYC 1389-597-1	mV:	10.77
25	TYC	8 h	16 m	13.62 s		21 °	16 '	23.8 "	*						TYC 1389-1921-1	mV:	11.11
26	TYC	8 h	15 m	31.63 s		20 °	55 '	37 "	*						TYC 1389-1819-1	mV:	11.12
27	TYC	8 h	16 m	35.80 s		20 °	43 '	21.1 "	*						TYC 1390-1192-1	mV:	11.24

Fig. 6.14 Data downloaded from Cartes du Ciel and sorted, in this case on star brightness. (Image by the author)

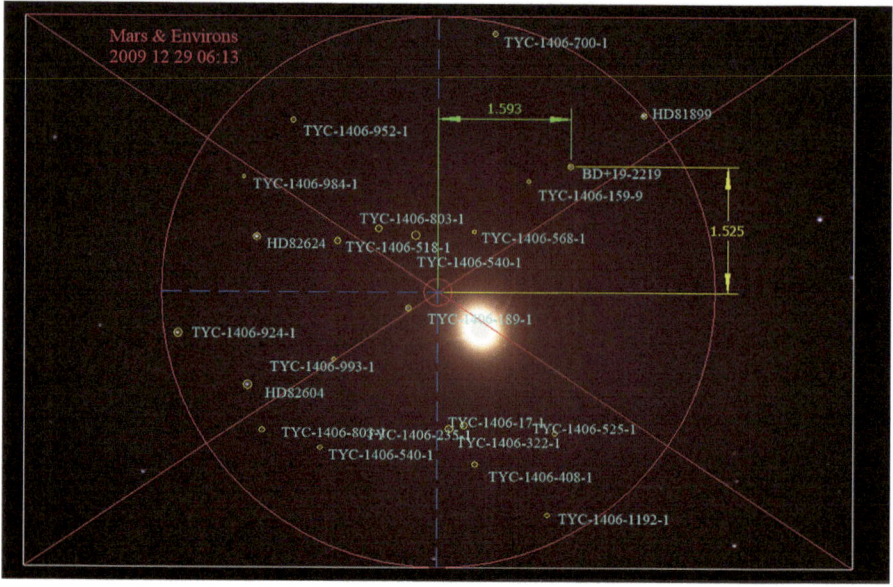

Fig. 6.15 Showing how the coordinates of one star are measured using the dimensioning tools in the CAD system. (Image by the author)

one quadrant of my circle and then made the layer containing those data invisible. That provided a reasonable compromise between speed and clogging the screen up with too much information.

Figure 6.16 shows how I added the x and y coordinates of the stars to my spreadsheet shown in Fig. 6.14. This had to be done manually. This was quite labour intensive. The best way to do it was to do it all in one hit, if you possibly can, because you get into a rhythm and soon start to rattle through the photos. Expect it to take several days.

I would have liked to try to automate this process. I have no doubt that it could be done, but in the end I felt that I didn't have enough data to repay the "overhead" cost of learning to write image analysis software. It might be a fun post-retirement project in a few years.

Of course you can buy software to do this stuff. My feeling is that I really don't like using "black box" software for data analysis. I don't know what it has done. For me that takes the fun out of it. You, dear reader, are fully entitled to differ.

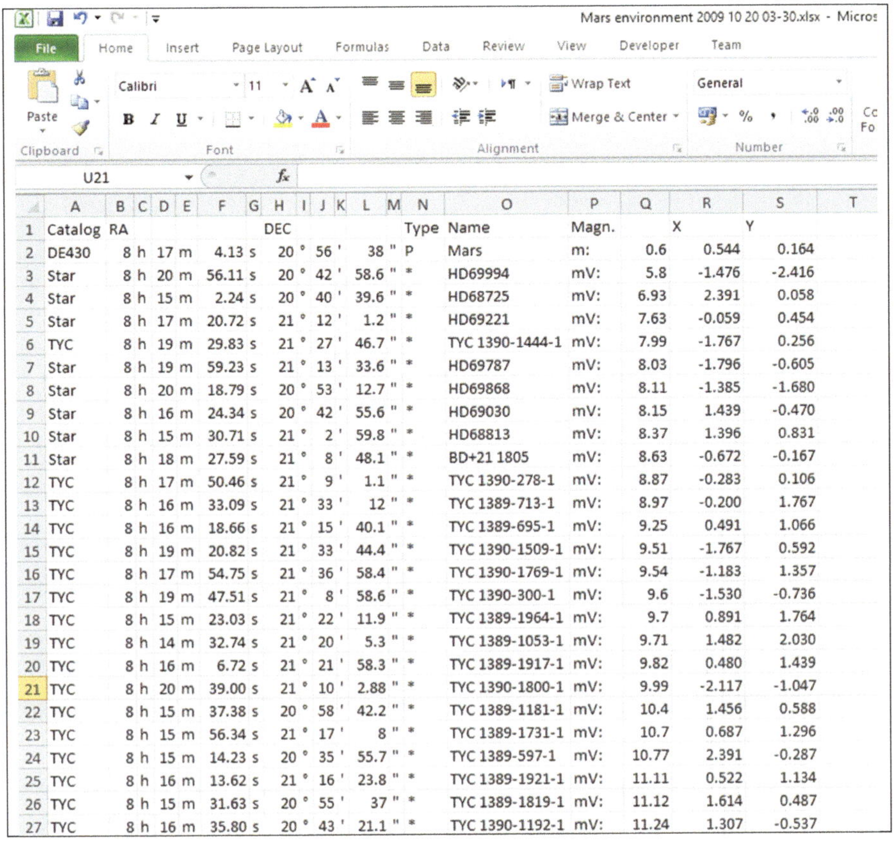

	A	B	C	D	E	F	G	H	I	J	K	L	M	N	O	P	Q	R	S	T
1	Catalog	RA					DEC							Type	Name	Magn.		X	Y	
2	DE430	8 h	17 m	4.13 s		20 °	56 '	38 "	P						Mars	m:		0.6	0.544	0.164
3	Star	8 h	20 m	56.11 s		20 °	42 '	58.6 "	*						HD69994	mV:		5.8	-1.476	-2.416
4	Star	8 h	15 m	2.24 s		20 °	40 '	39.6 "	*						HD68725	mV:		6.93	2.391	0.058
5	Star	8 h	17 m	20.72 s		21 °	12 '	1.2 "	*						HD69221	mV:		7.63	-0.059	0.454
6	TYC	8 h	19 m	29.83 s		21 °	27 '	46.7 "	*						TYC 1390-1444-1	mV:		7.99	-1.767	0.256
7	Star	8 h	19 m	59.23 s		21 °	13 '	33.6 "	*						HD69787	mV:		8.03	-1.796	-0.605
8	Star	8 h	20 m	18.79 s		20 °	53 '	12.7 "	*						HD69868	mV:		8.11	-1.385	-1.680
9	Star	8 h	16 m	24.34 s		20 °	42 '	55.6 "	*						HD69030	mV:		8.15	1.439	-0.470
10	Star	8 h	15 m	30.71 s		21 °	2 '	59.8 "	*						HD68813	mV:		8.37	1.396	0.831
11	Star	8 h	18 m	27.59 s		21 °	8 '	48.1 "	*						BD+21 1805	mV:		8.63	-0.672	-0.167
12	TYC	8 h	17 m	50.46 s		21 °	9 '	1.1 "	*						TYC 1390-278-1	mV:		8.87	-0.283	0.106
13	TYC	8 h	16 m	33.09 s		21 °	33 '	12 "	*						TYC 1389-713-1	mV:		8.97	-0.200	1.767
14	TYC	8 h	16 m	18.66 s		21 °	15 '	40.1 "	*						TYC 1389-695-1	mV:		9.25	0.491	1.066
15	TYC	8 h	19 m	20.82 s		21 °	33 '	44.4 "	*						TYC 1390-1509-1	mV:		9.51	-1.767	0.592
16	TYC	8 h	17 m	54.75 s		21 °	36 '	58.4 "	*						TYC 1390-1769-1	mV:		9.54	-1.183	1.357
17	TYC	8 h	19 m	47.51 s		21 °	8 '	58.6 "	*						TYC 1390-300-1	mV:		9.6	-1.530	-0.736
18	TYC	8 h	15 m	23.03 s		21 °	22 '	11.9 "	*						TYC 1389-1964-1	mV:		9.7	0.891	1.764
19	TYC	8 h	14 m	32.74 s		21 °	20 '	5.3 "	*						TYC 1389-1053-1	mV:		9.71	1.482	2.030
20	TYC	8 h	16 m	6.72 s		21 °	21 '	58.3 "	*						TYC 1389-1917-1	mV:		9.82	0.480	1.439
21	TYC	8 h	20 m	39.00 s		21 °	10 '	2.88 "	*						TYC 1390-1800-1	mV:		9.99	-2.117	-1.047
22	TYC	8 h	15 m	37.38 s		20 °	58 '	42.2 "	*						TYC 1389-1181-1	mV:		10.4	1.456	0.588
23	TYC	8 h	15 m	56.34 s		21 °	17 '	8 "	*						TYC 1389-1731-1	mV:		10.7	0.687	1.296
24	TYC	8 h	15 m	14.23 s		20 °	35 '	55.7 "	*						TYC 1389-597-1	mV:		10.77	2.391	-0.287
25	TYC	8 h	16 m	13.62 s		21 °	16 '	23.8 "	*						TYC 1389-1921-1	mV:		11.11	0.522	1.134
26	TYC	8 h	15 m	31.63 s		20 °	55 '	37 "	*						TYC 1389-1819-1	mV:		11.12	1.614	0.487
27	TYC	8 h	16 m	35.80 s		20 °	43 '	21.1 "	*						TYC 1390-1192-1	mV:		11.24	1.307	-0.537

Fig. 6.16 Showing how I added x and y coordinate data to my spreadsheet. (Image by the author)

There is one thing I have not done, which is to show you how to locate our target Mars. This was done by taking a one-second exposure immediately before the four-minute exposure photo we have been analysing. Mars would not have moved significantly relative to my field of view, especially as I was guiding on it, not a star. Figs. 6.17 and 6.18 show how I did this. I would zoom right in and draw a circle round the image of the planet. I usually used the feature to draw a circle through three points, manually selected by mouse clicks. I took several attempts until I was satisfied with the result. Getting this right is important. Your measurement is only as good as your judgement of where Mars is.

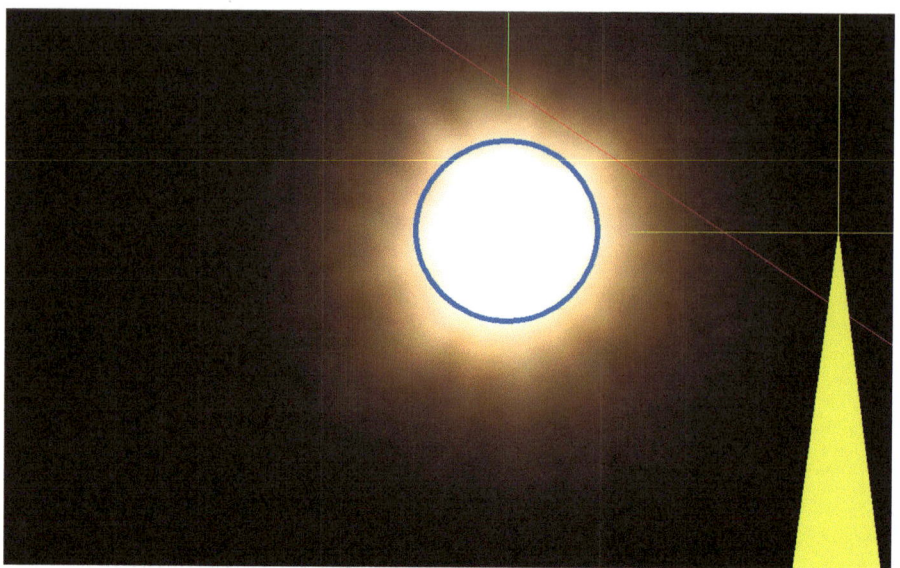

Fig. 6.17 A one-second exposure to get the position of Mars is taken immediately before the four-minute exposure. The planet does not then saturate so many pixels that you can't see where it is. This very zoomed-in picture shows how I drew a circle round the image of the planet. This usually took a few attempts until I was satisfied with the result. As with so much else, practice makes perfect – or at least adequate. (Image by the author)

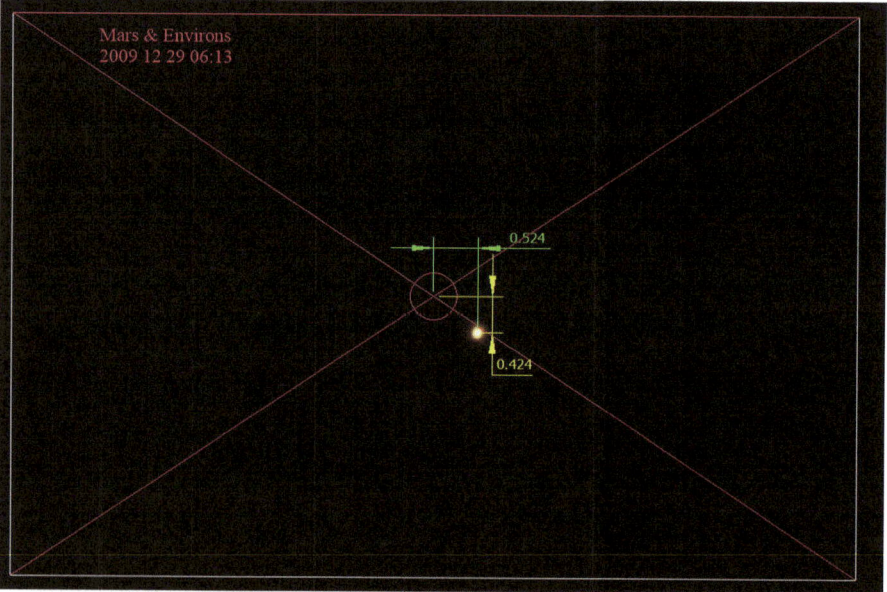

Fig. 6.18 Measuring the position of Mars from the one-second-exposure picture. (Image by the author)

How to Obtain the Position of Mars from the Star Positios

By the time I was done, I had 39 sets of data for the 2009–2010 apparition. Most of these were good data sets. I re-measured two or three others after I had plotted them up because they appeared to be anomalous (in the usual sense of the word). In each case I found and corrected errors. The only one I completely rejected as incorrigible to a credible answer was the first set.

One of the weaknesses of the scientific method is that there is a great temptation to stop once you get the answer you were looking for. I was looking for a smooth curve showing the apparent travel of Mars against the starry background. Once I got one, I suppose I am guilty of ceasing to look for further errors.

One method of triangulating planet positions from stars is given by Clark [1]. The method I actually used is not more accurate than this method, which was chosen for conceptual simplicity. I used it because it is more convenient. In particular, it gives an error estimate for the influence of each star. This influence is on a least-squares fit to the best position of the unknown object. This is important because in practice I did find myself misidentifying stars. I concluded that it is quicker to do this and lose up to five of my twenty stars than to bust my gut getting the identifications right in the first place. I rarely lost as many as five. Sometimes I didn't even lose any, but that was also rare.

I have adapted the method given in Chap. 12 of Montenbruck and Pfleger's Astronomy on the Personal Computer [110]. In their second edition, they give a program in a language called Pascal, which is now as dead as Latin. Dr. Jeannette Fine and I translated this into Fortran, a language which the computer science profession has been trying to kill since I was a grad student forty years ago, if not for longer. This translation is not quite straightforward, because Pascal handles input and output quite differently from Fortran. If truth be told, it handles it more elegantly and flexibly than Fortran.

They have failed miserably to kill Fortran, partly because there's a lot of legacy software written in Fortran and partly because Fortran has evolved and improved to the point where it is harder to criticize. There is also a lot of Fortran expertise kicking around in old timers like me. And you can get free Fortran compilers.

These authors have also released a fourth edition of their book [111], with everything now in C++, which seems to be a computer language with a future. Again, you can get free compilers. So my first thought was to use Montenbruck and Pfleger's C++ software. I soon had another thought coming. The software was developed under Microsoft's Visual C++ version 5.0. I quickly discovered that the source code won't run under recent versions of that compiler. The executables run, but you don't know what they are doing and can't customize them to read your data the way you want to read it. While fully accepting that I really ought to invest the time in learning C++ well enough to deal with little setbacks like that, the reality is that the time has to come from somewhere and I already know enough Fortran to dig myself out of trouble.

So I have stuck with the devil I know and will think about C++ as a project for when I retire.

The software I first used is the PGI Fortran compiler [110], with Code::Blocks [111] as the user interface. For this you have to load versions of Microsoft Visual Studio [112] and C++. They are both free if you choose the "Community" option, but both PGI Fortran and Code::Blocks are picky about which versions you should use. The free "Community" version of PGI Fortran comes with a license that expires on the following April 30th. When this happened to me in 2019, I was forced to do quite a lot of upgrading to get it to work again. But it all worked with Windows 10 64 bit, and it tolerates my reasonably modern but low-cost AMD quad core processor. (I spent my money on getting 64GB of RAM. That speeds things up far better than a fancy CPU ever would.) Eventually, in 2020, PGI Fortran became unavailable, so I had to stop and translate everything into a dialect compatible with the GNU Fortran (MinGW64) compiler.

I did have to tweak some compiler options to get the code to work. I was soon enough able to sort this out and find an alternative by setting a non-default compiler option. That's what I mean about being properly knowledgeable about the computer language. You're always going to encounter hiccups like this.

Before going into the Fortran, I need to show you the underlying calculation.

Figure 6.19 illustrates the challenge we face with geometric astrometry. We imagine objects projected onto a camera sensor which records their positions digitally. The challenge is to work out their right ascension and declination from the image.

The orientation of the sensor should be treated as unknown. In the case at hand, I piggybacked the DSLR onto a telescope in a convenient place where it did not get in the way of the finder scope. Its position along the telescope axis was as close as I could get the jubilee clip in Fig. 6.4 to the mount to minimize the amount of unbalance introduced. Its position in the circumferential direction was frankly random. Knowing that I would be able to correct for it later, I did not worry my little head about it.

The centre of the sensor would ideally be along the optical axis of my telephoto lens, but there is no need to bank on this being true. The correction of an off-axis sensor is straightforward. The different coordinate

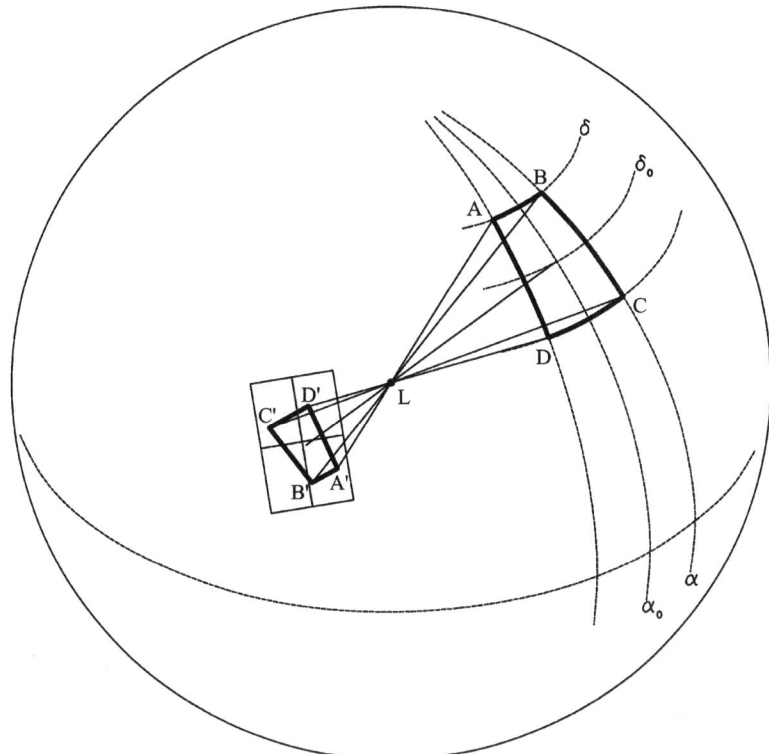

Fig. 6.19 This diagram illustrates the basic geometric astrometry problem. Imagine a region ABCD on the celestial sphere, bounded by great circles of constant Right Ascension α (BC and AD) or circles of constant declination δ (AB and CD). The lens L projects an image A'B'C'D' of ABCD onto a camera sensor. The centre of A'B'C'D' would ideally be at the centre of the sensor, but it might not be: we must allow for this. The orientation of the sensor is almost certainly not even close to the orientation of A'B'C'D'. Our task is to work out the coordinates of ABCD from the image recorded by the sensor. For clarity, the distance from L to the focal plane is greatly exaggerated in this diagram. (Image by the author)

systems for the projection of the celestial sphere, and the sensor chip, are shown in Fig. 6.20. In this diagram, a target object, whose coordinates are to be obtained, is shown. Reference stars with known positions are also shown.

By the time we have made the measurements described in Figs. 6.15 and 6.18, we will know the positions of these objects on the camera sensor. The units in which they have been measured were entirely arbitrary. This does not matter.

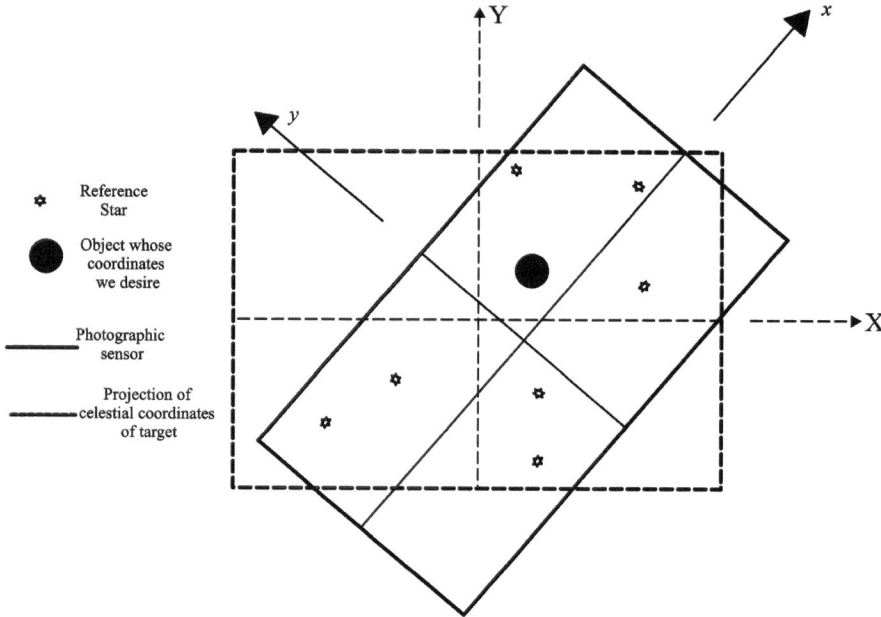

Fig. 6.20 The celestial coordinates of the region of the celestial sphere that we are examining are projected onto the plane of our photographic sensor. We imagine that the celestial coordinate system XY has origin (α_0, δ_0) (Fig. 6.19). The camera sensor has a coordinate system xy with origin at its centre and axes along its centre lines. Also shown are reference stars whose coordinates in the $\alpha\delta$ (RA and declination (Dec.)) system are known. A target, whose RA and Dec. $(\alpha T, \delta T)$ we wish to determine, is also shown. (Image by the author)

The coordinates of the ith reference stars in $\alpha\delta$ space are given by

$$X_i = ax_i + by_i + c,$$
$$Y_i = dx_i + ey_i + f. \tag{6.1}$$

We could use three reference stars, using their three α_i and three δ_i to give six equations for our six unknowns a, b, c, d, e and f. This would be fine if we knew their (x, y) positions exactly and knew that the stars had been correctly identified. In fact the coordinates I measured are not exact and there is a significant risk of misidentification: roughly 800 stars had their positions measured on photographs. That's hard on anyone's concentration span.

Instead I used more than three stars and obtained a least-squares fit to the unknowns a, b, c, d, e and f for each photograph.

The equations in the X and Y directions are independent. So we solve for a, b and c separately from the operations to solve for d, e and f.

Let us write the set of equations in a more abstract form which we can use to solve for both the X and Y directions.

$$
\begin{aligned}
t_i &\approx A_{i1}S_1 + \cdots + A_{im}S_m \quad \text{for } i = 1 \ldots n \\
&= \sum_{j=1}^{m} A_{ij} s_j \quad \text{for } i \ldots n,
\end{aligned}
\tag{6.2}
$$

where the s_i might be our x_i, y_i, or indeed other variables, and the A_{ij} could be such variables as a, b,…etc., in Eq. (6.1).

We will, however, need some coordinate transformations to accomplish our task. Right ascension (RA) and declination (Dec.) are spherical polar coordinates. The radius is usually assumed to have some mystical high value: after all the "celestial sphere" is a geometric idealization, not something physically real. It is just another instance of how old theories never quite die off. They survive in the jargon of the replacement theories. You see this all the time. For example, there's a lot of alchemical jargon in metallurgy, even though the alchemists would barely recognize the modern form of that art, transformed as it was by the invention of the electron microscope.

Anyway, back to coordinates. Please let me define a Cartesian coordinate system for the celestial sphere, with components ξ, η and ζ to avoid confusion with the coordinate systems defined on the camera sensor. The Cartesian coordinates of a unit vector with RA α and Dec. δ are

$$
\hat{e} = \begin{pmatrix} \cos\delta \cos(\alpha - \alpha_0) \\ \cos\delta \sin(\alpha - \alpha_0) \\ \sin\delta \end{pmatrix}.
\tag{6.3}
$$

Notice that the conventional angle of latitude used for right-handed spherical polar coordinates is the angle with respect to the North Pole, not the equator. This is $(90° - \delta)$. Of course

$$
\begin{aligned}
\cos(90° - \delta) &= \sin\delta, \\
\sin(90° - \delta) &= \cos\delta.
\end{aligned}
\tag{6.4}
$$

These formulae were used to write Eq. (6.3). When

$$\alpha = \alpha_0,$$
$$\delta = \delta_0, \tag{6.5}$$

in Fig. 6.19, let us call the unit vector

$$\hat{\mathbf{e}}_0 = \begin{pmatrix} \cos \delta_0 \\ 0 \\ \sin \delta_0 \end{pmatrix}. \tag{6.6}$$

Now let us imagine a star at (α, δ) in Fig. 6.19. Let the angle between this point and (α_0, δ_0) be ϕ. Then

$$\begin{aligned} \cos \phi &= & \hat{e} \cdot \hat{e}_0 \\ &= & \cos \delta_0 \cos \delta \cos (\alpha - \alpha_0) + \sin \delta_0 \sin \delta. \end{aligned} \tag{6.7}$$

Within the plane of the sensor, the following unit vectors, one north-south, the other east-west, define a coordinate system.

$$\hat{\mathbf{e}}_X = \begin{pmatrix} 0 \\ 1 \\ 0 \end{pmatrix}, \tag{6.8}$$

$$\hat{\mathbf{e}}_Y = \begin{pmatrix} \sin \delta_0 \\ 0 \\ -\cos \delta_0 \end{pmatrix}.$$

Let us now consider the point A' from Fig. (6.19). This is the projection of the point A at RA & Dec. (α, δ). Let us also imagine that the lens L in Fig. (6.19) projects the point (α_0, δ_0) onto the sensor plane along a vector **F**. It has to be true that $|\mathbf{F}| = F$ is the focal length of the lens if the image is in focus. We assume that it is and neglect any focusing error. Let the vector from the lens L to A' be **P**. Then

$$\mathbf{P} = \mathbf{F} + XF\hat{\mathbf{e}}_X + YF\hat{\mathbf{e}}_Y. \tag{6.9}$$

It follows that

$$|\mathbf{P}| = F\sqrt{1 + X^2 + Y^2}. \tag{6.10}$$

Also, since

$$F = |\mathbf{P}|\cos\phi, \tag{6.11}$$

from Eq. (6.7) it follows that

$$|\mathbf{P}| = \frac{F}{\cos\delta_0 \cos\delta \cos(\alpha - \alpha_0) + \sin\delta_0 \sin\delta}. \tag{6.12}$$

It is possible to rewrite Eq. (6.9) as follows using my ξ-η-ζ coordinates.

$$
\begin{aligned}
|\mathbf{P}| \begin{pmatrix} \cos\delta \cos(\alpha-\alpha_0) \\ \cos\delta \sin(\alpha-\alpha_0) \\ \sin\delta \end{pmatrix} &= F\begin{pmatrix} \cos\delta_0 \\ 0 \\ \sin\delta_0 \end{pmatrix} + F\begin{pmatrix} -Y\sin\delta_0 \\ -X \\ Y\cos\delta_0 \end{pmatrix} \\[2mm]
F\sqrt{1+X^2+Y^2}\begin{pmatrix} \cos\delta \cos(\alpha-\alpha_0) \\ \cos\delta \sin(\alpha-\alpha_0) \\ \sin\delta \end{pmatrix} &= F\begin{pmatrix} \cos\delta_0 \\ 0 \\ \sin\delta_0 \end{pmatrix} + F\begin{pmatrix} -Y\sin\delta_0 \\ -X \\ Y\cos\delta_0 \end{pmatrix} \\[2mm]
\sqrt{1+X^2+Y^2}\begin{pmatrix} \cos\delta \cos(\alpha-\alpha_0) \\ \cos\delta \sin(\alpha-\alpha_0) \\ \sin\delta \end{pmatrix} &= \begin{pmatrix} \cos\delta_0 \\ 0 \\ \sin\delta_0 \end{pmatrix} + \begin{pmatrix} -Y\sin\delta_0 \\ -X \\ Y\cos\delta_0 \end{pmatrix} \\[2mm]
\frac{1}{\cos\delta_0 \cos\delta \cos(\alpha-\alpha_0)+\sin\delta_0 \sin\delta}\begin{pmatrix} \cos\delta \cos(\alpha-\alpha_0) \\ \cos\delta \sin(\alpha-\alpha_0) \\ \sin\delta \end{pmatrix} &= \begin{pmatrix} \cos\delta_0 \\ 0 \\ \sin\delta_0 \end{pmatrix} + \begin{pmatrix} -Y\sin\delta_0 \\ -X \\ Y\cos\delta_0 \end{pmatrix}.
\end{aligned}
\tag{6.13}
$$

Equations (6.10) and (6.12) have been used in Eq. (6.13). Solving along the η axis for X gives

$$X = -\frac{\cos\delta \sin(\alpha-\alpha_0)}{\cos\delta_0 \cos\delta \cos(\alpha-\alpha_0)+\sin\delta_0 \sin\delta}. \tag{6.14}$$

Solving along the ξ axis for Y gives

$$
\begin{aligned}
\frac{\cos\delta \cos(\alpha-\alpha_0)}{\cos\delta_0 \cos\delta \cos(\alpha-\alpha_0)+\sin\delta_0 \sin\delta} &= \cos\delta_0 - Y\sin\delta_0 \\[2mm]
\frac{\cos\delta \cos(\alpha-\alpha_0)}{\cos\delta_0 \cos\delta \cos(\alpha-\alpha_0)+\sin\delta_0 \sin\delta} &- \frac{\cos\delta_0\left(\cos\delta_0 \cos\delta \cos(\alpha-\alpha_0)+\sin\delta_0 \sin\delta\right)}{\cos\delta_0 \cos\delta \cos(\alpha-\alpha_0)+\sin\delta_0 \sin\delta} \\
&= -Y\sin\delta_0 \\[2mm]
\frac{\cos\delta \cos(\alpha-\alpha_0)-\cos\delta_0\left(\cos\delta_0 \cos\delta \cos(\alpha-\alpha_0)+\sin\delta_0 \sin\delta\right)}{\cos\delta_0 \cos\delta \cos(\alpha-\alpha_0)+\sin\delta_0 \sin\delta} &= -Y\sin\delta_0
\end{aligned}
\tag{6.15}
$$

$$Y\sin\delta_0 = \frac{\cos\delta_0\left(\cos\delta_0\cos\delta\cos(\alpha-\alpha_0)+\sin\delta_0\sin\delta\right)+\cos\delta\cos(\alpha-\alpha_0)}{\cos\delta_0\cos\delta\cos(\alpha-\alpha_0)+\sin\delta_0\sin\delta}$$

$$= \frac{\left(1-\sin^2\delta_0\right)\cos\delta\cos(\alpha-\alpha_0)+\cos\delta_0\sin\delta_0\sin\delta-\cos\delta\cos(\alpha-\alpha_0)}{\cos\delta_0\cos\delta\cos(\alpha-\alpha_0)+\sin\delta_0\sin\delta}$$

$$= \frac{-\sin^2\delta_0\cos\delta\cos(\alpha-\alpha_0)+\cos\delta_0\sin\delta_0\sin\delta}{\cos\delta_0\cos\delta\cos(\alpha-\alpha_0)+\sin\delta_0\sin\delta} \qquad (6.16)$$

$$Y = -\frac{\sin\delta_0\cos\delta\cos(\alpha-\alpha_0)-\cos\delta_0\sin\delta}{\cos\delta_0\cos\delta\cos(\alpha-\alpha_0)+\sin\delta_0\sin\delta}$$

It is also possible to solve for α and δ. In fact, these are the quantities we really want.

Dividing the η component of Eq. (6.13) by the ξ component gives

$$\frac{\cos\delta\sin(\alpha-\alpha_0)}{\cos\delta\cos(\alpha-\alpha_0)} = \tan(\alpha-\alpha_0)$$

$$= \frac{-X}{\cos\delta_o-Y\sin\delta_0} \qquad (6.17)$$

Therefore,

$$\alpha = \alpha_0+\tan^{-1}\left(\frac{-X}{\cos\delta_o-Y\sin\delta_0}\right). \qquad (6.18)$$

Using the ζ component of the third line of Eq. (6.13) gives

$$\sqrt{1+X^2+Y^2}\sin\delta = \sin\delta_0+Y\cos\delta_0$$

$$\delta = \sin^{-1}\left(\frac{\sin\delta_0+Y\cos\delta_0}{\sqrt{1+X^2+Y^2}}\right). \qquad (6.19)$$

The program used is listed below. In a few places, lines have spilled over onto the following lines. I set the default line width to 132 columns. I daresay it might be better programming practice to use shorter lines. I would never do anything so preposterous as to claim that my programming is elegant, efficient or anything other than functional. I do my best to debug it, then, once it seems to give correct answers, I leave well alone – unless and until I trip over another bug.

There is a main program, which you have to have in Fortran programs. The calculation work is farmed out to subroutines. The input, output and subroutine management are controlled from the main program. Line numbers are given.

Main Program

(Listed with the permission of Jeannette M. Fine.)

```fortran
 1 !
 2 !
 3 program main
 4 !
 5 !
 6 ! FUNCTIONS:
 7 ! Console1 - Entry point of application.
 8 !
 9 !*********************************************************************
*********************
10 !
11 ! PROGRAM: Main
12 !
13 ! PURPOSE: Entry point for the application.
14 !
15 !*********************************************************************
*********************
16 !
17 ! Declarations NB These may very from version to version of
Fortran.
18 !
19 INTEGER I,J,K,M,N,i4,j4,IN,IM,I_STARCOUNT,ITHREE,IHR_RA,MIN_
RA,IDEG_DEG,MIN_DEG
20 INTEGER IYEAR,IMONTH,IDAY,IUTHOURS,IUTMINUTES
21 double precision AMAT(6,5),P,Q,H,H2,H3,S(4),EPS,XX_BODY,YY_
BODY,XX_CENTRE,YY_CENTRE
22 CHARACTER*30 NAMES(1000),NAME_CENTRE,NAME_BODY,CHARTEXT
23 CHARACTER*8 CHAR1,CHAR2,CHAR3,CHUNITS(21)
24 CHARACTER*10 CHAR10
25 CHARACTER*100 LONGTEXT
26 CHARACTER*35 IPFILE,OPFILE
27 INTEGER RA_CENTRE_H,DEC_CENTRE_D,RA_BODY_H,DEC_BODY_D,RA_
CENTRE_D,RA_BODY_D
28 INTEGER RA_CENTRE_M,DEC_CENTRE_M,RA_BODY_M,DEC_BODY_M
29 integer irad,iram
30 DOUBLE PRECISION RA_DEG(1000),DEC_DEG(1000),RA_OBSV(1000),DEC_
OBSV(1000),DELTA(1000)
31 DOUBLE PRECISION RA_CENTRE_DEG,DEC_CENTRE_DEG,X_BODY,Y_BODY,RA_
BODY_OBSV,DEC_BODY_OBSV
32 DOUBLE PRECISION RA_CENTRE_S,DEC_CENTRE_S,RA_BODY_S,DEC_
BODY_S,RA_CENTRE_OBSV,DEC_CENTRE_OBSV
33 DOUBLE PRECISION XX(1000),YY(1000),DET,FOC_LEN,SCALE,ARC
34 INTEGER RA_H(1000),DEC_D(1000),RA_M(1000),DEC_M(1000)
35 DOUBLE PRECISION RA_S(1000),DEC_S(1000),SEC_RA,SEC_DEG
36 DOUBLE PRECISION A,B,C,D,E,F,AA(1000,8),X(1000),Y(1000)
37 double precision D_RA,D_DEG,AA1(1000,8),RA_OBS,DEC_OBS
38 !
39 ! Most input and output is done in this Main program.
40 ! fort.26.txt is for debugging purposes only.
41 ! Most stuff written to it is normally commented out.
```

```
42 ! The file io_files_text (Unit 15) enables the input of variable
file names without recompiling.
43 ! The main reading is done from Unit 25, the main output from
Unit 36.
44 !
45 OPEN(UNIT=15,FILE='io_files.txt')
46 read(15,'(A35)')IPFILE
47 read(15,'(A35)')OPFILE
48 OPEN(UNIT=26,FILE='fort26.txt')
49 OPEN(UNIT=36,FILE=OPFILE)
50 OPEN(UNIT=25,FILE=IPFILE)
51 !
52 WRITE(26,'(/,6X,"FOTO: astrometric analysis of photographic
plates")')
53 WRITE(26,'(6X,"Based on a Pascal Program by Thomas Pfleger and
Oliver Montenbruck",/)')
54 WRITE(26,'(6X,"Input data file: fotoinp.csv",/)')
55 !READ DATE AND TIME
56 READ(25,*)IYEAR,IMONTH,IDAY,IUTHOURS,IUTMINUTES
57 WRITE(26,'(1X,"DATE:",1X,I4,1X,I2,1X,I2,2X,"TIME:",1X,I2,":"
,I2,//)')IYEAR,IMONTH,IDAY,IUTHOURS,IUTMINUTES
58 WRITE(36,'(1X,"ASTROMETRY METHOD OF MONTENBRUCK & PFLEGER FOR
LOCATING SOLAR SYSTEM OBJECTS ETC.")')
59 WRITE(36,'(1X,"ASTRONOMY ON THE PERSONAL COMPUTER - TR. DUNLOP -
2ND ED - 1994 - SPRINGER-VERLAG")')
60 WRITE(36,'(1X,"REPROGRAMMED IN FORTRAN BY J. M. FINE AND J. D.
CLARK (C) 2009",/)')
61 WRITE(36,'(1X,"DATE:",1X,I4,1X,I2,1X,I2,2X,"TIME:",1X,I2,":",I2,1
X,"UT",//)')IYEAR,IMONTH,IDAY,IUTHOURS,IUTMINUTES
62
63 ! We're going use a CSV file to format the input data.
64 ! We're not going to split this off as a subroutine.
65 ! Read coordinates of the plate center
66 READ(25,*)NAME_CENTRE,RA_CENTRE_H,RA_CENTRE_M,RA_CENTRE_S,DEC_
CENTRE_D,DEC_CENTRE_M,DEC_CENTRE_S
67 ! Change stuff into desired units
68 CALL DMS2DEG(DEC_CENTRE_D,DEC_CENTRE_M,DEC_CENTRE_S,DEC_CENTRE_
DEG)
69 CALL HMS2DEG(RA_CENTRE_H,RA_CENTRE_M,RA_CENTRE_S,RA_CENTRE_DEG)
70 CALL DEG2HMS(RA_CENTRE_DEG,IHR_RA,MIN_RA,SEC_RA)
71 CALL DEG2DMS(DEC_CENTRE_DEG,IDEG_DEG,MIN_DEG,SEC_DEG)
72 WRITE(26,'(1X,2(I4,1X),F9.4)')IHR_RA,MIN_RA,SEC_RA
73 WRITE(26,'(1X,2(I4,1X),F9.4,/)')IDEG_DEG,MIN_DEG,SEC_DEG
74 WRITE(26,601)NAME_CENTRE,RA_CENTRE_H,RA_CENTRE_M,RA_CENTRE_S,DEC_
CENTRE_D,DEC_CENTRE_M,DEC_CENTRE_S,RA_CENTRE_DEG,DEC_CENTRE_DEG
75 READ(25,*)NAME_BODY,X_BODY,Y_BODY
76 WRITE(26,604)NAME_BODY,X_BODY,Y_BODY
77 !
78 RA_CENTRE_D=15*RA_CENTRE_H
79 RA_BODY_D=15*RA_BODY_H
80 ! Read the data from the CSV file prepared in MS Excel
81 I_STARCOUNT=0
82 DO I=1,1000
```

```
 83 READ(25,*)NAMES(I),X(I),Y(I),RA_H(I),RA_M(I),RA_S(I),DEC_D(I),DE
C_M(I),DEC_S(I)
 84 ! WRITE(26,600)NAMES(I),X(I),Y(I),RA_H(I),RA_M(I),RA_S(I),DEC_D(I
),DEC_M(I),DEC_S(I)
 85 ! RA_D(I)=15*RA_H(I)
 86 CALL DMS2DEG(DEC_D(I),DEC_M(I),DEC_S(I),DEC_DEG(I))
 87 ! WRITE(26,603)DEC_DEG(I)
 88 CALL HMS2DEG(RA_H(I),RA_M(I),RA_S(I),RA_DEG(I))
 89 ! CALL DEG2DMS(DEC_DEG,IRAD,IRAM,RAS)
 90 ! WRITE(26,602)IRAD,IRAM,RAS,DEC_DEG
 91 ! CALL DEG2HMS(RA_DEG,IRAD,IRAM,RAS)
 92 ! WRITE(26,602)IRAD,IRAM,RAS,RA_DEG
 93 WRITE(26,607)NAMES(I),X(I),Y(I),RA_H(I),RA_M(I),RA_S(I),DEC_D(I),
DEC_M(I),DEC_S(I),RA_DEG(I),DEC_DEG(I)
 94
 95 IF(NAMES(I).EQ."END") GOTO 100
 96 I_STARCOUNT=I_STARCOUNT+1
 97 END DO
 98 close (unit=25)
 99 600 FORMAT(1X,a20,2(1X,F9.4),2(1X,I3),1X,F9.4,2(1X,I3),1X,F12.4)
100 601 FORMAT(1X,a20,20X,2(1X,I3),1X,F9.4,2(1X,I3),3(1X,F12.4))
101 602 FORMAT(1X,2(1X,I4),1X,F10.4,1X,F15.4)
102 603 FORMAT(1X,F15.4)
103 604 FORMAT(1X,a20,1X,1P,E13.6,1X,E13.6)
104 605 FORMAT(1X,F9.4,1X,F9.4,1X,F9.4)
105 606 FORMAT(8(1X,F9.4))
106 607 FORMAT(1X,a20,2(1X,F9.4),2(1X,I3),1X,F9.4,2(1X,I3),3(1X
,F12.4))
107 100 CONTINUE
108 CHAR10="STARS"
109 WRITE(26,'(/,1X,I4,1X,A5,/)')I_STARCOUNT,CHAR10
110 !
111 ! WRITE(26,*)'BEFORE SET AA'
112 ! Prepare data for Givens' Transformation.
113 DO J=1,I_STARCOUNT
114 ! WRITE(26,'(1X,4(F9.4,1X))')RA_CENTRE_DEG,DEC_CENTRE_DEG,RA_
DEG(J),DEC_DEG(J)
115 CALL EQUSTD(RA_CENTRE_DEG,DEC_CENTRE_DEG,RA_DEG(J),DEC_
DEG(J),XX(J),YY(J))
116 AA(J,1)=X(J)
117 AA(J,2)=Y(J)
118 AA(J,3)=1.0D0
119 AA(J,4)=+XX(J)
120 AA(J,5)=+YY(J)
121 ! WRITE(26,606)AA(J,1),AA(J,2),AA(J,3),AA(J,4),AA(J,5)
122 AA1(J,1)=X(J)
123 AA1(J,2)=Y(J)
124 AA1(J,3)=1.0D0
125 AA1(J,4)=+YY(J)
126
127 END DO
128 !
129 ! Never send numbers into subroutines as parameters.
130 ! If that goes wrong it goes HORRIBLY wrong,
```

```
131 ! e.g. overwriting the stored value of 3.
132 ITHREE=3
133 !300 WRITE(26,*)"REACHED 300"
134 ! Do the least squares fit fr the X direction parameters
135 CALL LSQFIT(AA,I_STARCOUNT,ITHREE,S)
136 !301 WRITE(26,*)"REACHED 301"
137 A=S(1)
138 B=S(2)
139 C=S(3)
140 !
141 ! CALCULATE COEFFICIENTS A,B,C,D,E,F
142 !
143 !WRITE(26,605)A,B,C
144 !
145 ! Do the least squares fit fr the Y direction parameters
146 CALL LSQFIT(AA1,I_STARCOUNT,ITHREE,S)
147 !302 WRITE(26,*)"REACHED 302"
148 D=S(1)
149 E=S(2)
150 F=S(3)
151 !WRITE(26,605)D,E,F
152 !303 WRITE(26,*)"REACHED 303"
153 !
154 ! Calculate equatorial coordinates (and errors for reference
stars)
155
156 DO I=1,I_STARCOUNT ! DO WE ALSO NEED TO DO FOR PLANET and CENTRE?
PROBABLY
157 XX(I) = A*X(I)+B*Y(I)+C
158 YY(I) = D*X(I)+E*Y(I)+F
159 CALL STDEQU(RA_CENTRE_DEG,DEC_CENTRE_DEG,XX(I),YY(I),RA_OBS,DEC_
OBS)
160 ! WRITE(26,*)RA_CENTRE_DEG,DEC_CENTRE_DEG,XX(I),YY(I),RA_OBS,DEC_
OBS
161 D_RA = (RA_OBS-RA_DEG(I))*DCOSD(DEC_DEG(I))
162 D_DEG = (DEC_OBS-DEC_DEG(I))
163 DELTA(I) = 3600.0D0 * DSQRT ( D_RA*D_RA + D_DEG*D_DEG )
164 ! WRITE(26,*)RA_OBS,DEC_OBS
165 RA_OBSV(I) = RA_OBS
166 DEC_OBSV(I) = DEC_OBS
167 END DO
168 !304 WRITE(26,*)"REACHED 304"
169 XX_BODY = A*X_BODY+B*Y_BODY+C
170 YY_BODY = D*X_BODY+E*Y_BODY+F
171 CALL STDEQU(RA_CENTRE_DEG,DEC_CENTRE_DEG,XX_BODY,YY_BODY,RA_BODY_
OBSV,DEC_BODY_OBSV)
172 XX_CENTRE = A*X_CENTRE+B*Y_CENTRE+C
173 YY_CENTRE = D*X_CENTRE+E*Y_CENTRE+F
174 CALLSTDEQU(RA_CENTRE_DEG,DEC_CENTRE_DEG,XX_CENTRE,YY_CENTRE,RA_
CENTRE_OBSV,DEC_CENTRE_OBSV)
175 !305 WRITE(26,*)"REACHED 305"
176 !
177 !focal length
178 !
```

```
179 DET=A*E-D*B
180 FOC_LEN=1.0D0/DSQRT(DABS(DET))
181 ARC=206264.8D0 !arcseconds per radian
182 SCALE=ARC/FOC_LEN
183 !306 WRITE(26,*)"REACHED 306"
184 !
185 ! Output results to unit 36. Change units where desired.
186 !
187 CHARTEXT="Plate constants:"
188 WRITE(36,'(1X,A20,/)')CHARTEXT
189 CHAR1="a = "
190 CHAR2="b = "
191 CHAR3="c = "
192 WRITE(36,652)CHAR1,A,CHAR2,B,CHAR3,C
193 652 FORMAT(1X,A4,1P,E13.6,1X,A4,E13.6,1X,A4,E13.6)
194 CHAR1="d = "
195 CHAR2="e = "
196 CHAR3="f = "
197 WRITE(36,653)CHAR1,D,CHAR2,E,CHAR3,F
198 653 FORMAT(1X,A4,1P,E13.6,1X,A4,E13.6,1X,A4,E13.6,/)
199 LONGTEXT="Effective focal length and image scale:"
200 WRITE(36,'(1X,A50,/)')LONGTEXT
201 CHAR1="F = "
202 CHAR2=" mm"
203 WRITE(36,'(1X,A4,1P,E13.6,1X,A4)')CHAR1,FOC_LEN,CHAR2
204 CHAR1="m = "
205 CHARTEXT="sec/mm"
206 WRITE(36,'(1X,A4,1P,E13.6,1X,A7,/)')CHAR1,SCALE,CHARTEXT
207 CHARTEXT="Coordinates:"
208 WRITE(36,'(1X,A20,/)')CHARTEXT
209 CHUNITS(1)="NAME"
210 CHUNITS(2)="x"
211 CHUNITS(3)="y"
212 CHUNITS(4)="X"
213 CHUNITS(5)="Y"
214 CHUNITS(6)="RA"
215 CHUNITS(7)="Dec"
216 CHARTEXT="Error"
217 WRITE(36,650)(CHUNITS(I),I=1,7),CHARTEXT
218 650 FORMAT(1X,A4,19X,A1,9X,A1,10X,A1,9X,A1,8X,A2,17X,A3,17X,A5)
219 CHUNITS(1)="mm"
220 CHUNITS(2)="mm"
221 CHUNITS(3)="mm"
222 CHUNITS(4)="mm"
223 CHUNITS(5)="h"
224 CHUNITS(6)="m"
225 CHUNITS(7)="s"
226 CHUNITS(8)="deg"
227 CHUNITS(9)="m"
228 CHUNITS(10)="s"
229 CHUNITS(11)="s"
230 WRITE(36,651)(CHUNITS(I),I=1,11)
231 651 FORMAT(24X,A2,8X,A2,8X,A2,9X,A2,7X,A1,4X,A1,3X,A1,9X,A3,2X,A
1,3X,A1,10X,A1)
```

```
232 CALL DEG2HMS(RA_BODY_OBSV,IHR_RA,MIN_RA,SEC_RA)
233 CALL DEG2DMS(DEC_BODY_OBSV,IDEG_DEG,MIN_DEG,SEC_DEG)
234 WRITE(36,655)NAME_BODY,X_BODY,Y_BODY,XX_BODY,YY_BODY,IHR_RA,MIN_
RA,SEC_RA,IDEG_DEG,MIN_DEG,SEC_DEG
235 655 FORMAT(1X,A20,4(1X,F9.4),0P,2(2(1X,I4),2X,F7.4))
236 CALL DEG2HMS(RA_CENTRE_OBSV,IHR_RA,MIN_RA,SEC_RA)
237 ! WRITE(26,*)RA_CENTRE_OBSV,IHR_RA,MIN_RA
238 CALL DEG2DMS(DEC_CENTRE_OBSV,IDEG_DEG,MIN_DEG,SEC_DEG)
239 WRITE(36,655)NAME_CENTRE,X_CENTRE,Y_CENTRE,XX_CENTRE,YY_
CENTRE,IHR_RA,MIN_RA,SEC_RA,IDEG_DEG,MIN_DEG,SEC_DEG
240 !
241 DO I=1,I_STARCOUNT
242 CALL DEG2HMS(RA_OBSV(I),IHR_RA,MIN_RA,SEC_RA)
243 WRITE(26,*)I,RA_OBSV(I),IHR_RA,MIN_RA
244 CALL DEG2DMS(DEC_OBSV(I),IDEG_DEG,MIN_DEG,SEC_DEG)
245 WRITE(36,656)NAMES(I),X(I),Y(I),XX(I),YY(I),IHR_RA,MIN_RA,SEC_
RA,IDEG_DEG,MIN_DEG,SEC_DEG,DELTA(I)
246 END DO
247 656 FORMAT(1X,A20,4(1X,F9.4),2(2(1X,I4),2X,0P
,F7.4,0P),2X,0P,F12.4)
248 !
249 close(unit=26)
250 CLOSE(unit=36)
251 STOP
252 END PROGRAM main
```

Angle Unit Changing Subroutines

```
253 !
254 SUBROUTINE DMS2DEG(IDEG,MIN,SEC, DECIMAL_DEG)
255 ! Change Deg Min Sec to Decimal Degrees
256 IF(IDEG.NE.0) THEN
257 ISIGN=IDEG/IABS(IDEG)
258 ELSE
259 ISIGN=1
260 ENDIF
261 DECIMAL_DEG=DFLOAT(IABS(IDEG))+(DFLOAT(MIN)/60.0D0)+(SEC/3600.
0D0)
262 DECIMAL_DEG=DECIMAL_DEG*ISIGN
263 RETURN
264 END SUBROUTINE DMS2DEG
265 !
266 SUBROUTINE HMS2DEG(IHR,MIN,SEC,DECIMAL_DEG)
267 ! Change Hour Min Sec to Decimal Degrees
268 DECIMAL_HR=DFLOAT(IHR)+(DFLOAT(MIN)/60.0D0)+(SEC/3600.0D0)
269 DECIMAL_DEG=DECIMAL_HR*360.0D0/24.0D0
270 RETURN
271 END SUBROUTINE HMS2DEG
272 !
273 SUBROUTINE DEG2DMS(DECIMAL_DEG,IDEG,MIN,SEC)
274 ! Change Decimal Degrees to Deg Min Sec
```

```
275 ANGLE=DABS(DECIMAL_DEG)
276 IDEG=ANGLE
277 ! write(26,*)'ideg=',ideg,'angle=',angle,'decimal_deg=',DECIMAL_
DEG
278 REMAINDER=ANGLE-IDEG
279 ! write(26,*)'remainder=',remainder
280 RMNDR60=REMAINDER*60.0D0
281 ! write(26,*)'RMNDR60=',RMNDR60
282 MIN=RMNDR60
283 ! write(26,*)'min=',min
284 REMAINDER=RMNDR60-MIN
285 ! write(26,*)'remainder=',remainder
286 SEC=REMAINDER*60.0D0
287 ! write(26,*)'sec=',sec
288 ! write(26,*)'DECIMAL_DEG=',DECIMAL_DEG
289 IF(DECIMAL_DEG.LT.0.0D0) IDEG=-IDEG
290 ! write(26,*)'ideg=',ideg
291 RETURN
292 END SUBROUTINE DEG2DMS
293 !
294 SUBROUTINE DEG2HMS(DECIMAL_DEG,IHR,MIN,SEC)
295 ! Change Decimal Degrees to Hour Min Sec
296 DECIMAL_HR=DECIMAL_DEG*24.0D0/360.0D0
297 ANGLE=DABS(DECIMAL_HR)
298 IHR=ANGLE
299 REMAINDER=DECIMAL_HR-IHR
300 RMNDR60=REMAINDER*60.0D0
301 MIN=RMNDR60
302 REMAINDER=RMNDR60-MIN
303 SEC=REMAINDER*60.0D0
304 IF(DECIMAL_HR.LT.0.0D0) IHR=-IHR
305 RETURN
306 END SUBROUTINE DEG2HMS
307 !
```

Least Squares Fitting Subroutine LSQFIT Has to Be Added

This subroutine is described in Chap. 4.

```
336 SUBROUTINE LSQFIT (A,N,M,S)
337 !-------------------------------------------------------------
---------------------
338 ! LSQFIT: Per Equations 3.1 through 3.13 above
339 ! solution of an overdetermined system of linear equations
340 ! A(i,1)*s(1)+...A(i,m)*s(m) - A(i,m+1) = 0 (i=1,..,n)
341 ! according to the method of least squares using Givens rotations
342 ! A: matrix of coefficients
343 ! N: number of equations (rows of A)
344 ! M: number of unknowns (M+1=columns of A, M=elements of S)
345 ! S: solution vector
```

```
346 !--------------------------------------------------------------------
--------------------
347 INTEGER I,J,K,M,N
348 double precision A(1000,8),P,Q,H,H2,H3,S(M),EPS
349
350 EPS = 1.0D-10
351 DO J=1,M ! loop over columns 1...M of A
352 ! eliminate matrix elements A(i,j) with i>j from column j
353 DO I=J+1,N
354 IF (A(I,J).NE.0.0D0)THEN
355 ! calculate p, q and new A(j,j); set A(i,j)=0
356 IF (DABS(A(J,J)).LT.EPS*DABS(A(I,J))) THEN
357 P=0.0D0
358 Q=1.0D0
359 A(J,J)=-A(I,J)
360 A(I,J)=0.0D0
361 ELSE
362 H=DSQRT(A(J,J)*A(J,J)+A(I,J)*A(I,J))
363 IF (A(J,J).LT.0.0D0) H=-H
364 P=A(J,J)/H
365 Q=-A(I,J)/H
366 A(J,J)=H
367 A(I,J)=0.0D0
368 END IF
369 ! calculate rest of the line
370 DO K=J+1,M+1
371 H2 = P*A(J,K) - Q*A(I,K)
372 A(I,K) = Q*A(J,K) + P*A(I,K)
373 A(J,K) = H2
374 END DO
375 END IF
376 END DO
377 END DO
378 ! backsubstitution
379
380 DO I = M,1,-1
381 H3=A(I,M+1)
382 DO K=I+1,M
383 H3=H3-A(I,K)*S(K)
384 END DO
385 S(I) = H3/A(I,I)
386 END DO
387 !
388 RETURN
389 END SUBROUTINE LSQFIT
```

Distance Unit Changing Subroutines

```
308 SUBROUTINE EQUSTD (RA0,DEC0,RA,DEC,XX,YY)
309 !--------------------------------------------------------------------
--------------------
310 ! EQUSTD: transformation of equatorial coordinates into
```

```
311 ! standard coordinates Equations 5.26 & 5.28 above
312 ! RA0,DEC0: right ascension and declination of the optical axis
(deg)
313 ! RA,DEC: right ascension and declination (deg)
314 ! XX,YY: standard coordinates
315 !------------------------------------------------------------------
----------------------
316 ! IMPLICIT DOUBLE PRECISION A-H,O-Z
317 ! REAL*8 C,DEC0,DEC,RA,RA0,XX,YY
318 C = DCOSD(DEC0)*DCOSD(DEC)*DCOSD(RA-RA0)+DSIND(DEC0)*DSIND(DEC);
319 XX = - ( DCOSD(DEC)*DSIND(RA-RA0) ) / C;
320 YY = - ( DSIND(DEC0)*DCOSD(DEC)*DCOSD    (RA-RA0)-
DCOSD(DEC0)*DSIND(DEC) ) / C;
321 END SUBROUTINE EQUSTD
322 !
323 SUBROUTINE STDEQU(RA0,DEC0,XX,YY,RA,DEC)
324 !------------------------------------------------------------------
----------------------
325 ! STDEQU: transformation from standard coordinates into
326 ! equatorial coordinates Equations 5.30 & 5.31 above
327 ! RA0,DEC0: right ascension and declination of the optical axis
(deg)
328 ! XX,YY: standard coordinates
329 ! RA,DEC: right ascension and declination (deg)
330 !------------------------------------------------------------------
----------------------
331 ! IMPLICIT DOUBLE PRECISION A-H,O-Z
332 RA = RA0 + DATAND ( -XX / (DCOSD(DEC0)-YY*DSIND(DEC0)) )
333 DEC = DASIND ( (DSIND(DEC0)+YY*DCOSD(DEC0))/DSQRT(1.0D0+XX*XX+YY
*YY) )
334 END SUBROUTINE STDEQU
335 !
```

An example of the data input to the program listed above is shown in Fig. 6.21. These data are stored and read as a comma separated variable (.csv) file. Microsoft Excel is able to create and edit these files very easily and efficiently. Excel 2010 offers Macintosh and MS-DOS-type .csv files as alternative formats to the default. I use the default .csv format in my Windows 10 environment without problems.

Results

A typical set of results appears in Table 6.1. It is the results from the data shown in Fig. 6.21. This is a reassuring set of results. The worst error associated with any star is 3.1490 arcsec for the star TYC-1406-408-1.

The results as-measured were not so reassuring.

The first attempt is shown in Table 6.2. Now we are seeing errors in the hundreds of arcseconds. I deal with this by throwing away the data from the

▲	A	B	C	D	E	F	G	H	I	J
1	2009	12	29	6	13					
2	CENTRE	9	31	50.9844	18	36	36.712			
3	MARS	0.524	-0.424	9	31	13.93	18	31	30.7	
4	HD82624	-2.169	0.667	9	34	14.11	18	41	21.5	
5	HD81899	2.473	2.149	9	29	36.41	19	14	31.2	
6	TYC-1406-924-1	-3.123	-0.504	9	35	2.12	18	21	31	
7	BD+19-2219	1.593	1.525	9	30	25.76	19	3	9.6	
8	TYC-1406-189-1	-0.354	-0.194	9	32	11.19	18	32	54.2	
9	TYC-1406-700-1	0.679	3.128	9	31	39.65	19	24	52.6	
10	BD+18-2213	0.136	-1.654	9	31	25.75	18	12	17.2	
11	TYC-1406-322-1	0.31	-1.606	9	31	15.3	18	13	23.6	
12	TYC-1406-190-1	-1.236	-0.812	9	33	0.63	18	21	32.3	
13	TYC-1406-952-1	-1.74	2.082	9	34	1.83	19	3	28.1	
14	TYC-1406-540-1	-1.401	-1.892	9	32	59.85	18	5	0.2	
15	TYC-1406-518-1	-1.203	0.624	9	33	12.84	18	43	3.9	
16	TYC-1406-408-1	0.443	-2.084	9	31	1.93	18	6	33.5	
17	TYC-1406-593-1	-0.267	0.692	9	32	14.75	18	46	18.8	
18	TYC-1406-568-1	0.43	0.734	9	31	31.27	18	48	38.1	
19	TYC-1406-159-1	1.083	1.348	9	30	56.02	18	59	18	
20	TYC-1406-984-1	-2.327	1.389	9	34	31.67	18	51	41.7	
21	TYC-1406-1192-1	1.319	-2.701	9	30	1.08	17	59	25.7	
22	TYC-1406-603-1	-0.715	0.773	9	32	43.76	18	46	24.6	
23	END	0	0	0	0	0	0	0	0	
24										

Fig. 6.21 Using Excel to inspect and edit the comma separated variable (.csv) file used as input to the program listed above. Notice that I populated the fields following the word "END" with zeros to prevent a fatal input error when reading that line. The fact that I populated them with zeros is neither here nor there. Any integers would have done

two worst stars and trying again. The stars in question are HD82604 and TYC-1406-803-1. The results are shown in Table 6.3.

Now there is one star with a much larger area than the others, TYC-1406-12-1. So let's drop that star and see what happens. In fact, we now recover the results in Table 6.1. At this point, I am going to deem the results to be good enough.

I also notice that the absolute values of coefficients c and f, which correspond to the centre offsets between the chip and celestial coordinate systems, are less than or equal to 0.91% of the absolute values of any of a, b, d or e. So I have detected a small offset. It is not enough to cause me loss of sleep.

Table 6.1 A typical set of results from the program above

```
ASTROMETRY METHOD OF MONTENBRUCK & PFLEGER FOR LOCATING SOLAR SYSTEM OBJECTS ETC.
ASTRONOMY ON THE PERSONAL COMPUTER - TR. DUNLOP - 2ND ED - 1994 - SPRINGER-VERLAG
REPROGRAMMED IN FORTRAN BY J. M. FINE AND J. D. CLARK (C) 2009

DATE: 2009 12 29  TIME:  6:13 UT


Plate constants:


a =  4.339115E-03 b = -6.948212E-04 c =  6.294252E-06
d =  6.948916E-04 e =  4.339904E-03 f =  1.463978E-06

Effective focal length and image scale:

F =  2.275422E+02  mm
m =  9.064902E+02 sec/mm

Coordinates:
```

NAME	x	y	X	Y	RA			Dec			Error
	mm	mm	mm	mm	h	m	s	deg	m	s	s
MARS	0.5240	-0.4240	0.0026	-0.0015	9	31	13.6465	18	31	32.3391	
CENTRE	0.0000	0.0000	0.0000	0.0000	9	31	50.8931	18	36	37.0140	
HD82624	-2.1690	0.6670	-0.0099	0.0014	9	34	14.2377	18	41	19.8079	2.4814
HD81899	2.4730	2.1490	0.0092	0.0110	9	29	36.3641	19	14	32.0240	1.0499
TYC-1406-924-1	-3.1230	-0.5040	-0.0132	-0.0044	9	35	2.1406	18	21	32.2712	1.3047
BD+19-2219	1.5930	1.5250	0.0059	0.0077	9	30	25.7530	19	3	9.2204	0.3925
TYC-1406-189-1	-0.3540	-0.1940	-0.0014	-0.0011	9	32	11.2174	18	32	52.5444	1.7007
TYC-1406-700-1	0.6790	3.1280	0.0008	0.0140	9	31	39.6254	19	24	54.2134	1.6505
BD+18-2213	0.1360	-1.6540	0.0017	-0.0071	9	31	25.7159	18	12	15.8179	1.4650
TYC-1406-322-1	0.3100	-1.6060	0.0025	-0.0068	9	31	15.2659	18	13	23.6192	0.4856
TYC-1406-190-1	-1.2360	-0.8120	-0.0048	-0.0044	9	33	0.4213	18	21	32.1981	2.9735
TYC-1406-952-1	-1.7400	2.0820	-0.0090	0.0078	9	34	1.7728	19	3	28.4471	0.8827
TYC-1406-540-1	-1.4010	-1.8920	-0.0048	-0.0092	9	32	59.8110	18	5	1.8366	1.7286
TYC-1406-518-1	-1.2030	0.6240	-0.0056	0.0019	9	33	12.9746	18	43	2.0569	2.6561
TYC-1406-408-1	0.4430	-2.0840	0.0034	-0.0087	9	31	2.1359	18	6	34.6382	3.1490
TYC-1406-593-1	-0.2670	0.6920	-0.0016	0.0028	9	32	14.7021	18	46	18.1066	0.9709
TYC-1406-568-1	0.4300	0.7340	0.0014	0.0035	9	31	31.1973	18	48	35.6329	2.6742
TYC-1406-159-1	1.0830	1.3480	0.0038	0.0066	9	30	56.1768	18	59	18.4067	2.2603
TYC-1406-984-1	-2.3270	1.3890	-0.0111	0.0044	9	34	31.6338	18	51	42.5591	1.0011
TYC-1406-1192-1	1.3190	-2.7010	0.0076	-0.0108	9	30	1.0219	17	59	26.3651	1.0629
TYC-1406-603-1	-0.7150	0.7730	-0.0036	0.0029	9	32	43.7525	18	46	26.0334	1.4373

One other thing I should mention is that the program calculates the position in the $\alpha\delta$ system projected onto of the centre of the camera sensor. It is not important to know this. I obtained the value by manually adjusting a guessed input value until the xy and XY values are all zero. The value calculated makes no difference to the calculated position of Mars.

The 2009–2010 Apparition of Mars

Table 6.4 lists the coordinates calculated for the 2009–2010 apparition of Mars from my photographs.

Plotting these data is, as always, instructive. I was able to track down a couple of errors that way because the positions on certain dates looked out of kilter with the others. I never did manage to satisfy myself that I had good data for 20 October 2009, so I rejected the data for that night.

Table 6.2 The original results for 29 December 2019

```
ASTRONOMY ON THE PERSONAL COMPUTER - TR. DUNLOP - 2ND ED - 1994 - SPRINGER-VERLAG
REPROGRAMMED IN FORTRAN BY J. M. FINE AND J. D. CLARK (C) 2009

DATE: 2009 12 29  TIME:  6:13 UT


Plate constants:

a =  4.195261E-03 b = -7.479143E-04 c =  5.500457E-05
d =  6.017891E-04 e =  4.271421E-03 f =  8.576549E-05

Effective focal length and image scale:

F =  2.333177E+02  mm
m =  8.840513E+02 sec/mm

Coordinates:
```

NAME	x	y	X	Y	RA			Dec			Error
	mm	mm	mm	mm	h	m	s	deg	m	s	s
MARS	0.5240	-0.4240	0.0026	-0.0014	9	31	13.7060	18	31	45.6547	
CENTRE	0.0000	0.0000	0.0001	0.0001	9	31	50.1863	18	36	54.4023	
HD82624	-2.1690	0.6670	-0.0095	0.0016	9	34	9.5270	18	42	9.6451	80.9850
HD81899	2.4730	2.1490	0.0088	0.0108	9	29	42.5096	19	13	31.8491	104.8072
TYC-1406-924-1	-3.1230	-0.5040	-0.0127	-0.0039	9	34	54.5649	18	22	57.2077	137.8432
HD82604	-2.0970	-1.6770	-0.0075	-0.0083	9	33	39.3293	18	7	54.7421	504.6158
BD+19-2219	1.5930	1.5250	0.0056	0.0076	9	30	29.5604	19	2	34.5849	64.2610
TYC-1406-189-1	-0.3540	-0.1940	-0.0013	-0.0010	9	32	9.6236	18	33	19.4815	33.6946
TYC-1406-700-1	0.6790	3.1280	0.0006	0.0139	9	31	42.7610	19	24	14.3956	58.2800
BD+18-2213	0.1360	-1.6540	0.0019	-0.0069	9	31	24.0212	18	12	53.9423	44.2362
TYC-1406-322-1	0.3100	-1.6060	0.0026	-0.0066	9	31	13.9699	18	13	57.7234	39.0324
TYC-1406-190-1	-1.2360	-0.8120	-0.0045	-0.0041	9	32	56.5206	18	22	24.8769	78.6580
TYC-1406-952-1	-1.7400	2.0820	-0.0088	0.0079	9	33	59.0355	19	3	49.9576	45.2490
TYC-1406-540-1	-1.4010	-1.8920	-0.0044	-0.0088	9	32	54.7456	18	6	12.9570	102.9133
TYC-1406-803-1	-2.9700	-1.6770	-0.0112	-0.0089	9	34	32.2880	18	6	4.0889	665.0780
TYC-1406-518-1	-1.2030	0.6240	-0.0055	0.0020	9	33	10.2400	18	43	33.8055	47.5256
TYC-1406-408-1	0.4430	-2.0840	0.0035	-0.0085	9	31	0.7494	18	7	12.9321	42.8745
TYC-1406-593-1	-0.2670	0.6920	-0.0016	0.0029	9	32	13.9709	18	46	30.8530	16.3613
TYC-1406-568-1	0.4300	0.7340	0.0013	0.0035	9	31	31.9545	18	48	34.4003	10.3991
TYC-1406-159-1	1.0830	1.3480	0.0036	0.0065	9	30	58.7766	18	58	56.0038	44.8618
TYC-1406-984-1	-2.3270	1.3890	-0.0107	0.0046	9	34	27.1447	18	52	25.2483	77.6048
TYC-1406-1192-1	1.3190	-2.7010	0.0076	-0.0107	9	30	0.9822	17	59	56.5717	30.9032
TYC-1406-603-1	-0.7150	0.7730	-0.0035	0.0030	9	32	42.1490	18	46	46.2608	31.5065
TYC-1406-12-1	1.4100	-1.7080	0.0072	-0.0064	9	30	6.0486	18	14	42.8319	20.7210

The result of all my labour is shown in Fig. 6.22. There are other ways to present the same data. Fig. 6.23 shows the epicycle plotted relative to the ecliptic. I tend to use the term epicycle for the period of apparent back-and-forth motion, even though nobody now thinks epicycles are anything but an illusion. It should be emphasized that both Figs. 6.22 and 6.23 are planar projections of the celestial sphere: for example the ecliptic is actually a great circle, but it comes out as a curve, not a line, in Fig. 6.22. To Southern Hemisphere observers, these two images would appear to be upside down. I vividly remember experiencing the opposite phenomenon during my only visit to Australia. No matter how many times I told myself that it was I who was inverted compared to what for me was normal, and that the sky had not turned around, in my heart I found this very hard to believe. At the time, from Australia, Saturn was just below Spica in the sky. As the tour guide at Sydney Observatory explained this, I piped up and said that from my back

Table 6.3 Better results for 29 December 2019

```
| ASTROMETRY METHOD OF MONTENBRUCK & PFLEGER FOR LOCATING SOLAR SYSTEM OBJECTS ETC.
  ASTRONOMY ON THE PERSONAL COMPUTER - TR. DUNLOP - 2ND ED - 1994 - SPRINGER-VERLAG
  REPROGRAMMED IN FORTRAN BY J. M. FINE AND J. D. CLARK (C) 2009

DATE: 2009 12 29  TIME:  6:13 UT


Plate constants:

a =  4.335178E-03 b = -6.912920E-04 c =  1.314600E-09
d =  6.939782E-04 e =  4.340722E-03 f =  4.275716E-09


Effective focal length and image scale:

F =  2.276402E+02  mm
m =  9.061000E+02 sec/mm


Coordinates:
```

NAME	x	y	X	Y	RA			Dec			Error
	mm	mm	mm	mm	h	m	s	deg	m	s	s
MARS	0.5240	-0.4240	0.0026	-0.0015	9	31	13.7894	18	31	31.8694	
CENTRE	0.0000	0.0000	0.0000	0.0000	9	31	50.9844	18	36	36.7129	
HD82624	-2.1690	0.6670	-0.0099	0.0014	9	34	14.1710	18	41	20.0312	1.7054
HD81899	2.4730	2.1490	0.0092	0.0110	9	29	36.4871	19	14	31.6255	1.1725
TYC-1406-924-1	-3.1230	-0.5040	-0.0132	-0.0044	9	35	2.0795	18	21	32.4772	1.5858
BD+19-2219	1.5930	1.5250	0.0059	0.0077	9	30	25.8575	19	3	8.8798	1.5589
TYC-1406-189-1	-0.3540	-0.1940	-0.0014	-0.0011	9	32	11.2983	18	32	52.2767	2.4642
TYC-1406-700-1	0.6790	3.1280	0.0008	0.0140	9	31	39.5952	19	24	54.3125	1.8798
BD+18-2213	0.1360	-1.6540	0.0017	-0.0071	9	31	25.8993	18	12	15.2135	2.9103
TYC-1406-322-1	0.3100	-1.6060	0.0025	-0.0068	9	31	15.4568	18	13	22.9908	2.3157
TYC-1406-190-1	-1.2360	-0.8120	-0.0048	-0.0044	9	33	0.4834	18	21	31.9914	2.1096
TYC-1406-952-1	-1.7400	2.0820	-0.0090	0.0078	9	34	1.6578	19	3	28.8304	2.5479
TYC-1406-540-1	-1.4010	-1.8920	-0.0048	-0.0092	9	32	59.9188	18	5	1.4776	1.6105
TYC-1406-518-1	-1.2030	0.6240	-0.0056	0.0019	9	33	12.9652	18	43	2.0880	2.5395
TYC-1406-408-1	0.4430	-2.0840	0.0034	-0.0087	9	31	2.3587	18	6	33.9053	6.1251
TYC-1406-593-1	-0.2670	0.6920	-0.0016	0.0028	9	32	14.7428	18	46	17.9724	0.8339
TYC-1406-568-1	0.4300	0.7340	0.0014	0.0035	9	31	31.2757	18	48	35.3752	2.7260
TYC-1406-159-1	1.0830	1.3480	0.0038	0.0066	9	30	56.2611	18	59	18.1307	3.4225
TYC-1406-984-1	-2.3270	1.3890	-0.0110	0.0044	9	34	31.5210	18	51	42.9369	2.4504
TYC-1406-1192-1	1.3190	-2.7010	0.0076	-0.0108	9	30	1.3259	17	59	25.3703	3.5238
TYC-1406-603-1	-0.7150	0.7730	-0.0036	0.0029	9	32	43.7634	18	46	25.9974	1.3982
TYC-1406-12-1	1.4100	-1.7080	0.0073	-0.0064	9	30	5.3915	18	14	27.5176	17.5181

yard, it was above Spica. The tour guide gave me a funny look, thought for a second, and, presumably recognizing my accent, said, "Did you just come from England?" Once she realized that I was astronomically knowledge-able, she gave me a much more thorough tour of the southern sky than I would otherwise have gotten, all with a 16-inch telescope. Sometimes it pays to open your big mouth.

Anyway, to change RA and Dec. coordinates into ecliptic coordinates, I first transformed α and δ into Cartesian coordinates, using essentially Eq. (6.3). Then I rotated the coordinates about the ξ axis using a rotation matrix. (For readers who are not familiar with rotation matrices, there is a short section on rotations in Chap. 7, which will teach you what you need to know.)

$$\begin{pmatrix} \xi_{ecl} \\ \eta_{ecl} \\ \zeta_{ecl} \end{pmatrix} = \begin{pmatrix} 1 & 0 & 0 \\ 0 & \cos\varphi_{ax} & -\sin\varphi_{ax} \\ 0 & \sin\varphi_{ax} & \cos\varphi_{ax} \end{pmatrix} \begin{pmatrix} \xi \\ \eta \\ \zeta \end{pmatrix}, \tag{6.21}$$

Table 6.4 Coordinates calculated from my observations of the 2009–2010 apparition of Mars

Year	Month	Date	Time (UT)	RA (decimal hours)	DEC (Decimal Degrees)	RA (rad)	DEC (rad)	Ecliptic longitude λ	Ecliptic latitude β
2009	10	20	03:30:00	8.282	21.026	−2.168	0.367	0.821	0.692
2009	10	22	03:58:00	8.355	20.766	−2.187	0.362	0.800	0.682
2009	10	23	01:52:00	8.385	20.689	−2.195	0.361	0.792	0.678
2009	10	25	02:51:00	8.451	20.519	−2.213	0.358	0.773	0.671
2009	10	29	01:52:00	8.578	20.180	−2.246	0.352	0.738	0.655
2009	11	2	02:29:00	8.701	19.840	−2.278	0.346	0.704	0.640
2009	11	7	02:29:00	8.845	19.421	−2.316	0.339	0.666	0.622
2009	11	8	02:21:00	8.873	19.351	−2.323	0.338	0.658	0.618
2009	11	13	03:51:00	9.006	18.966	−2.358	0.331	0.623	0.601
2009	11	15	02:36:00	9.055	18.827	−2.371	0.329	0.611	0.594
2009	11	17	02:17:00	9.102	18.691	−2.383	0.326	0.598	0.588
2009	11	22	02:27:00	9.213	18.386	−2.412	0.321	0.570	0.573
2009	12	11	01:03:00	9.499	17.811	−2.487	0.311	0.496	0.538
2009	12	16	01:25:00	9.533	17.874	−2.496	0.312	0.487	0.536
2009	12	23	01:10:00	9.546	18.127	−2.499	0.316	0.481	0.539
2009	12	29	06:13:00	9.520	18.526	−2.492	0.323	0.485	0.548
2010	1	1	02:50:00	9.497	18.761	−2.486	0.327	0.489	0.554
2010	1	8	02:17:00	9.409	19.460	−2.463	0.340	0.508	0.573
2010	1	18	01:51:00	9.210	20.684	−2.411	0.361	0.553	0.611
2010	1	26	22:22:00	8.983	21.809	−2.352	0.381	0.609	0.649
2010	2	1	00:15:00	8.843	22.393	−2.315	0.391	0.645	0.671
2010	2	14	21:21:00	8.492	23.517	−2.223	0.410	0.740	0.717
2010	2	17	19:13:00	8.432	23.648	−2.208	0.413	0.757	0.724
2010	2	21	02:45:00	8.372	23.749	−2.192	0.415	0.774	0.730
2010	2	21	19:43:00	8.361	23.772	−2.189	0.415	0.778	0.731
2010	2	26	23:31:00	8.291	23.832	−2.170	0.416	0.799	0.738
2010	3	2	01:09:00	8.268	23.886	−2.165	0.417	0.805	0.740
2010	3	2	19:59:00	8.261	23.823	−2.163	0.416	0.808	0.740
2010	3	4	21:06:00	8.241	23.820	−2.158	0.416	0.814	0.741
2010	3	6	16:35:00	8.231	23.747	−2.155	0.414	0.818	0.741
2010	3	7	21:01:00	8.227	23.722	−2.154	0.414	0.819	0.740
2010	3	13	20:03:00	8.225	23.510	−2.153	0.410	0.821	0.737
2010	3	21	20:15:00	8.268	23.088	−2.165	0.403	0.811	0.727
2010	4	2	20:24:00	8.416	22.201	−2.203	0.387	0.772	0.701
2010	4	4	20:07:00	8.449	22.025	−2.212	0.384	0.764	0.696
2010	4	5	21:12:00	8.467	21.929	−2.217	0.383	0.759	0.693
2010	4	8	21:17:00	8.522	21.641	−2.231	0.378	0.744	0.684
2010	5	6	21:27:00	9.201	18.186	−2.409	0.317	0.575	0.571
2010	5	15	22:14:00	9.465	16.766	−2.478	0.293	0.513	0.524

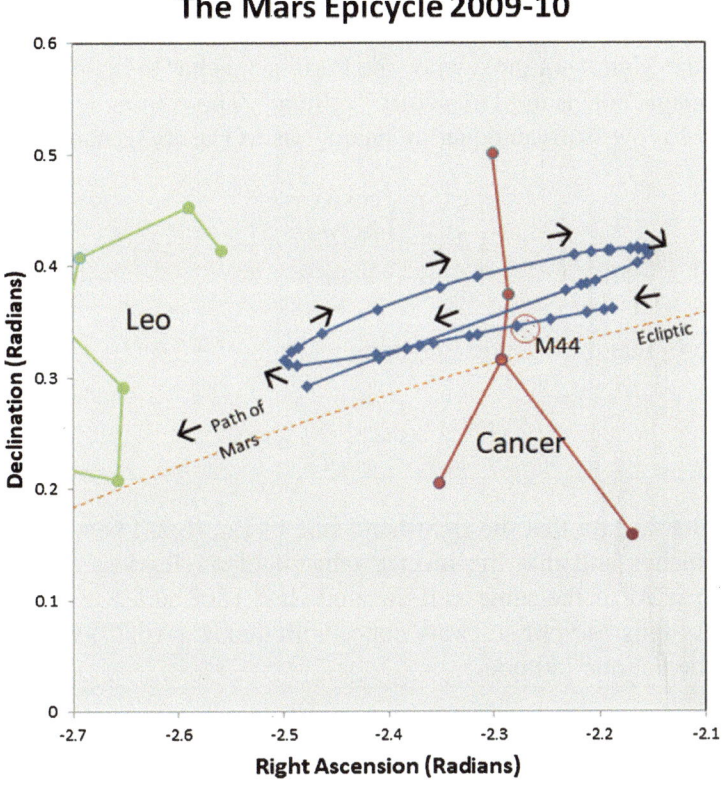

Fig. 6.22 The epicycle traced out by Mars plotted from the data in Table 6.4. The constellation Cancer and part of Leo are shown to add context

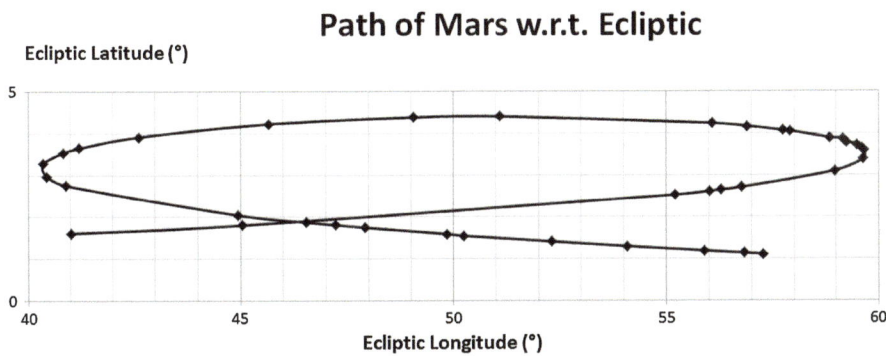

Fig. 6.23 The epicycle shown in Fig. 6.22 for the 2009–2010 apparition, presented in ecliptic co-ordinates, this time in degrees, helps the reader get a feeling for how far away from the ecliptic the epicycle is

Where φ_{ax} is my measured tilt of the Earth's axis given by Eq. (5.54), 23.6°. I knew that this must be a rotation about the ξ axis, because this axis points towards, the March equinox, when the Earth's axis has to lie in the $\eta\zeta$ plane. The subscript "ecl" is used to signify "ecliptic". The ecliptic longitude λ and latitude β follow from an equation analogous to Eq. (6.3), namely

$$\begin{pmatrix} \xi_{ecl} \\ \eta_{ecl} \\ \zeta_{ecl} \end{pmatrix} = \begin{pmatrix} \cos\beta\cos\lambda \\ \cos\beta\sin\lambda \\ \sin\beta \end{pmatrix}. \tag{6.21}$$

It follows that the ecliptic longitude

$$\lambda = \tan^{-1}\left(\frac{\xi_{ecl}}{\eta_{ecl}}\right), \tag{6.22}$$

taking suitable care that the right-hand side of Eq. (6.22) becomes singular if η_{ecl} vanishes and that the inverse tangent of angles less than −90° or greater than 90° is the same as if the angle had 180° added to or subtracted from it. As long as your software copes with that, Eq. (6.22) will be fine.

Also the ecliptic latitude

$$\beta = \sin^{-1}\zeta_{ecl}. \tag{6.23}$$

The ecliptic longitude and latitude for this apparition are plotted against time in Figs. 6.24 and 6.25 respectively.

From the data shown in Table 6.4 and in Figs. 6.22, 6.23, 6.24, and 6.25, our next task is to work out a preliminary orbit for Mars.

Later Observing Campaigns

I had two further observing campaigns for the Mars apparitions of 2011–2012 and 2020.

For the former, I used the same equipment as for the 2009–2010 campaign. In a fit of enthusiasm in 2013, I sold it and bought a 9.25" SCT on a used Celestron CGEM Mount. With hindsight I came to regret this slightly. I let myself be persuaded that an 11″ telescope was too heavy for a lady of a certain age, although the 9.25″ was a good scope. The problem was the mount. I never got on very well with it, and I half think I bought somebody else's trouble. I now have a Celestron CGX mount and I like it a lot.

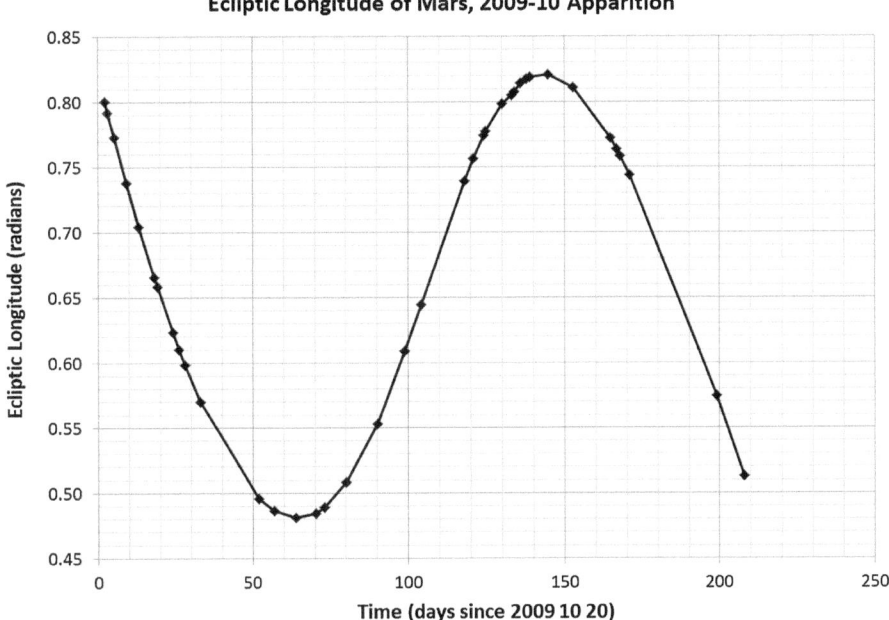

Fig. 6.24 The ecliptic longitude of Mars during the 2009–2010 apparition as a function of time

For my 2015 book on viewing and imaging the Solar System [105], I photographed a solar telescope at the house of my friend Trevor Taylor. He offered to buy a copy of the book if I included his observatory in the background. He had built it himself and was justly proud of the workmanship. Dear reader, like the razor salesman who bought the company, I was so impressed that I bought Trevor's observatory (Figs. 6.26 & 6.27).

The size of the dome opening constrains me to use a small secondary guide scope. In principle, this ought to be a problem. In practice I have not found it to be so. My experience is that by far the most crucial component of the guiding system is the quality of the camera. I have twice upgraded the camera. Both times the guiding improved greatly, largely because the exposure time went down and so the frequency of corrections went up. Also, I have not been disappointed by the results of guiding my photos to measure the position of the planet by following stars rather than Mars. The result is much better pictures.

For guiding I have switched to PHD2. The Guidemaster system I used to use appears to be no longer maintained. There is a strong community supporting this powerful software, and there are many YouTube training videos

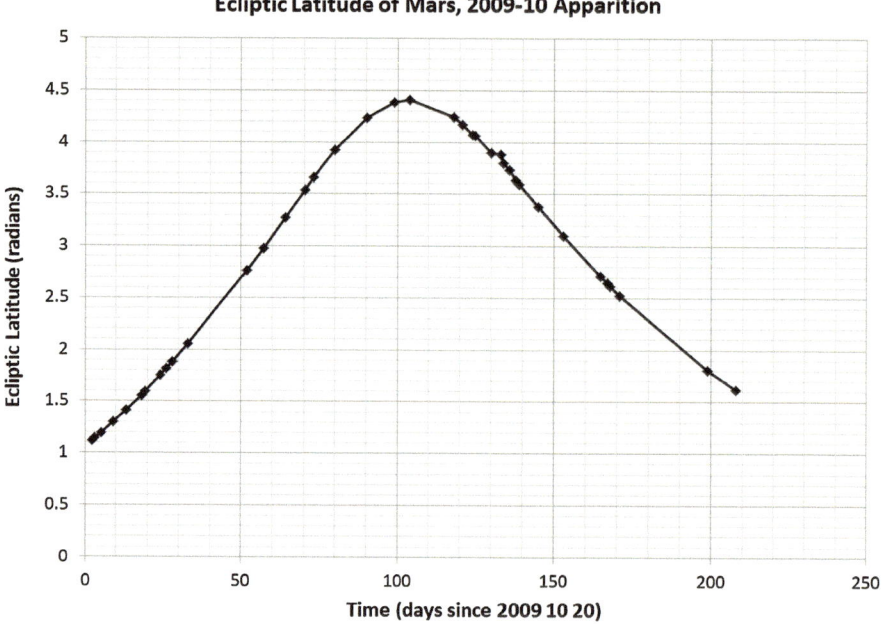

Fig. 6.25 The ecliptic latitude of Mars during the 2009–2010 apparition as a function of time

Fig. 6.26 The author in her observatory next to her 11" SCT. The mount holding the piggybacked DSLR can just be seen to the right of the telescope. (Image by Pauline Thomas, used with permission)

Fig. 6.27 My observatory building in my back yard. (Image: Author)

to get you started. I am the first to admit that I needed them. At first, PHD2 left me a very confused little bunny.

Balanced against the disadvantage of a finite-width dome slit is the fact that the dome does keep out stray light from neighbours who don't use bathroom curtains and the like. I'm on a steep hillside. The observatory is a long way below street level. The houses shield it from street lights. Behind me is a social club, which does not shine stray lights, and behind that is a sixty-foot cliff. Nobody will be able to build houses on it. So I do pretty well for light pollution despite being in a suburban location. Most of the houses are in the valley below. There is one factory down below. This is a source of light pollution and of turbulent air from its chimneys. The seeing is most certainly not helped (Fig. 6.28).

Fig. 6.28 My worst light pollution problem. The chimneys also make the air turbulent on otherwise still nights. It is due south of my observatory. (Image: Author)

Before I list the results of my later observing campaigns, I will tell you of one mistake I made in 2020. I bought a 300-mm telephoto lens for my DSLR and used it for many of my shots to position Mars. At first sight, the quality of the photographs was very impressive (Fig. 6.29).

Alas, it turned out that the only un-distorted region was around the centre of the image: I had to throw most of the data away. I was getting about ±6 arcsec agreement with Cartes du Ciel compared to ±2 arcsec or less with my 500-mm catadioptric lens (Fig. 6.30).

There are two morals of this little tale. First, process the images as you go, not in a batch at the end of the campaign. You then have time to revert to Plan A. Second, don't be deceived by pretty pictures. All that glitters is not gold.

Fig. 6.29 A picture taken with my 300-mm lens on my DSLR. 4 minutes at f/5.6, processed with Deep Sky Stacker and Gimp. A didymium filter was used to reduce sodium light pollution. It is a nice-looking picture, with a very dark sky, but as explained in the text, it was not so good for data analysis. 26 October 2020. (Image: Author)

Fig. 6.30 A picture taken with my 500-mm lens on my DSLR. 4 minutes at f/8, processed with Deep Sky Stacker and Gimp. A didymium filter was used to reduce
sodium light pollution. This picture is a lot less pretty than the one in Fig. 6.29, but
the data are more usable and more accurate. 20 August 2020. (Image: Author)

Data from 2011 to 2012 (Table 6.5 and Fig. 6.31)

Table 6.5 Coordinates calculated from my observations of the 2011–2012 apparition of Mars

Year	Month	Day	Time (UT)	RA (decimal hours)	Dec (decimal degrees)	RA (Rad)	Dec (Rad)	Ecliptic longitude λ (rad)	Ecliptic latitude β (rad)
2011	10	2	05:06:00	8.709	19.389	−2.280	0.338	−2.233	0.018
2011	11	25	06:21:00	10.620	10.937	−2.780	0.191	−2.736	0.035
2012	1	6	06:48:00	11.547	6.297	−3.023	0.110	−2.989	0.053
2012	1	16	02:47:00	11.643	5.987	−3.048	0.104	−3.014	0.058
2012	1	17	01:47:00	11.649	5.980	−3.050	0.104	−3.016	0.059
2012	2	2	05:45:00	11.646	6.500	−3.049	0.113	−3.011	0.067
2012	2	4	04:25:00	11.632	6.646	−3.045	0.116	−3.007	0.068
2012	2	11	05:04:00	11.554	7.319	−3.025	0.128	−2.984	0.070
2012	2	19	23:13:00	11.403	8.379	−2.985	0.146	−2.940	0.072
2012	2	26	02:15:00	11.269	9.297	−2.950	0.162	−2.902	0.073
2012	3	4	20:58:00	11.079	10.422	−2.900	0.182	−2.849	0.072
2012	3	11	22:06:00	10.905	11.348	−2.855	0.198	−2.801	0.069
2012	3	19	20:14:00	10.734	12.065	−2.810	0.211	−2.756	0.064
2012	3	23	22:02:00	10.646	12.480	−2.787	0.218	−2.732	0.062
2012	3	27	21:01:00	10.580	12.700	−2.770	0.222	−2.715	0.060
2012	3	30	21:30:00	10.538	12.810	−2.759	0.224	−2.705	0.057
2012	4	10	21:06:00	10.448	12.824	−2.735	0.224	−2.683	0.049
2012	4	12	21:44:00	10.442	12.766	−2.734	0.223	−2.682	0.048
2012	5	14	23:04:00	10.729	9.778	−2.809	0.171	−2.770	0.027

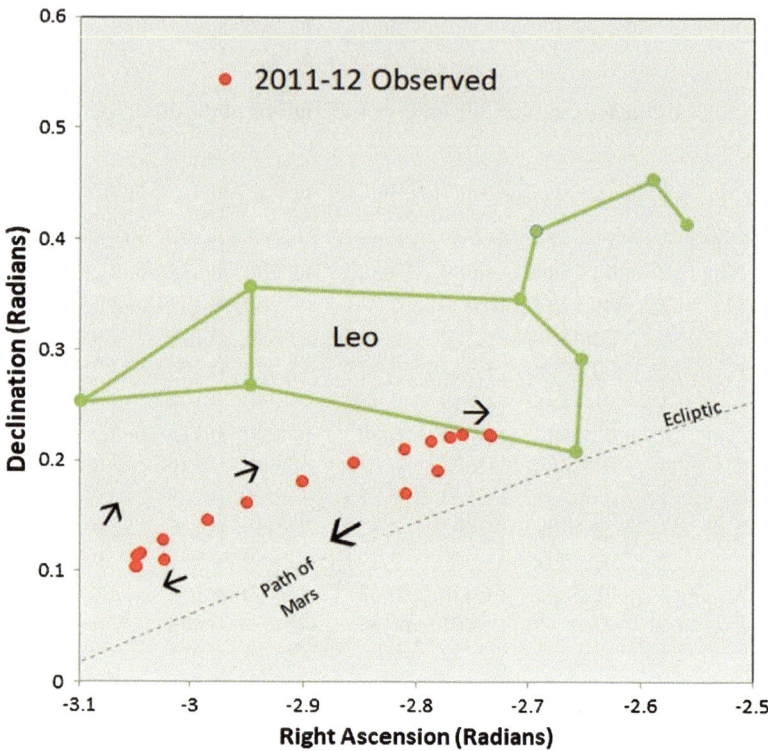

Fig. 6.31 The data from Table 6.5 plotted next to the constellation Leo to show where the points were. (Image: Author)

Data from 2020 (Table 6.6 and Fig. 6.32)

Table 6.6 Coordinates calculated from my observations of the 2020 apparition of Mars

Year	Month	Date	Time (UT)	RA (decimal hours)	Dec (decimal degrees)	RA (rad)	Dec (Rad)	Ecliptic longitude λ	Ecliptic latitude β
2020	8	20	02:53:00	1.658	5.864	−0.434	0.102	0.363	0.264
2020	9	2	02:53:00	1.817	6.674	−0.476	0.116	0.397	0.293
2020	9	5	02:23:00	1.834	6.770	−0.480	0.118	0.401	0.296
2020	9	17	23:38:00	1.812	6.784	−0.474	0.118	0.395	0.294
2020	9	30	00:24:00	1.656	6.290	−0.433	0.110	0.359	0.271
2020	10	9	23:37:00	1.459	5.687	−0.382	0.099	0.314	0.242
2020	10	12	22:54:00	1.395	5.502	−0.365	0.096	0.300	0.232
2020	10	15	01:41:00	1.350	5.378	−0.353	0.094	0.289	0.226
2020	10	26	21:15:00	1.124	4.886	−0.294	0.085	0.237	0.195
2020	11	3	21:27:00	1.018	4.832	−0.266	0.084	0.212	0.183
2020	11	14	22:23:00	0.958	5.214	−0.251	0.091	0.195	0.183
2020	11	19	18:33:00	0.964	5.545	−0.252	0.097	0.194	0.189
2020	11	26	20:37:00	1.007	6.190	−0.264	0.108	0.200	0.204

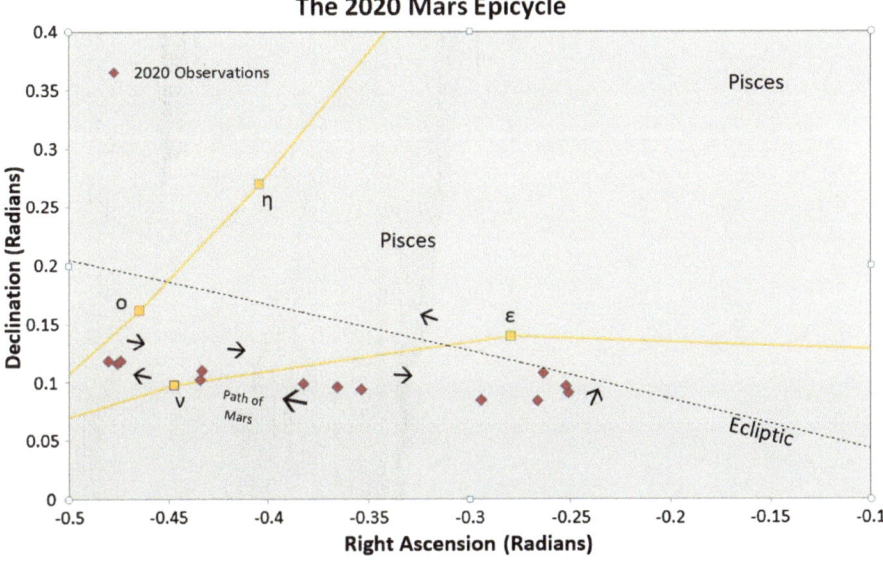

Fig. 6.32 The data from Table 6.6 plotted next to the constellation Pisces to show where the points were. (Image: Author)

Chapter 7

First Shot at the Orbit of Mars

You will need to have read, or at least to understand, Chaps. 2 and 3 before attempting this chapter. Knowing what's in Chap. 4 won't hurt either, though you could at a pinch catch that knowledge up later.

I tried several methods to find a preliminary orbit of Mars. (Yes I did look behind the sofa and underneath the cushions. Alas, no method did I find.) For orbits clearly inclined to that of the Earth, there are well-documented methods due to Laplace and Gauss [113]. The problem with these methods is that they break down for coplanar or nearly coplanar orbits.

Gauss' method requires three observations. It very cunningly exploits Kepler's laws that the celestial body orbits in an ellipse which is planar and that equal areas are swept out in equal times (Fig. 7.1).

By some careful jiggery-pokery, these two facts can be used to estimate the distance to the sun from the middle observation and the velocity of the body at that point. As we shall see, that's enough information to determine the entire ellipse along which the orbit plies.

Gauss became very famous because he developed this method to determine the orbit of the newly discovered asteroid Ceres, which was about to disappear behind the Sun. He predicted where it would reappear from behind the Sun. He had almost certainly refined this method by the time he wrote it up [114], so there is considerable uncertainty about how he actually first solved the problem.

J. Clark, *Calculate the Orbit of Mars!*,
https://doi.org/10.1007/978-3-030-78267-2_7,
© Springer Nature Switzerland AG 2021

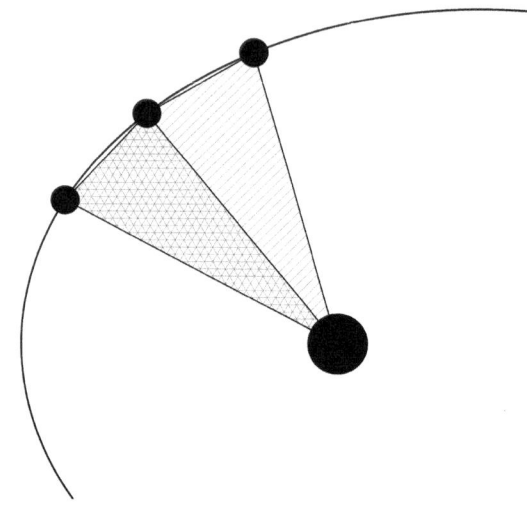

Fig. 7.1 Two areas swept out by an orbiting body, which are coplanar, and whose magnitudes are proportional to the time intervals of the observations

I did in fact try Gauss' method, but I concluded that it was basically a random number generator for the case of Mars. The results were not consistent for different threesomes of observations.

So began a quest to find a method that could work. At first, I thought I had found one, due to Bauschinger, in a 1906 book written in German [115]. Bauschinger came from a distinguished scientific family. His father discovered the Bauschinger effect, well known to students of the mechanics of steels. Anyway, fortunately, my German is just about good enough to follow Bauschinger Junior. His method is essentially a variant of Gauss' method with two "middle" observations, not one. Two sets of simultaneous Gauss-like equations are posed and solved. The real problem was that Bauschinger used Victorian mathematical methods. Since the celebrated thermodynamicist Gibbs' vector notation was published at about the same time as Bauschinger's book, he did not use vector notation, but instead used spherical trigonometry. Nowadays that's a lost art because we have something better: vector methods. So I gave up on Bauschinger.

I also found, in English, a series of papers by Karimi and Mortari [116, 117, 118, 119, 120, 121] developing methods for nearly coplanar orbits. These papers are around a decade old.

I was trying the methods of Karimi and Mortari simultaneously with Bauschinger's method and a 1965 method due to Herget [122]. I suppose trying three methods at once is a good way to drive yourself insane and take forever to get nowhere. But it's what I did.

Eventually Herget's method won. I got it to work, and it gave me a pre-liminary orbit. I was able to say "auf Wiedersehen" to Bauschinger and "goodbye" to Karimi and Mortari.

Before plunging into Herget's method, the time has come to look at how to orient an orbit in a plane other than the plane of our coordinate system.

We now turn our attention to the problem of how to work out the elliptical orbit of Mars when we see it from the surface of another planet which also orbits the Sun in an ellipse, and which does some other inconvenient things, because it is a rotating globe of roughly spherical shape, orbiting about the centre of mass of itself and its unusually large moon.

Non-coplanar Orbits Facing in Random Directions

The planets as they orbit roughly follow a line across the sky called the "ecliptic". The strict definition of the ecliptic is that it is the path the Sun appears to follow through the sky. We now know that this is equivalent to saying that the ecliptic is a projection of the Earth's orbital plane.

But please note the word "roughly" in the above paragraph. The apparent positions of the planets don't follow the ecliptic exactly, and we need to allow for this.

Figure 7.2 shows how to envisage the relative positions of the Sun, the Earth and Mars at any time. The vectors joining them obey the equation

$$\mathbf{r} + \mathbf{R} = \boldsymbol{\rho}, \tag{7.1}$$

where $\mathbf{r}$ is the Sun-to-Mars vector, $\mathbf{R}$ is the Sun-to-Earth vector and $\boldsymbol{\rho}$ is the Earth-to-Mars vector. Some writers, such as Danby [123], point R from Earth towards the Sun, but other writers, such as Curtis [124], writing about the orbits of satellites about the Earth, points R from the largest body (in his case the Earth) towards its satellite. Unfortunately, you can't be compatible with everyone.

I have chosen to point R from the Earth towards the Sun, as in Fig. 7.2.

Because of the finite speed of light, an observer on Earth wouldn't actually see Mars at the same position as the observer of Fig. 7.2 when Earth is at the position shown in Fig. 7.2. Light would take almost half-an-hour to cross this diagram. I expect an error of up to 10 minutes in the times of my measured Mars positions to result. I propose to neglect such effects in what follows. Most other modern writers on this subject do [123,124], despite being well aware that relativistic corrections to the Newtonian picture exist.

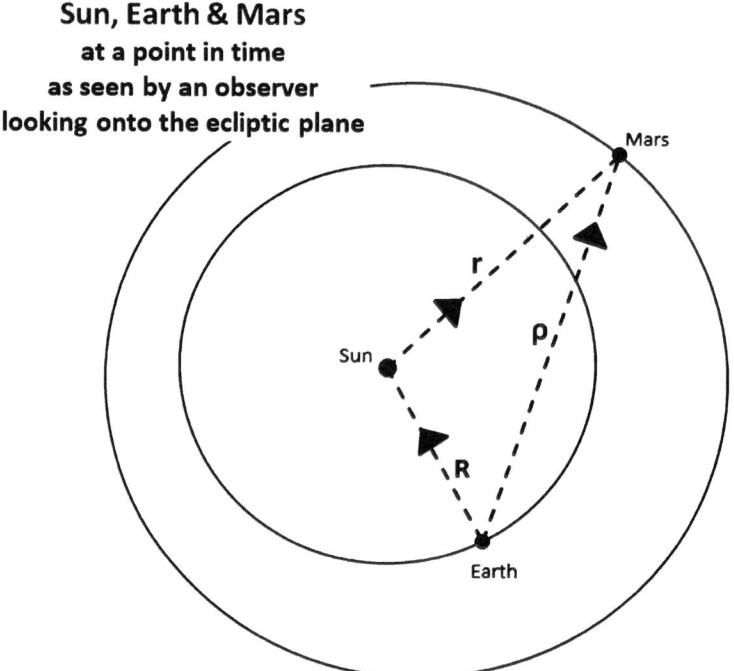

Fig. 7.2 A diagram defining the vectors r, R and ρ relating the positions of the Sun, the Earth and Mars. The triangle SEM joining the Sun, Earth and Mars is not necessarily coplanar with the ecliptic, the plane of the orbit of the Earth. (Image by the author)

It should be emphasized at this point that the orbits drawn in Fig. 7.2 are unknown to us. They might not look as shown. In this chapter, we will work out the orbit of the Earth. In later chapters, we will work out that of Mars.

The first re-orientation we could have is a rotation within the plane of the ecliptic, by an angle customarily designated ϖ. Fig. 7.3 shows an example.

The next possible re-orientation is that the ellipse could be tilted relative to the plane of the ecliptic. One tilt is enough to define all possible ellipses which still have one focus at the Sun. Fig. 7.4 shows a tilt axis that allows this. This tilt axis could of course have any orientation. Two numbers are required to describe the tilt: the orientation of the axis and the amount of tilt.

The amount of tilt is commonly given the very reasonable name of *inclination*. The name of the angle of the tilt axis is also perfectly reasonable, but a little less obvious. The line crosses the orbit at two points, seen in

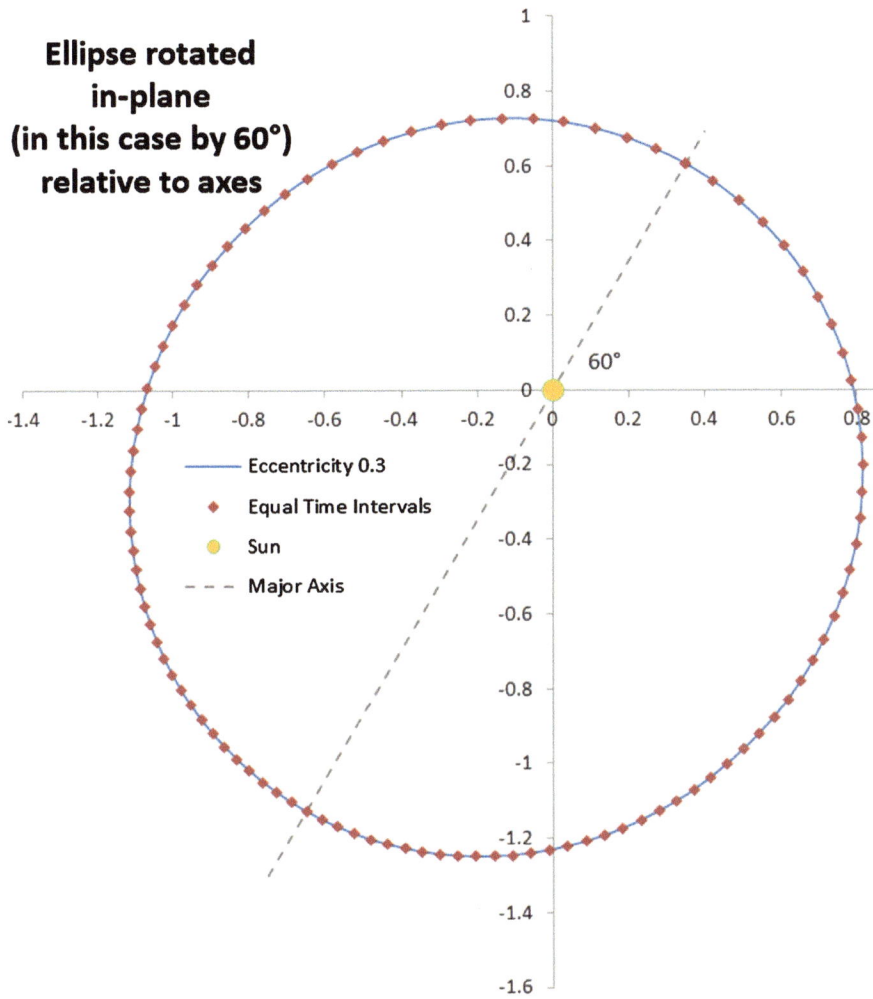

Fig. 7.3 Our ellipse could be rotated in plane by an angle ϖ. In this case, $\varpi = 60° = \pi/3$ radians

Fig. 7.4. Conventionally we pick the one where the planet is moving in the +z direction, assuming right-handed coordinates, and call the angle from the x-axis to this point the *longitude of the ascending node*. The longitude of the ascending node is normally given the symbol Ω. There is less consensus about the symbol for the inclination. I have seen the Greek letter iota (ι) or the English letter i and its upper case equivalent I. I'm afraid I have been a bit inconsistent. Sometimes I use I and sometimes i.

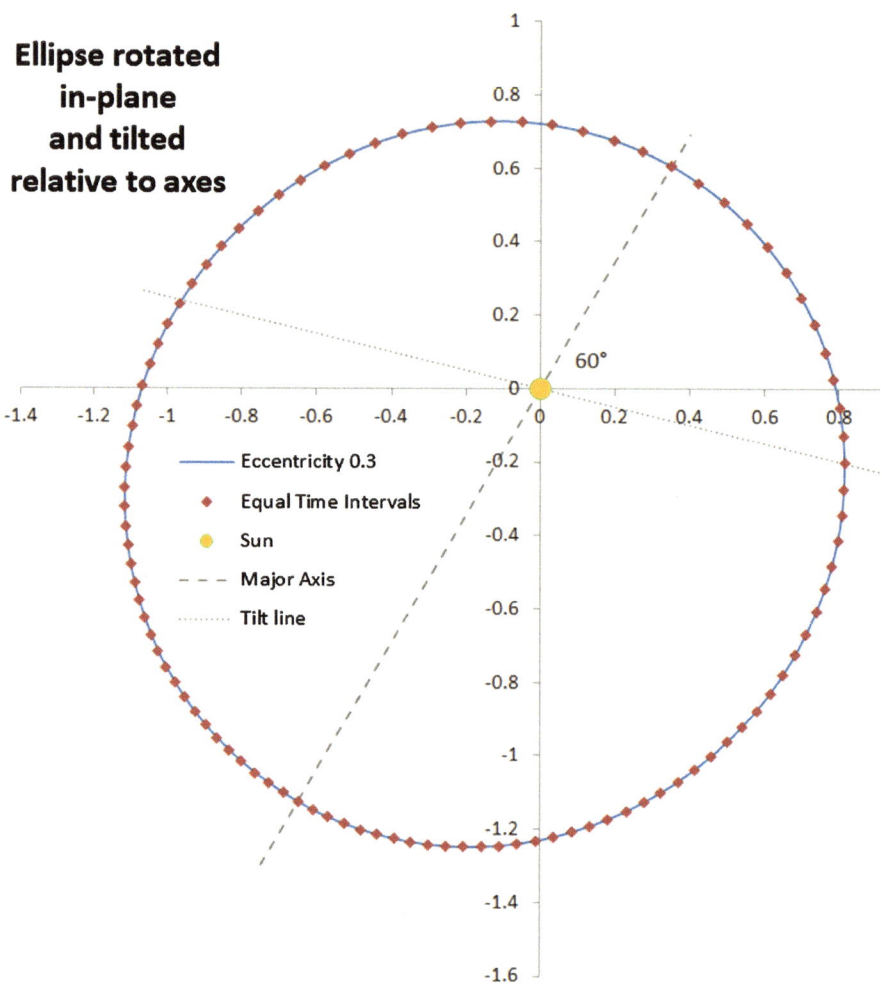

Fig. 7.4 Our ellipse could be rotated in plane by an angle ϖ. In this case, ϖ = 60° = π/3 radians. There is also a line about which the orbit is tilted

Rotations

There is no pre-ordained reason why the orbit of Mars would have its apsis line from perihelion to aphelion parallel to that of the Earth, or indeed, why it should lie in the same plane as that of the Earth. We already know that it is not quite coplanar with the Earth's axis: see Fig. 6.22, where the path of Mars around the 2009–10 opposition is slightly north of the ecliptic. Our first task then is to work out a way to express an orbit in an arbitrary orientation.

A rotation can be considered to be a transformation between coordinate system where the position of the origin does not change. This rather abstract mathematical idea can be made physically relevant by associating one coordinate system with a reference frame and another with an object which is rotated.

This object could be a body. For example, we could have a reference frame associated with the centre of the Earth, which co-rotates with the Earth. We could have another frame with origin at the centre of the Earth, which does not rotate relative to the International Celestial Reference Frame [125], a quantitative and intellectually respectable version of that frame colloquially, but vaguely, known as the "fixed stars".

The object associated with the frame of reference could also be a more abstract entity such as the path of an orbit. Indeed, there is no reason whatsoever why the orbit of Mars would lie exactly in the plane of the ecliptic. Nor is there any reason why its perihelion should lie along the major axis of the Earth's orbit.

And finally, of course, the true anomaly, the measure of where the planet is along the path of its orbit, is not at all likely to be the same as that of the Earth. In fact, there are systems where the true anomalies are linked, due to small second-order satellite-to-satellite gravitational interactions, the most spectacular example being the link, or "resonance" of the orbits of Jupiter's moons Io, Europa and Ganymede, which hold each other in a pattern with orbital period ratios of 1:2:4 [126, 83]. But the assumption I am making to determine the orbit of Mars is that there are not going to be any such shortcuts.

Any three-dimensional rotation can be broken down into successive rotations, each about an axis. Or, to put it another way, by successively rotating about convenient axes, we can rotate to any orientation.

So let us consider a coordinate system with perpendicular x and y axes and another with equally perpendicular x' and y' axes. The x' and y' axes are rotated by an angle θ with respect to the x and y axes, as shown in Fig. 7.5. A point P is shown, with Cartesian coordinates (x_P, y_P).

These coordinates may be expressed as polar coordinates via the relations

$$x_P = r_P \cos \vartheta$$
$$y_P = r_P \sin \vartheta$$

(7.2)

The polar coordinates in the x', y' frame of reference are

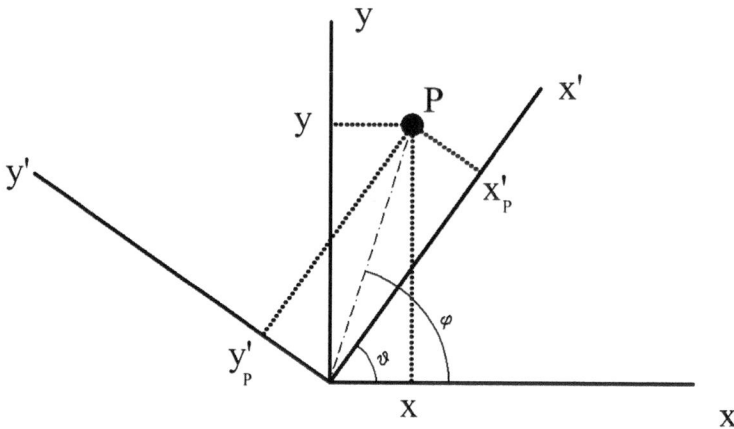

Fig. 7.5 Two coordinate systems in a plane

$$x'_P = r_P \cos(\varphi - \vartheta)$$
$$y'_P = r_P \sin(\varphi - \vartheta)$$
$$(7.3)$$

From Eq. (7.3), we can use the trigonometric addition formulae to obtain

$$
\begin{aligned}
x'_P &= r_P \cos\varphi \cos\vartheta + r_P \sin\varphi \sin\vartheta \\
&= \quad x_P \cos\vartheta + y_P \sin\vartheta \\
y'_P &= r_P \sin\varphi \cos\vartheta - r_P \sin\vartheta \cos\varphi \\
&= \quad y_P \cos\vartheta - x_P \sin\vartheta
\end{aligned}
\qquad (7.4)
$$

We can write Eq. (7.4) in matrix form:

$$
\begin{pmatrix} x'_P \\ y'_P \end{pmatrix} = \begin{pmatrix} \cos\vartheta & \sin\vartheta \\ -\sin\vartheta & \cos\vartheta \end{pmatrix} \begin{pmatrix} x_P \\ y_P \end{pmatrix}.
\qquad (7.5)
$$

To get from the (x,y) system to the (x', y') system, we rotate in the reverse direction. Since $\cos(-\vartheta) = \cos\vartheta$ and $\sin(-\vartheta) = -\sin\vartheta$, we have

$$
\begin{pmatrix} x_P \\ y_P \end{pmatrix} = \begin{pmatrix} \cos\vartheta & -\sin\vartheta \\ \sin\vartheta & \cos\vartheta \end{pmatrix} \begin{pmatrix} x'_P \\ y'_P \end{pmatrix}.
\qquad (7.6)
$$

The square matrix in Eq. (7.6) is the both inverse and the transpose of the square matrix in Eq. (7.5). That it is the transpose can be seen by inspection. That it is the inverse can be seen by multiplying the two matrices.

$$\begin{pmatrix} \cos\vartheta & \sin\vartheta \\ -\sin\vartheta & \cos\vartheta \end{pmatrix}\begin{pmatrix} \cos\vartheta & -\sin\vartheta \\ \sin\vartheta & \cos\vartheta \end{pmatrix} = \begin{pmatrix} \cos^2\vartheta + \sin^2\vartheta & -\sin\vartheta\cos\vartheta + \sin\vartheta\cos\vartheta \\ -\sin\vartheta\cos\vartheta + \sin\vartheta\cos\vartheta & \sin^2\vartheta + \cos^2\vartheta \end{pmatrix}$$

$$= \begin{pmatrix} 1 & 0 \\ 0 & 1 \end{pmatrix}. \tag{7.7}$$

We could add the third dimension as follows. In right-handed coordinate systems, the usual kind, the z and z' axes coincide and point out of the page. We use the usual formula for matrix multiplication that the *ij*th element of the product P of two 3×3 matrices A and B is given by

$$P_{ij} = \sum_{k=1}^{3} A_{ik} B_{kj} \tag{7.8}$$

to rewrite Eq. (7.6) as

$$\begin{pmatrix} x_P \\ y_P \\ z_P \end{pmatrix} = \begin{pmatrix} \cos\vartheta & -\sin\vartheta & 0 \\ \sin\vartheta & \cos\vartheta & 0 \\ 0 & 0 & 1 \end{pmatrix}\begin{pmatrix} x'_P \\ y'_P \\ z'_P \end{pmatrix}. \tag{7.9}$$

This formula enables us to rotate about the *z*-axis to get from a "new" to an "old" coordinate system.

It is worth pointing out that the rotation does not change the magnitude of the vector (x_P', y_P', z_P'). We now show this.

$$\begin{pmatrix} x_P \\ y_P \\ z_P \end{pmatrix} = \begin{pmatrix} \cos\vartheta & -\sin\vartheta & 0 \\ \sin\vartheta & \cos\vartheta & 0 \\ 0 & 0 & 1 \end{pmatrix}\begin{pmatrix} x'_P \\ y'_P \\ z'_P \end{pmatrix}$$

$$= \begin{pmatrix} x'_P \cos\vartheta - y'_P \sin\vartheta \\ x'_P \cos\vartheta - y'_P \sin\vartheta \\ z'_P \end{pmatrix}. \tag{7.10}$$

The magnitude of the rotated vector

$$\sqrt{x'^2 + y'^2 + z'^2} = \sqrt{\begin{array}{l} x'^2_P \cos^2\vartheta + y'^2_P \sin^2\vartheta - 2x_P y_P \sin\vartheta\cos\vartheta + x'^2_P \sin^2\vartheta \\ + y'^2_P \cos^2\vartheta + 2x_P y_P \sin\vartheta\cos\vartheta + z'^2_P \end{array}}$$

$$= \sqrt{x'^2_P\left(\cos^2\vartheta + \sin^2\vartheta\right) + y'^2_P\left(\sin^2\vartheta + \cos^2\vartheta\right) + z'^2_P} \tag{7.11}$$

$$= \sqrt{x'^2_P + y'^2_P + z'^2_P}$$

The 3×3 matrix in Eq. (7.9) also has the property that its transpose is its inverse. We now show this by multiplying this matrix by its transpose.

$$\begin{pmatrix} \cos\vartheta & \sin\vartheta & 0 \\ -\sin\vartheta & \cos\vartheta & 0 \\ 0 & 0 & 1 \end{pmatrix}\begin{pmatrix} \cos\vartheta & -\sin\vartheta & 0 \\ \sin\vartheta & \cos\vartheta & 0 \\ 0 & 0 & 1 \end{pmatrix} = \begin{pmatrix} \cos^2\vartheta + \sin^2\vartheta + 0 & -\sin\vartheta\cos\vartheta + \sin\vartheta\cos\vartheta + 0 & 0 \\ -\sin\vartheta\cos\vartheta + \sin\vartheta\cos\vartheta + 0 & \cos^2\vartheta + \sin^2\vartheta + 0 & 0 \\ 0 & 0 & 1 \end{pmatrix}$$

$$= \begin{pmatrix} 1 & 0 & 0 \\ 0 & 1 & 0 \\ 0 & 0 & 1 \end{pmatrix}. \tag{7.12}$$

Using the inverse matrix corresponds to rotating back to the original frame of reference. Unsurprisingly, therefore, matrices possessing the property that their transpose is also their inverse, and that do not change the magnitudes of vectors that they rotate, are called rotation matrices.

We now consider the case of rotation about the x-axis. In Fig. 7.6, we can work out that in a right-handed coordinate system x points out of the page, and the analogue of Eq. (7.9) is

$$\begin{pmatrix} x_P \\ y_P \\ z_P \end{pmatrix} = \begin{pmatrix} 1 & 0 & 0 \\ 0 & \cos\psi & -\sin\psi \\ 0 & \sin\psi & \cos\psi \end{pmatrix}\begin{pmatrix} x'_P \\ y'_P \\ z'_P \end{pmatrix}. \tag{7.13}$$

Consider the matrices

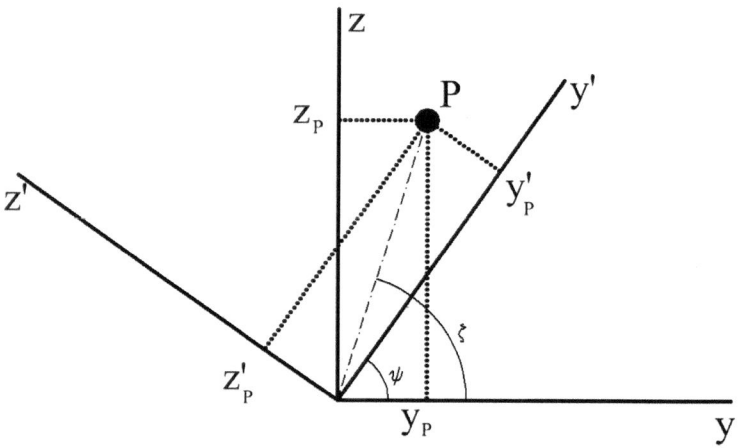

Fig. 7.6 Two coordinate systems in a plane

$$R_\omega = \begin{pmatrix} \cos\omega & -\sin\omega & 0 \\ \sin\omega & \cos\omega & 0 \\ 0 & 0 & 1 \end{pmatrix}$$

$$R_I = \begin{pmatrix} 1 & 0 & 0 \\ 0 & \cos I & -\sin I \\ 0 & \sin I & \cos I \end{pmatrix} \qquad (7.14)$$

We have

$$
R_\omega R_I = \begin{pmatrix} \cos\omega & -\sin\omega & 0 \\ \sin\omega & \cos\omega & 0 \\ 0 & 0 & 1 \end{pmatrix} \begin{pmatrix} 1 & 0 & 0 \\ 0 & \cos I & -\sin I \\ 0 & \sin I & \cos I \end{pmatrix}
$$

$$
= \begin{pmatrix} \cos\omega & -\sin\omega\cos I & \sin\omega\sin I \\ \sin\omega & \cos\omega\cos I & -\cos\omega\cos I \\ 0 & \sin I & \cos I \end{pmatrix} \qquad (7.15)
$$

whereas

$$
R_I R_\omega = \begin{pmatrix} 1 & 0 & 0 \\ 0 & \cos I & -\sin I \\ 0 & \sin I & \cos I \end{pmatrix} \begin{pmatrix} \cos\omega & -\sin\omega & 0 \\ \sin\omega & \cos\omega & 0 \\ 0 & 0 & 1 \end{pmatrix}
$$

$$
= \begin{pmatrix} \cos\omega & -\sin\omega & 0 \\ \sin\omega & \sin\omega\cos I & \cos\omega\cos I \\ \sin\omega\sin I & \cos\omega\sin I & \cos I \end{pmatrix} \qquad (7.16)
$$

$$
\neq R_\omega R_I.
$$

In mathematics, you can prove something by disproving the opposite. Here we have proved that the order in which rotation matrices are multiplied affects the result. This corresponds to something physical. If you take a die, as in the singular of dice, with faces numbered from one to six, and rotate it twice, about different axes, then do the rotations the opposite way around, you will find that the number on the top face is different.

It is important to understand that the order in which you perform rotations matters.

Here is another example. In Fig. 7.7, a book is rotated twice, through 90° clockwise about the +z-axis, then clockwise about the +y-axis. It ends up with the spine facing in the $+z$ direction and the front cover facing in the $+x$ direction.

In Fig. 7.8, the book is rotated twice, through 90° clockwise about the $+y$-axis, then clockwise about the $+z$-axis. In other words, the rotation operations are the same as those shown in Fig. 7.7, but in the reverse order. Now the book ends up with the spine facing in the $-x$ direction and the front cover facing in the $-y$ direction.

Reversing the order of the rotations changes the outcome.

To these matrices, we add the rotation matrix

$$R_\Omega = \begin{pmatrix} \cos\Omega & -\sin\Omega & 0 \\ -\sin\Omega & \cos\Omega & 0 \\ 0 & 0 & 1 \end{pmatrix}. \tag{7.17}$$

It is no coincidence that I have chosen the symbols ω, I and Ω.

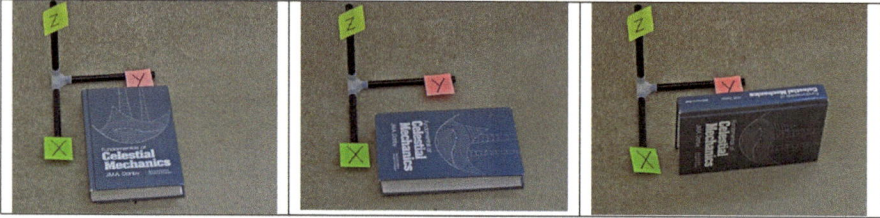

Fig. 7.7 A book is rotated through 90° clockwise about the $+z$-axis, then clockwise about the $+y$-axis. (Images by the author)

Fig. 7.8 A book is rotated through 90° clockwise about the $+y$-axis, then clockwise about the $+z$-axis. The rotations are the same ones shown in Fig. 7.7, but performed in the reverse order. After the rotations, the book ends up facing a different way than after the rotations in Fig. 7.7. (Images by the author. (the book is Danby's celestial Mechanics [89]. I wonder if Charles Dickens might have said, "Experiment performed by Mrs. Do-as-you-Would-be-Danby?". Sorry.))

The closed-form expressions of the matrix product $R_\omega R_I R_\Omega$ is too complicated to be of much practical use. In practice, such rotations will be done numerically.

It is fortunately straightforward to do this in Microsoft Excel. There is a function MMULT that does it for you. There is a bit of a trick to this, but let us take for demonstration purposes the case where

$$\omega = 340^\circ,$$
$$I = 2^\circ, \& \tag{7.18}$$
$$\Omega = 50^\circ.$$

Then

$$R_\omega = \begin{pmatrix} 0.940 & 0.342 & 0 \\ -0.342 & 0.940 & 0 \\ 0 & 0 & 1 \end{pmatrix}$$

$$R_I = \begin{pmatrix} 1 & 1 & 0 \\ 0 & 0.999 & -0.035 \\ 0 & 0.035 & 0.999 \end{pmatrix} \tag{7.19}$$

$$R_\Omega = \begin{pmatrix} 0.643 & -0.766 & 0 \\ 0.766 & 0.643 & 0 \\ 0 & 0 & 1 \end{pmatrix}$$

We multiply them in two stages, first to obtain $R_\omega R_I$ and then secondly to obtain $(R_\omega R_I) R_\Omega$.

Figure 7.9 shows the first stage multiplication. The arguments of the function MMULT are the two matrices E5:G7, outlined in blue, and E9:G11, outlined in green. You have to have selected the range I5:K7 into which to deposit the answer. Then there is a twiddly bit. If you just hit ENTER, the function won't work. You have to hit CTRL-SHIFT-ENTER instead

In step two, we multiply the product of the first two matrices by the third. The result is shown in Fig. 7.10.

It should be pointed out that it does not matter where in our spreadsheet the matrices are. By selecting regions in the MMULT arguments, we tell it where the input matrices are. It could also be pointed out that I have rounded the displays of the matrices to show three decimal places. As can be seen by looking at the intermediate matrix I5:K7 in Fig. 7.10, or by comparing the matrix I9:K11 in Figs. 7.9 and 7.10, the spreadsheet actually does the calculation to machine accuracy. The end result in our example is

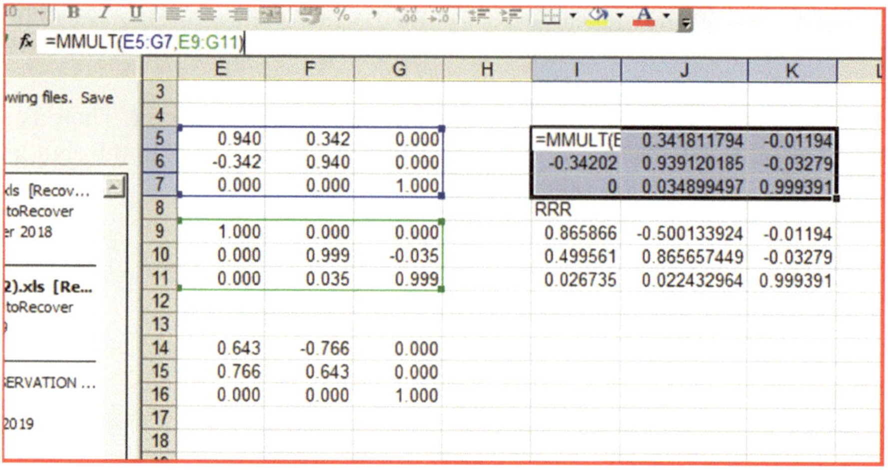

Fig. 7.9 Showing the matrix multiplication operation MMULT in Microsoft Excel

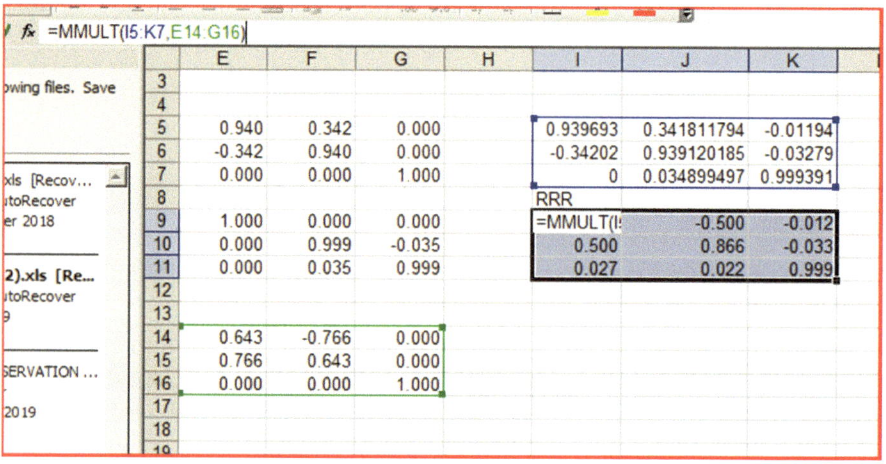

Fig. 7.10 Showing the second matrix multiplication operation with MMULT in Microsoft Excel

$$R_\omega R_I R_\Omega = \begin{pmatrix} 0.866 & -0.500 & -0.012 \\ 0.500 & 0.866 & -0.033 \\ 0.027 & 0.022 & 0.999 \end{pmatrix}. \tag{7.20}$$

Microsoft Excel can equally easily multiply a 3 x 3 matrix by a 3 x 1 matrix (also called a column vector) (Fig. 7.11).

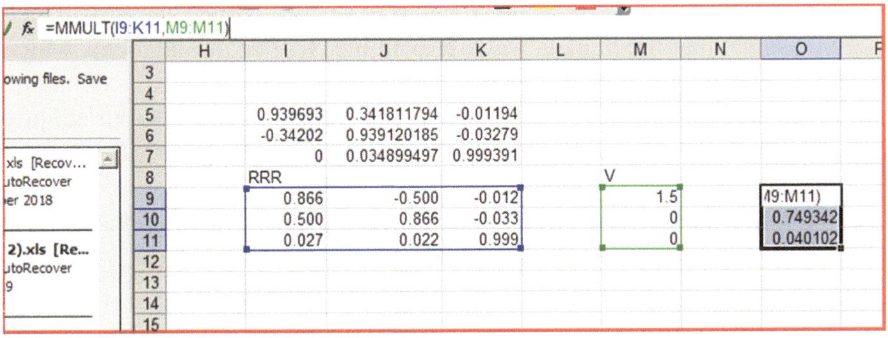

Fig. 7.11 Showing the matrix multiplication operation MMULT in Microsoft Excel. We have multiplied the 3 × 3 matrix I9:K11 by the column vector M9:M11 to give the resulting column vector in O9:O11

The calculation performed is, to three decimal places,

$$R_\omega R_I R_\Omega \begin{pmatrix} 1.5 \\ 0 \\ 0 \end{pmatrix} = \begin{pmatrix} 0.866 & -0.500 & -0.012 \\ 0.500 & 0.866 & -0.033 \\ 0.027 & 0.022 & 0.999 \end{pmatrix} \begin{pmatrix} 1.5 \\ 0 \\ 0 \end{pmatrix} = \begin{pmatrix} 1.299 \\ 0.749 \\ 0.040 \end{pmatrix}. \qquad (7.21)$$

We can check that this rotation has not changed the size of the column vector:

$$\sqrt{1.5^2 + 0^2 + 0^2} = \sqrt{1.299^2 + 0.749^2 + 0.040^2} = 1.500, \qquad (7.22)$$

to three decimal places.

What These Angles Are Used for

These angles are needed to orient an elliptical orbit. I'm going to illustrate a hypothetical system where we are observing from a blue planet whose orbit has eccentricity 0.2 and a semi-major axis given by

$$e_{\text{blue}} = 0.2,$$
$$a_{\text{blue}} = 1\text{AU}. \qquad (7.23)$$

We can take it as read that the star in this system, which is much, much more massive than the planets we will examine, will be at one focus, which gives us an origin.

The plane in which our planet orbits can be considered to be our ecliptic whose inclination is by definition zero. There are therefore no ascending or descending nodes for our planet.

This is not the case for the planet, which we call the red planet, whose orbit we wish to determine. It has

$$e_{red} = 0.3,$$
$$a_{red} = 1.5 \text{AU}. \tag{7.24}$$

The characteristics of these planets may remind you of somewhere we know, but the orbits have been made more eccentric to help us to see what is going on. From a space craft high above the ecliptic in this system, the view in Fig. 7.12 is what we might see.

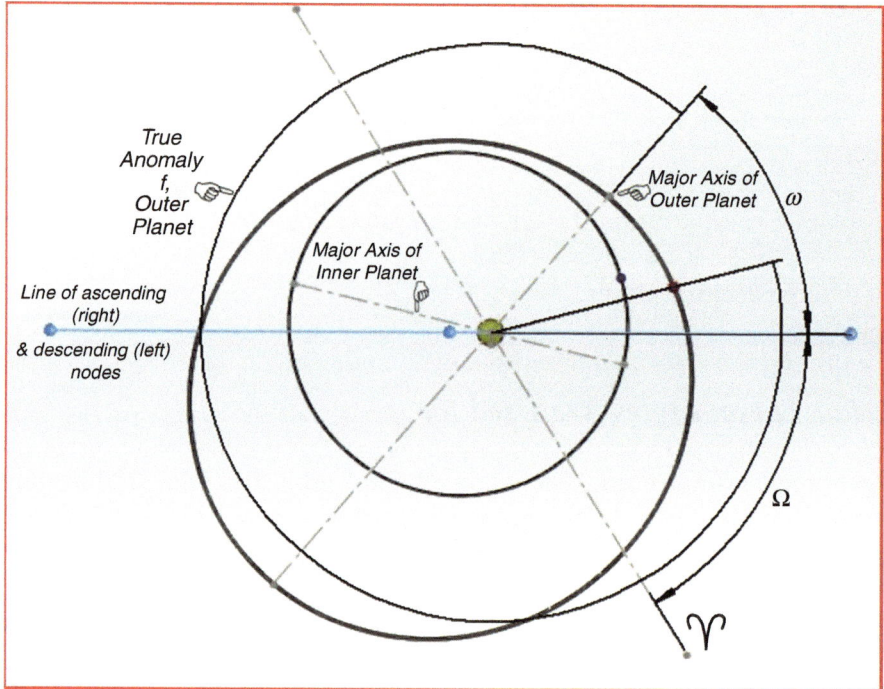

Fig. 7.12 A view of the orbits of our inner (blue) planet and our outer (red) planet. The symbols are explained in the text

The major axes of orbits of the two planets are not collinear. Were they to orbit with only negligible mutual gravitational attraction, there is absolutely no reason why they should be aligned.

Incidentally, one of the benefits of drawing Fig. 7.12 from scratch, rather than stealing a diagram from Wikipedia or wherever, is that I began to notice things that I would never otherwise have noticed. Were my blue and red planets to have mass comparable to those of the Earth and Mars, they would significantly attract one another gravitationally, especially during close approaches, and the orbits I have shown would be perturbed. That is one reason why we don't actually observe such planetary orbits in our solar system: they would not be stable. There are all sorts of evidence that the Earth's orbit has been stable for billions of years, from radioactive dating to the fact that the biosphere has only been able to thrive and prosper with temperatures broadly in the range 0 ± 40 °C, with most of the surface being in the upper half of this range for at least some of the time. Asteroids, on the other hand, can have more eccentric orbits, presumably because their mutual gravitation attraction is much lower.

Anyway, I digress. Fig. 7.13 shows the inclination angle I of the orbits of our "red" and "blue" planets.

One consequence of the nonzero value of I is that the angles Ω and ω in Fig. 7.12 are not coplanar. Sometimes when I is small, people add them anyway.

Figure 7.14 shows out two-planet system from various angles, to give you a sense of how they look

If I felt mischievous, I could also show another two-planet system, identical except that I have pointed the major axis of the outer, "red", planet in a different direction. In Fig. 7.15, I have done just that.

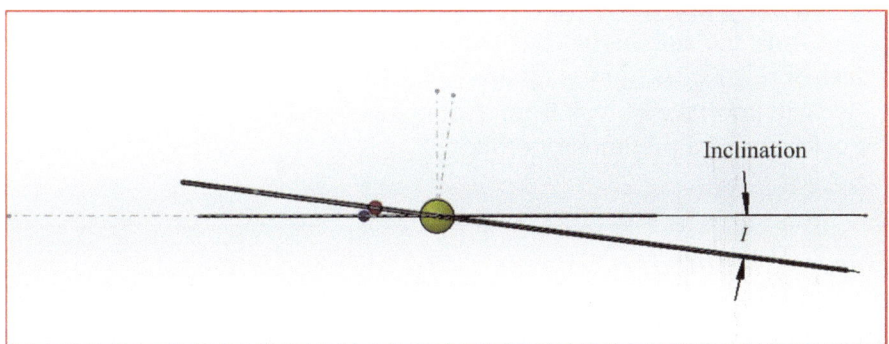

Fig. 7.13 Showing the angle of inclination of orbit of the "red" planet to that of the "blue" one

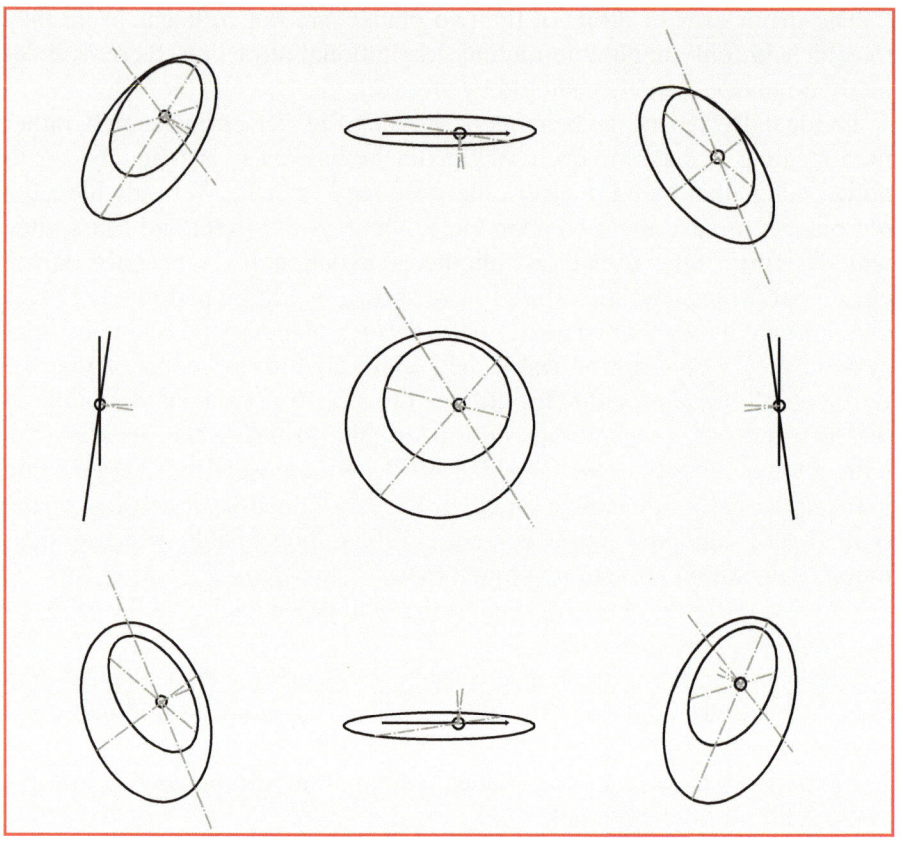

Fig. 7.14 Views of our two-planet system from various angles

From our point of view, if we wish to determine the orbit of the "red" planet from the surface of the "blue" planet, our task is to determine the values of a_{red}, e_{red}, Ω_{red}, ω_{red}, I_{red} and f_{red}.

We will now take a break from rotations and look at the method of Herget for determining a preliminary orbit.

Herget's Method

This is taken from Herget's 1965 paper [122]. This paper cannot easily be read without reference to Herget's earlier book [127], which he published privately, because he was concerned that textbooks cost too much for graduate students. Ironically, I paid well over $100 to obtain a used copy. An account of the method is also given by Danby [89]. I didn't follow either author precisely, but repeatedly tried one, then the other, until I thought I

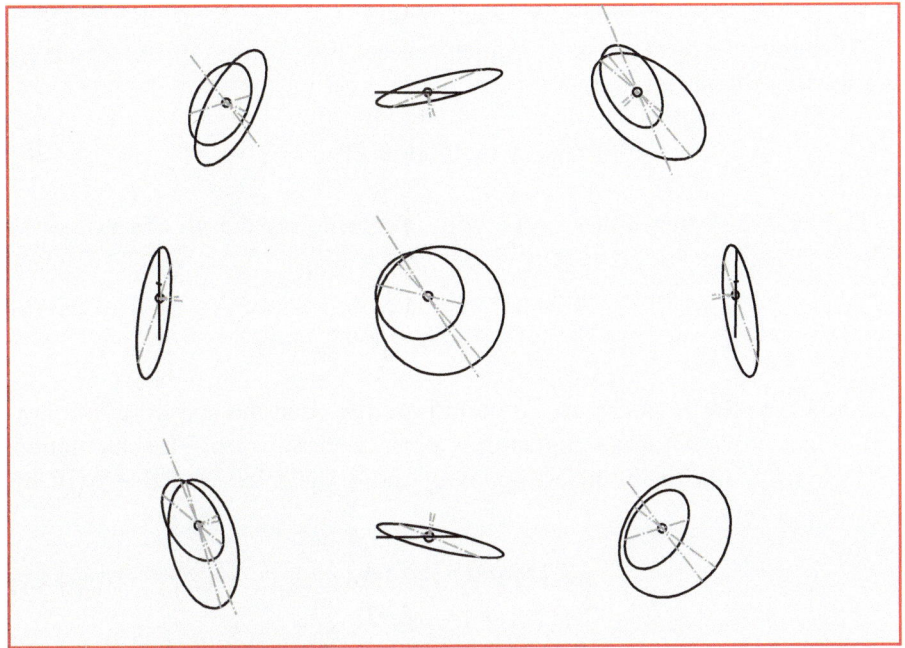

Fig. 7.15 Showing our imaginary planetary system, only with the major axis of the outer, "red" planet pointing in a different direction

understood the method. I believe I even made a small simplification of my own.

At various observation times t_i, Eq. (7.1) (see also Fig. 7.2) is

$$\mathbf{r}_i = \rho_i \hat{\rho}_i - \mathbf{R}_i \qquad (7.25)$$

We know the $\mathbf{R}_i$ and the unit vector $\hat{\rho}_i$, whose components contain the angular position we measured at the time of the *ith* observation. But we don't know the $\mathbf{r}_i$ or the magnitudes of the ρ_i. Herget's method is a way of finagling enough of the unknowns to obtain a preliminary orbit. This method is robust enough that it does not fail at low orbital inclinations to the ecliptic.

I say "finagling" because, despite the considerable algebraic and trigonometric manipulation required, we seed the method with guessed values of $\mathbf{r}$ for two of the observations. It can be any two, and it's just as easy to make them the first two and last two observations: we then guess $\mathbf{r}_1$ and $\mathbf{r}_n$. Since we know where the Sun is, we have by implication guessed a value of the difference in the true anomalies of the 1st and nth observations, $(f_n - f_1)$.

Everything else we will have to calculate.

By using spherical polar coordinates about the position of the observer, we see that the vector

$$\hat{\rho} = \left(\cos\beta \cos\lambda, \cos\beta \sin\lambda, \sin\beta \right). \tag{7.26}$$

Define two further unit vectors, with i representing the ith observation:

$$\begin{aligned} \hat{\mathbf{A}}_i &= \left(-\sin\lambda_i, \cos\lambda_i, 0 \right), \\ \hat{\mathbf{D}}_i &= \left(-\sin\beta_i \cos\lambda_i, -\sin\beta_i \cos\lambda_i, \cos\beta_i \right). \end{aligned} \tag{7.27}$$

The variables ρ_1 and ρ_n are to be imposed to seed the solution, but they are in fact unknown in an angles-only orbit determination. The calculation will be seeded with initial estimates of these variables, and they will be allowed to vary.

Let

$$\begin{aligned} P_i\left(\rho_1, \rho_n\right) &= \rho_i \hat{\rho}_i \bullet \hat{\mathbf{A}}_i, \\ Q_i\left(\rho_1, \rho_n\right) &= \rho_i \hat{\rho}_i \bullet \hat{\mathbf{D}}_i, \end{aligned} \tag{7.28}$$

It can be seen from Eqs. (7.26) and (7.27) that the two scalar products in Eq. (7.28) would vanish if the $\rho_i \hat{\rho}_i$ were perfectly known. Any nonzero value they have represents the error in the $\rho_i \hat{\rho}_i$. Part of our task is to find a way to minimize this error.

We will also write Eq. (7.28) as

$$\begin{aligned} P_i\left(\rho_1, \rho_n\right) &= \left(\mathbf{r}_i + \mathbf{R}_i \right) \bullet \hat{\mathbf{A}}_i, \\ Q_i\left(\rho_1, \rho_n\right) &= \left(\mathbf{r}_i + \mathbf{R}_i \right) \bullet \hat{\mathbf{D}}_i, \end{aligned} \tag{7.29}$$

because we know the $\mathbf{R}_i$ and will be using trial solutions of some of the $\mathbf{r}_i$. We also know $\hat{\rho}_i$ because the β_i and λ_i are my observations, which determine $\hat{\rho}_i$ in Eq. (7.26). We do not however know a priori the scalar magnitudes ρ_i. These will have to be calculated.

We start with estimated, and therefore imposed, values for the first and last quantities ρ_1 and ρ_n, but not for the other values of ρ_i. Let us call these first estimates ρ_1^e and ρ_n^e. Let the values of corrections to these estimates be $\Delta\rho_1$ and $\Delta\rho_n$. Although it would be lovely if

$$\begin{aligned} P_i\left(\rho_1^e + \Delta\rho_1, \rho_n^e + \Delta\rho_n\right) &= \quad 0 \text{ and} \\ Q_i\left(\rho_1^e + \Delta\rho_1, \rho_n^e + \Delta\rho_n\right) &= 0, \text{ for } i = 2, \ldots, n-1, \end{aligned} \tag{7.30}$$

we know in our hearts that that's not going to happen. A more realistic approximation would be

$$P_i\left(\rho_1^e,\rho_n^e\right)+\frac{\partial P_i}{\partial \rho_1}\Delta\rho_1+\frac{\partial P_i}{\partial \rho_n}\Delta\rho_n = \quad 0 \text{ and}$$

$$Q_i\left(\rho_1^e,\rho_n^e\right)+\frac{\partial Q_i}{\partial \rho_1}\Delta\rho_1+\frac{\partial Q_i}{\partial \rho_n}\Delta\rho_n = 0, \text{ for } i = 2,\ldots,n-1.$$

(7.31)

We can approximate the derivatives by

$$\frac{\partial P_i}{\partial \rho_1} \simeq \frac{P_i\left(\rho_1+\delta,\rho_n\right)-P_i\left(\rho_1-\delta,\rho_n\right)}{2\delta},$$

$$\frac{\partial P_i}{\partial \rho_n} \simeq \frac{P_i\left(\rho_1,\rho_n+\delta\right)-P_i\left(\rho_1,\rho_n-\delta\right)}{2\delta},$$

$$\frac{\partial Q_i}{\partial \rho_1} \simeq \frac{Q_i\left(\rho_1+\delta,\rho_n\right)-Q_i\left(\rho_1-\delta,\rho_n\right)}{2\delta},$$

$$\frac{\partial Q_i}{\partial \rho_n} \simeq \frac{Q_i\left(\rho_1,\rho_n+\delta\right)-Q_i\left(\rho_1,\rho_n-\delta\right)}{2\delta},$$

(7.32)

where δ is a small finite number which we will adjust to be as small as possible so that a computer program works.

We still have work to do before we can write down Eqs. (7.32). We have to calculate the P_i and the Q_i. Herget (1948) and Danby use one method, which involves hypergeometric functions. Escobal [128] avoids these and sticks to trigonometric functions. Therefore, I followed Escobal's method.

It never hurts to know where we are trying to get to. We are trying to calculate trial values of $\mathbf{r}_i$ that we can use in Eqs. (7.29) and (7.32) to calculate P_i, Q_i and their derivatives.

We will then use these to write out Eqs. (7.31) as a set of overdetermined equations, from which we will find least-squares fit solutions for the unknowns $\Delta\rho_1$ and $\Delta\rho_n$. This may or may not involve some trial and error in choosing initial estimates ρ_1^e and ρ_n^e. In fact, I quickly discovered that you need to make good initial guesses for the method to work.

We find ρ_1^e and ρ_n^e by thinking up plausible guesstimates for $\mathbf{r}_1^e$ and $\mathbf{r}_n^e$ and using Eq. (7.25) to calculate ρ_1^e and ρ_n^e. This is because there are methods to estimate the other $\mathbf{r}_i^e$ from $\mathbf{r}_1^e$ and $\mathbf{r}_n^e$.

To work out the terms on the right-hand side of Eq. (7.32), let us use Eq. (7.25). This is rearranged to

$$\rho_i \hat{\rho}_i = \mathbf{R}_i + \mathbf{r}_i \tag{7.33}$$

Therefore

$$\begin{aligned}
(\rho_i)\hat{\rho}_i \pm \delta\hat{\rho}_1 &= \mathbf{R}_i + \mathbf{r}_i \pm \delta\hat{\rho}_1, \\
(\rho_i)\hat{\rho}_i \pm \delta\hat{\rho}_n &= \mathbf{R}_i + \mathbf{r}_i \pm \delta\hat{\rho}_n.
\end{aligned} \tag{7.34}$$

So

$$\begin{aligned}
P_i(\rho_1 \pm \delta, \rho_n) &= (\mathbf{r}_i + \mathbf{R}_i \pm \delta\hat{\rho}_1) \bullet \hat{\mathbf{A}}_i, \\
Q_i(\rho_1 \pm \delta, \rho_n) &= (\mathbf{r}_i + \mathbf{R}_i \pm \delta\hat{\rho}_1) \bullet \hat{\mathbf{D}}_i, \\
P_i(\rho_1, \rho_n \pm \delta) &= (\mathbf{r}_i + \mathbf{R}_i \pm \delta\hat{\rho}_n) \bullet \hat{\mathbf{A}}_i, \\
Q_i(\rho_1, \rho_n \pm \delta) &= (\mathbf{r}_i + \mathbf{R}_i \pm \delta\hat{\rho}_n) \bullet \hat{\mathbf{D}}_i.
\end{aligned} \tag{7.35}$$

In practice, the $\mathbf{r}_i$ will be estimates at this point, so let us use the notation $\mathbf{r}_i^e$ as a reminder of this.

$$\begin{aligned}
P_i(\rho_1 \pm \delta, \rho_n) &= (\mathbf{r}_i^e + \mathbf{R}_i \pm \delta\hat{\rho}_1) \bullet \hat{\mathbf{A}}_i, \\
Q_i(\rho_1 \pm \delta, \rho_n) &= (\mathbf{r}_i^e + \mathbf{R}_i \pm \delta\hat{\rho}_1) \bullet \hat{\mathbf{D}}_i, \\
P_i(\rho_1, \rho_n \pm \delta) &= (\mathbf{r}_i^e + \mathbf{R}_i \pm \delta\hat{\rho}_n) \bullet \hat{\mathbf{A}}_i, \\
Q_i(\rho_1, \rho_n \pm \delta) &= (\mathbf{r}_i^e + \mathbf{R}_i \pm \delta\hat{\rho}_n) \bullet \hat{\mathbf{D}}_i.
\end{aligned} \tag{7.36}$$

$$\begin{aligned}
\frac{\partial P_i}{\partial \rho_1} &\simeq \frac{\left((\mathbf{r}_i^e + \mathbf{R}_i + \delta\hat{\rho}_1) \bullet \hat{\mathbf{A}}_i\right) - \left((\mathbf{r}_i^e + \mathbf{R}_i - \delta\hat{\rho}_1) \bullet \hat{\mathbf{A}}_i\right)}{2\delta}, \\
\frac{\partial P_i}{\partial \rho_n} &\simeq \frac{\left((\mathbf{r}_i^e + \mathbf{R}_i + \delta\hat{\rho}_n) \bullet \hat{\mathbf{A}}_i\right) - \left((\mathbf{r}_i^e + \mathbf{R}_i - \delta\hat{\rho}_n) \bullet \hat{\mathbf{A}}_i\right)}{2\delta}, \\
\frac{\partial Q_i}{\partial \rho_1} &\simeq \frac{\left((\mathbf{r}_i^e + \mathbf{R}_i + \delta\hat{\rho}_1) \bullet \hat{\mathbf{D}}_i\right) - \left((\mathbf{r}_i^e + \mathbf{R}_i - \delta\hat{\rho}_1) \bullet \hat{\mathbf{D}}_i\right)}{2\delta}, \\
\frac{\partial Q_i}{\partial \rho_n} &\simeq \frac{\left((\mathbf{r}_i^e + \mathbf{R}_i + \delta\hat{\rho}_n) \bullet \hat{\mathbf{D}}_i\right) - \left((\mathbf{r}_i^e + \mathbf{R}_i - \delta\hat{\rho}_n) \bullet \hat{\mathbf{D}}_i\right)}{2\delta}.
\end{aligned} \tag{7.37}$$

Hence

Although the textbooks don't mention it, this simplifies to

$$\frac{\partial P_i}{\partial \rho_1} \simeq \frac{2\delta \hat{\rho}_1 \cdot \hat{\mathbf{A}}_i}{2\delta} = \hat{\rho}_1 \cdot \hat{\mathbf{A}}_i,$$

$$\frac{\partial P_i}{\partial \rho_n} \simeq \frac{2\delta \hat{\rho}_n \cdot \hat{\mathbf{A}}_i}{2\delta} = \hat{\rho}_n \cdot \hat{\mathbf{A}}_i,$$

$$\frac{\partial Q_i}{\partial \rho_1} \simeq \frac{2\delta \hat{\rho}_1 \cdot \hat{\mathbf{D}}_i}{2\delta} = \hat{\rho}_1 \cdot \hat{\mathbf{D}}_i,$$ (7.38)

$$\frac{\partial Q_i}{\partial \rho_n} \simeq \frac{2\delta \hat{\rho}_n \cdot \hat{\mathbf{D}}_i}{2\delta} = \hat{\rho}_n \cdot \hat{\mathbf{D}}_i.$$

Thus, our set of overdetermined equations is

$$P_i\left(\rho_1^e, \rho_n^e\right) + \left(\hat{\rho}_1 \cdot \hat{\mathbf{A}}_i\right)\Delta\rho_1 + \left(\hat{\rho}_n \cdot \hat{\mathbf{A}}_i\right)\Delta\rho_n = \qquad 0 \text{ and}$$

$$Q_i\left(\rho_1^e, \rho_n^e\right) + \left(\hat{\rho}_1 \cdot \hat{\mathbf{D}}_i\right)\Delta\rho_1 + \left(\hat{\rho}_n \cdot \hat{\mathbf{D}}_i\right)\Delta\rho_n = 0, \text{ for } i = 2, \ldots, n-1.$$ (7.39)

Equations (7.31) actually have much the same form as Eq. (6.2). They can be solved by the same method due to Givens [96]. Recycling subroutines used there will be quite labour-saving.

We now have enough information to get the radial distances $\mathbf{r}_1$ and $\mathbf{r}_n$. We can't yet get a velocity at one of them, which we will need to turn the information we have into an orbit.

That will actually be quite a task. I needed to learn a few more tricks for manipulating ellipses. If you have reached this far, dear reader, you will already have the mathematical knowledge to keep going. All you will need is patience and perhaps a bit of stamina.

But first we need to do some stuff. At the moment, we are missing some vital ingredients. The most obvious of these is that we have not yet inserted any physics. All we have done is geometry. There is still a lot of geometry to do, notably using trigonometric methods, but in the process we will insert the constraint that Eq. (3.60) must be obeyed. This equation contains physics as well as geometry. We will discover that if the angular momentum per unit mass h is known, only one orbit can pass through two points $\mathbf{r}_1$ and $\mathbf{r}_n$ that we will have found by Herget's method. This will help us to determine the orbit.

Lagrange's f and g Functions

These functions enable us to satisfy the requirement that if the angular momentum per unit mass h is known, only one orbit can pass through two points $\mathbf{r}_i$ and $\mathbf{r}_j$ I'm going to introduce a coordinate system, based not on the centre of the ellipse, as in Chap. 1, but on the attracting focus S, that is, the one occupied by the Sun. This system is shown in Fig. 7.16.

We start by writing the coordinate $\mathbf{r}$ as

$$\mathbf{r} = x\hat{\mathbf{x}} + y\hat{\mathbf{y}} \tag{7.40}$$

and the velocity $\mathbf{v}$ as

$$\mathbf{v} = \dot{\mathbf{r}} = \dot{x}\hat{\mathbf{x}} + \dot{y}\hat{\mathbf{y}}. \tag{7.41}$$

Again the subscript 0 is used to indicate the value of a quantity at $t = 0$. Then

$$\mathbf{r}_0 = x_0\hat{\mathbf{x}} + y_0\hat{\mathbf{y}} \tag{7.42}$$

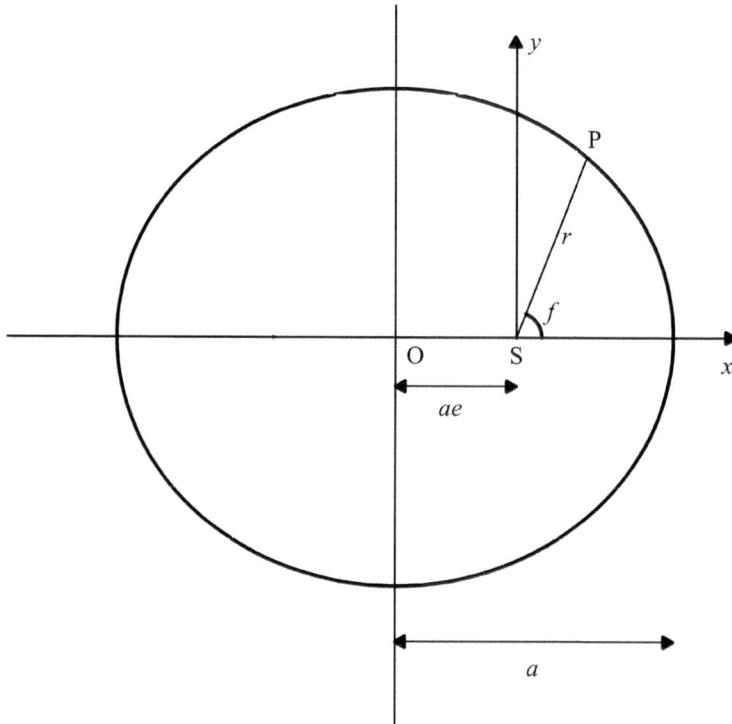

Fig. 7.16 Showing the xy coordinate system used in this section. Note that this system has its origin at the attracting focus of the ellipse not the centre

and

$$\mathbf{v}_0 = \dot{\mathbf{r}}_0 = \dot{x}_0\hat{\mathbf{x}} + \dot{y}_0\hat{\mathbf{y}}. \tag{7.43}$$

The (constant) angular momentum per unit mass may be expressed in terms of the initial conditions as

$$\mathbf{h} = \mathbf{r}_0 \times \mathbf{v}_0 = \begin{vmatrix} \hat{\mathbf{x}} & \hat{\mathbf{y}} & \hat{\mathbf{z}} \\ x_0 & y_0 & 0 \\ \dot{x}_0 & \dot{y}_0 & 0 \end{vmatrix} = (x_0\dot{y}_0 - y_0\dot{x}_0)\hat{\mathbf{z}} \tag{7.44}$$

Hence, the magnitude of the angular momentum per unit mass is

$$h = x_0\dot{y}_0 - y_0\dot{x}_0. \tag{7.45}$$

The unit vectors in the X and Y directions can be obtained from Eqs. (7.42) and (7.43):

$$\hat{\mathbf{y}} = \frac{\mathbf{r}_0 - x_0\hat{\mathbf{x}}_0}{y_o} \tag{7.46}$$

whence

$$\begin{aligned}
\mathbf{v}_0 &= \hat{\mathbf{x}}\dot{x}_0 + \dot{y}\left(\frac{\mathbf{r}_0 - x_0\hat{\mathbf{x}}_0}{y_o}\right) \\
&= \left(\frac{y_0\dot{x}_0 - x_0\dot{y}_0}{y_o}\right)\hat{\mathbf{x}} + \frac{\dot{y}_0}{y_0}\mathbf{r}_0 \\
&= -\left(\frac{h}{y_o}\right)\hat{\mathbf{x}} + \frac{\dot{y}_{00}}{y_o}\mathbf{r}_0.
\end{aligned} \tag{7.47}$$

Thus

$$\hat{\mathbf{x}} = \frac{\dot{y}_0}{h}\mathbf{r}_0 - \frac{y_0}{h}\mathbf{v}_0. \tag{7.48}$$

Substituting into Eq. (7.46) gives

$$\hat{\mathbf{y}} = \frac{1}{y_0}\mathbf{r}_0 - \frac{x_0}{y_0}\left(\frac{\dot{y}_0}{h}\mathbf{r}_0 - \frac{y_0}{h}\mathbf{v}_0\right)$$

$$= \frac{h - x_0\dot{y}_0}{y_0}\mathbf{r}_0 + \frac{x_0}{h}\mathbf{v}_0 \tag{7.49}$$

$$= -\frac{\dot{x}_0}{h}\mathbf{r}_0 + \frac{y_0}{h}\mathbf{v}_0.$$

In the last step, Eq. (7.45) has been used.

Now let us back-substitute these values of the unit vectors into Eqs. (7.40) and (7.41) at the later time t.

$$\mathbf{r} = X\left(\frac{\dot{y}_0}{h}\mathbf{r}_0 - \frac{y_0}{h}\mathbf{v}_0\right)Y\left(-\frac{\dot{y}_0}{h}\mathbf{r}_0 + \frac{x_0}{h}\mathbf{v}_0\right)$$

$$= \frac{x\dot{y}_0 - y\dot{x}_0}{h}\mathbf{r}_0 + \frac{-xy_0 + yx_0}{h}\mathbf{v}_0 \tag{7.50}$$

and

$$\mathbf{v} = \dot{x}\left(\frac{\dot{y}_0}{h}\mathbf{r}_0 - \frac{y_0}{h}\mathbf{v}_0\right)\dot{y}\left(-\frac{\dot{x}_0}{h}\mathbf{r}_0 + \frac{x_0}{h}\mathbf{v}_0\right)$$

$$= \frac{\dot{x}\dot{y}_0 - \dot{y}\dot{x}_0}{h}\mathbf{r}_0 + \frac{-\dot{x}y_0 + \dot{y}x_0}{h}\mathbf{v}_0. \tag{7.51}$$

We can set

$$f_{\text{Lag}} = \frac{x\dot{y}_0 - y\dot{x}_0}{h} \quad \text{and}$$

$$g_{\text{Lag}} = \frac{-xy_0 + yx_0}{h}. \tag{7.52}$$

The time derivatives of these functions are

$$\dot{f}_{\text{Lag}} = \frac{\dot{x}\dot{y}_0 - \dot{y}\dot{x}_0}{h} \quad \text{and}$$

$$\dot{g}_{\text{Lag}} = \frac{-\dot{x}y_0 + \dot{y}x_0}{h}. \tag{7.53}$$

Then

$$x(t) = f_{\text{Lag}}(t, t_0)x_0(t_0) + g_{\text{Lag}}(t, t_0)\dot{x}_0(t_0),$$
$$y(t) = f_{\text{Lag}}(t, t_0)y_0(t_0) + g_{\text{Lag}}(t, t_0)\dot{y}_0(t_0). \tag{7.54}$$

The time dependencies of the terms have been explicitly added for clarity.

So, if we can obtain values for f_{Lag} and g_{Lag} and their time derivatives, we can obtain the coordinates (x, y) of any point on the ellipse from a knowledge of the point (x_0, y_0) and the velocity $(\dot{x}_0, \dot{y}_0)$ at that time. This is exactly what we do to obtain a preliminary orbit: get a rough estimate of (x_0, y_0) and $(\dot{x}_0, \dot{y}_0)$ at the time of one observation.

Therefore, our next move is to find usable forms of f_{Lag} and g_{Lag}. We could do this in terms of the true anomaly (the angle f in Fig. 7.16). I made that mistake. We can do the derivation in terms of the eccentric anomaly E in about half as many pages. Therefore, I propose to take you along an easier path up this particular mountain, to save you finding out what I discovered the hard way (Fig. 7.17).

We need time derivatives $\dot{x}$ & $\dot{y}$.

At the point P,

$$x = r \cos f,$$
$$y = r \sin f. \tag{7.55}$$

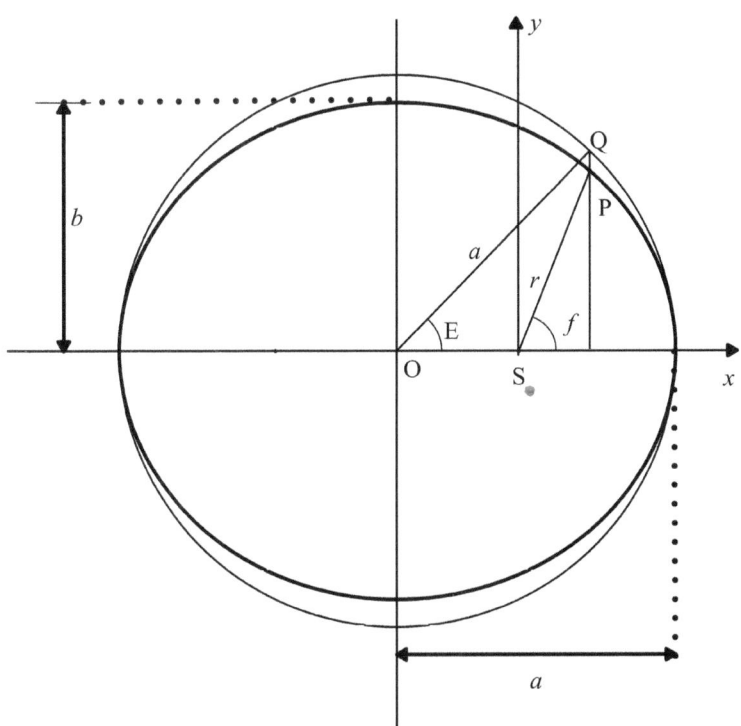

Fig. 7.17 Showing the geometric relation between eccentric anomaly E and true anomaly f

Because the distance OS is ae, it follows that

$$
\begin{aligned}
x &= a\cos E - ae \\
&= a(\cos E - e).
\end{aligned}
$$
(7.56)

Using Eqs. (2.6) and (3.66)

$$
\begin{aligned}
y &= r\sin f \\
&= b\sin E \\
&= a\sqrt{1-e^2}\,\sin E.
\end{aligned}
$$
(7.57)

Equation (3.68) tells us that

$$
r = x^2 + y^2 = a(1 - e\cos E).
$$
(3.68)

By differentiating Eqs. (7.56), (7.57) and (3.68) with respect to time, we obtain

$$
\dot{r} = ae\dot{E}\sin E.
$$
(7.58)

$$
\begin{aligned}
\dot{x} &= \dot{r}\cos f - r\dot{f}\sin f \\
&= -a\dot{E}\sin E.
\end{aligned}
$$
(7.59)

$$
\begin{aligned}
\dot{y} &= \dot{r}\sin f + r\dot{f}\cos f \\
&= a\dot{E}\sqrt{1-e^2}\,\cos E.
\end{aligned}
$$
(7.60)

The time derivative of the eccentric anomaly, $\dot{E}$, appears in each of Eqs. (7.58), (7.59) and (7.60). With a little bit of craftiness, we can obtain this from Kepler's equation, by remembering that precisely because the mean anomaly M is a mean, it is proportional to time, so it has a constant time derivative. I worked this value out in Eq. (3.90), namely

$$
\dot{M} = \sqrt{\mu a}^{-3/2}.
$$
(7.61)

But of course, from Kepler's equation (3.79)

$$
\begin{aligned}
\dot{M} &= \frac{d}{dt}(E - e\sin E) \\
&= \dot{E} - e\dot{E}\cos E \\
&= \dot{E}(1 - e\cos E)
\end{aligned}
$$
(7.62)

Therefore, using Eqs. (3.67) and (3.68), namely,

$$\hat{\mathbf{e}} = +\sqrt{e} \tag{7.63}$$

$$
\begin{aligned}
\sqrt{\mu}a^{-3/2} &= \dot{E}\left(1 - e\cos E\right) \\
&= \dot{E}\left(1 - e\cos E\right) \\
&= \dot{E}\frac{r}{a}, \tag{7.64}
\end{aligned}
$$

$$\text{so}\,\dot{E} = \frac{1}{r}\sqrt{\frac{\mu}{a}}.$$

We can now substitute Eqs. (7.56), (7.57), (7.59), (7.60) and (7.64) into Eqs. (7.52) and (7.53).

$$
\begin{aligned}
f_{\text{Lag}} &= \frac{x\dot{y}_0 - y\dot{x}_0}{h} \\[2mm]
&= \frac{\left(a(\cos E - e)\right)\left(a\,\dot{E}_0\sqrt{1-e^2}\cos E_0\right) - \left(a\sqrt{1-e^2}\sin E\right)\left(-a\,\dot{E}_0\sin E_0\right)}{h} \\[2mm]
&= \frac{a^2\sqrt{1-e^2}\,\dot{E}_0\left[(\cos E - e)(\cos E_0) - (\sin E)(-\sin E_0)\right]}{h} \\[2mm]
&= \frac{a^2\sqrt{1-e^2}\,\dot{E}_0\left[(\cos E_0\cos E - e\cos E_0) + (\sin E_0\sin E)\right]}{h} \\[2mm]
&= \frac{a^2\sqrt{1-e^2}\,\dot{E}_0\left[\left(\cos(E-E_0)\right) - e\cos E_0\right]}{h} \\[2mm]
&= \frac{a^{3/2}\sqrt{\mu\left(1-e^2\right)}\left[\left(\cos(E-E_0)\right) - e\cos E_0\right]}{r_0 h} \\[2mm]
&= \frac{a\sqrt{\mu a\left(1-e^2\right)}\left[\left(\cos(E-E_0)\right) - e\cos E_0\right]}{r_0 h} \\[2mm]
&= \frac{a\left[\left(\cos(E-E_0)\right) - e\cos E_0\right]}{r_0} \\[2mm]
&= \frac{-ae\cos E_0}{r_0} + \frac{a\left[\cos(E-E_0)\right]}{r_0} \\[2mm]
&= \frac{a\left(r_0/a - 1\right)}{r_0} + \frac{a\left[\cos(E-E_0)\right]}{r_0} \\[2mm]
&= 1 - \frac{a}{r_0}\left[1 - \cos(E-E_0)\right].
\end{aligned}
\tag{7.65}
$$

In deriving Eq. (7.65), Eqs. (3.68) and (3.86) were used.

$$
\begin{aligned}
g_{\text{Lag}} &= \frac{y x_0 - x y_0}{h} \\[2mm]
&= \frac{\left(a\sqrt{1-e^2}\sin E\right)\left(a\left(\cos E_0 - e\right)\right) - \left(a\left(\cos E - e\right)\right)\left(a\sqrt{1-e^2}\sin E_0\right)}{h} \\[2mm]
&= \frac{a^2\sqrt{1-e^2}\left[\left(\sin E\right)\left(\left(\cos E_0 - e\right)\right) - \left(\left(\cos E - e\right)\right)\left(\sin E_0\right)\right]}{h} \\[2mm]
&= \frac{a^2\sqrt{1-e^2}\left[\sin E\cos E_0 - e\sin E - \sin E_0\cos E + e\sin E_0\right]}{h} \\[2mm]
&= \frac{a^2\sqrt{1-e^2}\left[\sin E\cos E_0 - \sin E_0\cos E + e\sin E_0 - e\sin E\right]}{h} \\[2mm]
&= \frac{a^2\sqrt{1-e^2}\left[\sin\left(E-E_0\right) + E - E + E_0 - E_0 + e\sin E_0 - e\sin E\right]}{h} \\[2mm]
&= \frac{a^2\sqrt{1-e^2}\left[\sin\left(E-E_0\right) + \left(E - e\sin E\right) - E + E_0 - \left(E_0 - e\sin E_0\right)\right]}{h} \quad\quad (7.66) \\[2mm]
&= \frac{a^2\sqrt{1-e^2}\left[\sin\left(E-E_0\right) + M - E + E_0 - M_0\right]}{h} \\[2mm]
&= \frac{a^2\sqrt{1-e^2}\left[\sin\left(E-E_0\right) - \left(E-E_0\right) + \left(M - M_0\right)\right]}{\sqrt{\mu a\left(1-e^2\right)}} \\[2mm]
&= \frac{a^{3/2}\left(M - M_0\right)}{\sqrt{\mu}} - \frac{a^{3/2}\left[\left(E-E_0\right) - \sin\left(E-E_0\right)\right]}{\sqrt{\mu}} \\[2mm]
&= \frac{T\left(M - M_0\right)}{2\pi} - \frac{a^{3/2}\left[\left(E-E_0\right) - \sin\left(E-E_0\right)\right]}{\sqrt{\mu}} \\[2mm]
&= \left(t - t_0\right) - \frac{a^{3/2}\left[\left(E-E_0\right) - \sin\left(E-E_0\right)\right]}{\sqrt{\mu}}
\end{aligned}
$$

In deriving Eq. (7.66), Eqs. (3.68), (3.79) and (3.86) were used.

The quickest way to obtain the time of derivatives of f_{Lag} and g_{Lag} is to differentiate Eqs. (7.65) and (7.66).

Using Eq. (7.64)

$$\dot{f}_{\text{Lag}} = \frac{d}{dt}\left\{1 - \frac{a}{r_0}\left[1 - \cos\left(E - E_0\right)\right]\right\}$$

$$= \frac{a}{r_0}\dot{E}\sin\left(E - E_0\right)$$

$$= \frac{a}{r_0}\frac{1}{r}\sqrt{\frac{\mu}{a}}\sin\left(E - E_0\right)$$

$$= \frac{\sqrt{\mu a}}{r\ r_0}\sin\left(E - E_0\right). \tag{7.67}$$

Last but not least, and again using Eq. (7.64)

$$\dot{g}_{\text{Lag}} = \frac{d}{dt}\left\{(t - t_0) - \frac{a^{3/2}\left[\left(E - E_0\right) - \sin\left(E - E_0\right)\right]}{\sqrt{\mu}}\right\}$$

$$= 1 - \frac{a^{3/2}\left[\dot{E}\left(1 - \cos\left(E - E_0\right)\right)\right]}{\sqrt{\mu}}$$

$$= 1 - \frac{a^{3/2}\dot{E}\left(1 - \cos\left(E - E_0\right)\right)}{\sqrt{\mu}} \tag{7.68}$$

$$= 1 - \frac{a\sqrt{a}\sqrt{\mu}\left(1 - \cos\left(E - E_0\right)\right)}{r\sqrt{a}\sqrt{\mu}}$$

$$= 1 + \frac{a}{r}\left(\cos\left(E - E_0\right) - 1\right).$$

The Sector–Triangle Ratio

To use Herget's method, I need values of $\mathbf{r}$, the Sun–Mars distance. What I have is measurements of right ascension α and declination δ, conveniently re-expressed as ecliptic longitude λ and ecliptic latitude β. Thanks to my endeavours reported in Chap. 4, I also have $\mathbf{R}$ at the observation times.

The values of λ and β tell me the direction but not the magnitude of the vectors $\boldsymbol{\rho}$ at the observation times.

I largely follow the account of Chap. 6 of Escobal [128] because, although it is a little lengthy, it avoids the use of hypergeometric functions, with which I have guessed that most readers are unfamiliar.

To squeeze more information out of this situation will require some ingenuity. One mathematician who was not short of that commodity was Gauss.

I'm going to use his idea of working out the ratio of two areas: the sector and triangle described by two points in the orbit. These are illustrated in Figs. 7.18.

In Eq. (3.37), we already showed that the sector area $\mathbf{S}$ is

$$\mathbf{S}_S \;=\; \frac{\Delta t}{2}\,\mathbf{r}\times\dot{\mathbf{r}} \;=\; \frac{\Delta t}{2}\,\mathbf{h}\,. \tag{3.37}$$

From Fig. 7.19, it can be seen that the area of the triangle is, if the Sun–planet distances are $\mathbf{r}_1$ and $\mathbf{r}_2$,

$$\mathbf{A}_T = \frac{1}{2}\left(\mathbf{r}_1\times\mathbf{r}_2\right) = \frac{1}{2}r_1 r_2 \sin\left(f_2 - f_1\right)\hat{\mathbf{A}}_T. \tag{7.69}$$

The ratio of the two areas

$$Y = \frac{\left|\mathbf{S}_S\right|}{\left|\mathbf{A}_T\right|} = \frac{h\,t}{r_1\,r_2\sin\left(f_2 - f_1\right)}. \tag{7.70}$$

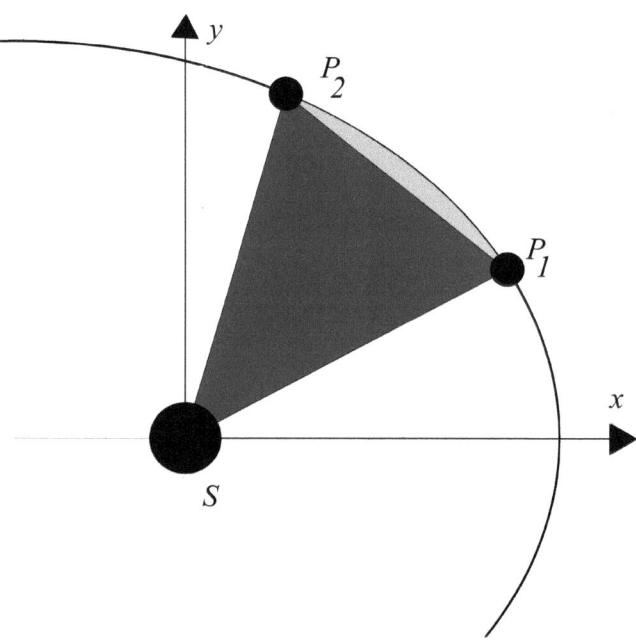

Fig. 7.18 The area of the triangle swept out by a planet orbiting from position P1 to position P2 is shown dark grey. The area of the sector swept out is the sum of the dark grey and light grey areas

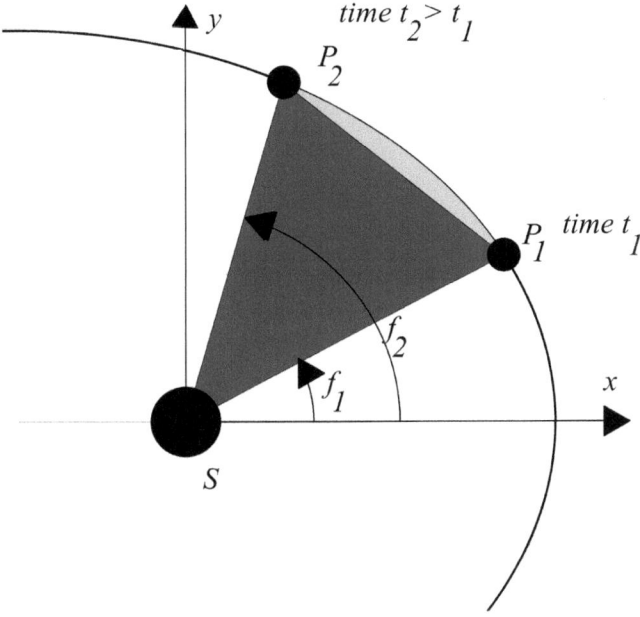

Fig. 7.19 Repeat of Fig. 7.18 with true anomalies drawn in. The orbit is in a counter clockwise direction. In the Solar System, this is equivalent to looking down from an imaginary spacecraft to the north of the Sun

It's a bit annoying that the letter Y is used for this ratio. Some authors call it y. Either way the ratio is too easily confused with spatial coordinates. I stuck with capital Y, used by Herget [122], to try to be marginally less confusing.

Equation (3.60) tells us that

$$r = \frac{h^2}{\mu\left[1 + e\cos(f)\right]} \tag{3.60}$$

At time t_1:

$$\left[1 + e\cos(f_1)\right] = \frac{h^2}{\mu\, r_1} \tag{7.71}$$

At a later time t_2:

$$\left[1 + e\cos(f_2)\right] = \frac{h^2}{\mu\, r_2} \tag{7.72}$$

Adding Eq. (7.72) to Eq. (7.71) gives

$$
\frac{h^2}{\mu}\left(\frac{1}{r_1}+\frac{1}{r_2}\right)= \quad 2+e\cos(f_1)+e\cos(f_2)
$$

$$
= 2+2e\cos\left(\frac{f_2+f_1}{2}\right)\cos\left(\frac{f_2-f_1}{2}\right).
$$

(7.73)

At this point, I don't know the eccentricity or the true anomalies, and I'm going to try to eliminate terms to get to a point where I can make progress. First, let's try to eliminate the term $e\cos[(f_1+f_2)/2]$.

Since

$$
\cos f = \cos\left(\frac{f+f}{2}\right)= \cos^2\frac{f}{2}-\sin^2\frac{f}{2}
$$

(7.74)

It follows that

$$
\sqrt{r}\cos\frac{f}{2}=\pm\sqrt{\frac{r(1+\cos f)}{2}},
$$

$$
\sqrt{r}\sin\frac{f}{2}=\pm\sqrt{\frac{r(1-\cos f)}{2}}.
$$

(7.75)

From Eqs. (3.67) and (3.78), namely,

$$
r\cos f = a(\cos E - e)
$$

(3.67)

and

$$
r = a(1-e\cos E)
$$

(3.68)

It follows that

$$\sqrt{r}\cos\frac{f}{2} = \sqrt{\frac{r+r\cos f}{2}}$$

$$= \sqrt{\frac{a(1-e\cos E)+a(\cos E-e)}{2}}$$

$$= \sqrt{\frac{a-ae\cos E+a\cos E-ae}{2}}$$

$$= \sqrt{\frac{a(1-e)+a(1-e)\cos E}{2}}$$

$$= \sqrt{\frac{a(1-e)(1+\cos E)}{2}}$$

$$= \sqrt{\frac{a(1-e)\left(1+\cos^2\frac{E}{2}-\sin^2\frac{E}{2}\right)}{2}}$$

$$= \sqrt{\frac{a(1-e)\left(2\cos^2\frac{E}{2}\right)}{2}}$$

$$= \sqrt{a(1-e)}\cos\frac{E}{2}. \tag{7.76}$$

Similarly

$$\sqrt{r}\sin\frac{f}{2} = \sqrt{\frac{r-r\cos f}{2}}$$

$$= \sqrt{\frac{a(1-e\cos E)-a(\cos E-e)}{2}}$$

$$= \sqrt{\frac{a-ae\cos E-a\cos E+ae}{2}}$$

$$= \sqrt{\frac{a(1+e)-a(1+e)\cos E}{2}}$$

$$= \sqrt{\frac{a(1+e)(1-\cos E)}{2}} \tag{7.77}$$

$$= \sqrt{\frac{a(1+e)\left(1-\cos^2\frac{E}{2}+\sin^2\frac{E}{2}\right)}{2}}$$

$$= \sqrt{\frac{a(1+e)\left(2\sin^2\frac{E}{2}\right)}{2}}$$

$$= \sqrt{a(1+e)}\sin\frac{E}{2}.$$

Now, the factors $\cos((f_2 - f_1)/2)$ and $\cos((f_2 + f_1)/2)$ can be multiplied by $\sqrt{(r_1 r_2)}$ and expanded as follows:

$$\sqrt{r_2 r_1}\cos\left(\frac{f_2 + f_1}{2}\right) = \left[\sqrt{r_2}\cos\frac{f_2}{2}\right]\left[\sqrt{r_1}\cos\frac{f_1}{2}\right] - \left[\sqrt{r_2}\sin\frac{f_2}{2}\right]\left[\sqrt{r_1}\sin\frac{f_1}{2}\right]$$

$$\sqrt{r_2 r_1}\cos\left(\frac{f_2 - f_1}{2}\right) = \left[\sqrt{r_2}\cos\frac{f_2}{2}\right]\left[\sqrt{r_1}\cos\frac{f_1}{2}\right] + \left[\sqrt{r_2}\sin\frac{f_2}{2}\right]\left[\sqrt{r_1}\sin\frac{f_1}{2}\right]$$

$$\sqrt{r_2 r_1}\cos\left(\frac{f_2 + f_1}{2}\right) = \left[\sqrt{a(1-e)}\cos\frac{E_2}{2}\right]\left[\sqrt{a(1-e)}\cos\frac{E_1}{2}\right] - \left[\sqrt{a(1+e)}\sin\frac{E_2}{2}\right]\left[\sqrt{a(1+e)}\sin\frac{E_1}{2}\right]$$

$$\sqrt{r_2 r_1}\cos\left(\frac{f_2 - f_1}{2}\right) = \left[\sqrt{a(1-e)}\cos\frac{E_2}{2}\right]\left[\sqrt{a(1-e)}\cos\frac{E_1}{2}\right] + \left[\sqrt{a(1+e)}\sin\frac{E_2}{2}\right]\left[\sqrt{a(1+e)}\sin\frac{E_1}{2}\right]$$

$$\sqrt{r_2 r_1}\cos\left(\frac{f_2 + f_1}{2}\right) = (a(1-e))\left[\cos\frac{E_2}{2}\right]\left[\cos\frac{E_1}{2}\right] - (a(1+e))\left[\sin\frac{E_2}{2}\right]\left[\sin\frac{E_1}{2}\right]$$

$$\sqrt{r_2 r_1}\cos\left(\frac{f_2 - f_1}{2}\right) = (a(1-e))\left[\cos\frac{E_2}{2}\right]\left[\cos\frac{E_1}{2}\right] + (a(1+e))\left[\sin\frac{E_2}{2}\right]\left[\sin\frac{E_1}{2}\right]$$

$$\sqrt{r_2 r_1}\cos\left(\frac{f_2 + f_1}{2}\right) = a\cos\left(\frac{E_2 - E_1}{2}\right) - ae\cos\left(\frac{E_2 + E_1}{2}\right)$$

$$\sqrt{r_2 r_1}\cos\left(\frac{f_2 - f_1}{2}\right) = -ae\cos\left(\frac{E_2 - E_1}{2}\right) + a\cos\left(\frac{E_2 + E_1}{2}\right)$$

(7.78)

Multiplying the last line of Eq. (7.78) by e and adding it to the penultimate line of Eq. (7.78) gives

$$\sqrt{r_2 r_1}\left[\cos\left(\frac{f_2 - f_1}{2}\right) + e\cos\left(\frac{f_2 + f_1}{2}\right)\right] = a\cos\left(\frac{E_2 - E_1}{2}\right) - ae^2\cos\left(\frac{E_2 - E_1}{2}\right).$$

(7.79)

Rearranging

$$e\cos\left(\frac{f_2 + f_1}{2}\right) = \frac{p}{\sqrt{r_2 r_1}}\cos\left(\frac{E_2 - E_1}{2}\right) - \cos\left(\frac{f_2 - f_1}{2}\right).$$

(7.80)

since from Eq. (3.62)

$$p = \frac{h^2}{\mu} = a\left(1 - e^2\right)$$

(7.81)

Plug Eq. (7.80) into Eq. (7.73):

$$p\left(\frac{1}{r_1}+\frac{1}{r_2}\right)=2+\frac{2p}{\sqrt{r_2 r_1}}\cos\left(\frac{E_2-E_1}{2}\right)\cos\left(\frac{f_2-f_1}{2}\right)-2\cos^2\left(\frac{f_2-f_1}{2}\right). \quad (7.82)$$

$$p\left[\left(\frac{1}{r_1}+\frac{1}{r_2}\right)-\frac{2}{\sqrt{r_2 r_1}}\cos\left(\frac{E_2-E_1}{2}\right)\cos\left(\frac{f_2-f_1}{2}\right)\right]=2-2\cos^2\left(\frac{f_2-f_1}{2}\right)$$

$$p=\frac{2-2\cos^2\left(\frac{f_2-f_1}{2}\right)}{\left[\left(\frac{1}{r_1}+\frac{1}{r_2}\right)-\frac{2}{\sqrt{r_2 r_1}}\cos\left(\frac{E_2-E_1}{2}\right)\cos\left(\frac{f_2-f_1}{2}\right)\right]}$$

$$=\frac{2\sin^2\left(\frac{f_2-f_1}{2}\right)}{\left[\left(\frac{1}{r_1}+\frac{1}{r_2}\right)-\frac{2}{\sqrt{r_2 r_1}}\cos\left(\frac{E_2-E_1}{2}\right)\cos\left(\frac{f_2-f_1}{2}\right)\right]} \quad (7.83)$$

$$=\frac{2\sin^2\left(\frac{f_2-f_1}{2}\right)}{\left[\left(\frac{r_2+r_1}{r_2 r_1}\right)-\frac{2}{\sqrt{r_2 r_1}}\cos\left(\frac{E_2-E_1}{2}\right)\cos\left(\frac{f_2-f_1}{2}\right)\right]}$$

$$=\frac{2 r_2 r_1 \sin^2\left(\frac{f_2-f_1}{2}\right)}{\left[(r_2+r_1)-2\sqrt{r_2 r_1}\cos\left(\frac{E_2-E_1}{2}\right)\cos\left(\frac{f_2-f_1}{2}\right)\right]}.$$

But from Eq. (7.70)

$$p=\frac{Y^2 r_2^2 r_1^2 \sin^2\left(f_2-f_1\right)}{\mu\,\Delta t^2}$$

$$=\frac{2 r_2 r_1 \sin^2\left(\frac{f_2-f_1}{2}\right)}{\left[(r_2+r_1)-2\sqrt{r_2 r_1}\cos\left(\frac{E_2-E_1}{2}\right)\cos\left(\frac{f_2-f_1}{2}\right)\right]}. \quad (7.84)$$

So

$$\gamma^2 \left(\sin\left(f_2 - f_1\right) \right)^2 = \frac{\mu\, \Delta t^2\, 2\sin^2\left(\dfrac{f_2 - f_1}{2}\right)}{r_2 r_1 \left[\left(r_2 + r_1\right) - 2\sqrt{r_2 r_1}\,\cos\left(\dfrac{E_2 - E_1}{2}\right)\cos\left(\dfrac{f_2 - f_1}{2}\right) \right]}$$

$$\gamma^2 \left(2\sin^2\left(\dfrac{f_2 - f_1}{2}\right)\cos^2\left(\dfrac{f_2 - f_1}{2}\right) \right) = \frac{2\mu\, \Delta t^2\, \sin^2\left(\dfrac{f_2 - f_1}{2}\right)}{r_2 r_1 \left[\left(r_2 + r_1\right) - 2\sqrt{r_2 r_1}\,\cos\left(\dfrac{E_2 - E_1}{2}\right)\cos\left(\dfrac{f_2 - f_1}{2}\right) \right]}$$

$$\gamma^2 = \frac{\mu\, \Delta t^2\, \sec^2\left(\dfrac{f_2 - f_1}{2}\right)}{r_2 r_1 \left[\left(r_2 + r_1\right) - 2\sqrt{r_2 r_1}\,\cos\left(\dfrac{E_2 - E_1}{2}\right)\cos\left(\dfrac{f_2 - f_1}{2}\right) \right]}.$$

$$(7.85)$$

This expression, Eq. (7.85), is obviously a bit cumbersome. We can simplify the next stage of manipulation by defining

$$l = \frac{r_1 + r_2}{4\sqrt{r_2 r_1}\,\cos\left(\dfrac{f_2 - f_1}{2}\right)} - \frac{1}{2}, \qquad (7.86)$$

$$m = \frac{\mu \Delta t^2}{\left[2\sqrt{r_2 r_1}\,\cos\left(\dfrac{f_2 - f_1}{2}\right) \right]^3}, \qquad (7.87)$$

and

$$\begin{aligned}
\xi &= \frac{1}{2}\left[1 - \cos\left(\dfrac{E_2 - E_1}{2}\right) \right] \\
&= \frac{1}{2}\left[1 - \cos^2\left(\dfrac{E_2 - E_1}{4}\right) + \sin^2\left(\dfrac{E_2 - E_1}{4}\right) \right] \qquad (7.88) \\
&= \sin^2\left(\dfrac{E_2 - E_1}{4}\right).
\end{aligned}$$

If we know r_1, r_2 and the angle between them, then l and m are both known. x, however, is not. Let us now examine the quantity

$$\frac{m}{l+\xi} = \frac{\mu \Delta t^2}{\left[2\sqrt{r_2 r_1}\,\cos\left(\dfrac{f_2 - f_1}{2}\right)\right]^3} \frac{1}{\dfrac{r_1 + r_2}{4\sqrt{r_2 r_1}\,\cos\left(\dfrac{f_2 - f_1}{2}\right)} - \dfrac{1}{2} + \dfrac{1}{2} - \dfrac{1}{2}\cos\left(\dfrac{E_2 - E_1}{2}\right)}$$

$$= \frac{\mu \Delta t^2 \sec^2\left(\dfrac{f_2 - f_1}{2}\right)}{2 r_2 r_1 \left[4\sqrt{r_2 r_1}\,\cos\left(\dfrac{f_2 - f_1}{2}\right)\right]} \frac{1}{\left[\dfrac{r_1 + r_2}{4\sqrt{r_2 r_1}\,\cos\left(\dfrac{f_2 - f_1}{2}\right)} - \dfrac{1}{2}\cos\left(\dfrac{E_2 - E_1}{2}\right)\right]}$$

$$= \frac{\mu \,\Delta t^2 \sec^2\left(\dfrac{f_2 - f_1}{2}\right)}{r_2 r_1 \left[(r_2 + r_1) - 2\sqrt{r_2 r_1}\,\cos\left(\dfrac{E_2 - E_1}{2}\right)\cos\left(\dfrac{f_2 - f_1}{2}\right)\right]}$$

$$= Y^2.$$

(7.89)

Introduction of Kepler's Equation

We have to do this in order to provide a relationship between Δt and Δf.
Kepler's equation (3.79) is that

$$M(t) = E(t) - e\sin E(t) \tag{3.79}$$

We can write the mean anomaly at time t after the perihelion time t_p as

$$M = \frac{2\pi}{T}(t - t_p). \tag{7.90}$$

To this we add the effect of Kepler's third law, Eq. (3.88)

$$\frac{a^3}{T^2} = \frac{\mu}{4\pi^2} \tag{3.88}$$

Therefore, we can write that

$$M(t - t_p) = \frac{\sqrt{\mu}}{a^{3/2}} = E(t) - e\sin E(t). \tag{7.91}$$

The difference in Kepler's equation at our two times is

$$
\frac{\sqrt{\mu}\Delta t}{a^{3/2}} = E(t_2) - E(t_1) + e\sin E(t_1) - e\sin E(t_2)
$$
$$
= E(t_2) - E(t_1) - 2e\sin\left(\frac{E_2 - E_1}{2}\right)\cos\left(\frac{E_2 + E_1}{2}\right). \tag{7.92}
$$

But by Eq. (7.78)

$$
e\cos\left(\frac{E_2 + E_1}{2}\right) = \cos\left(\frac{E_2 - E_1}{2}\right) - \frac{\sqrt{r_2 r_1}}{a}\cos\left(\frac{f_2 - f_1}{2}\right). \tag{7.93}
$$

Therefore, the difference between the Kepler's equation at the two times t_1 and t_2 is

$$
\frac{\sqrt{\mu}\Delta t}{a^{3/2}} = E(t_2) - E(t_1) - \sin(E_2 - E_1) + \frac{\sqrt{r_2 r_1}}{a}\sin\left(\frac{E_2 - E_1}{2}\right)\cos\left(\frac{f_2 - f_1}{2}\right). \tag{7.94}
$$

We can get three expressions for $\sin[(E_2 - E_1)/2]$, viz.

$$
\sin\left(\frac{E_2 - E_1}{2}\right) = \sin\frac{E_2}{2}\cos\frac{E_1}{2} - \cos\frac{E_2}{2}\sin\frac{E_1}{2}
$$
$$
= \frac{\sqrt{r_2 r_1}}{\sqrt{a(1-e)}\sqrt{a(1+e)}}\left(\sin\frac{f_2}{2}\cos\frac{f_1}{2} - \cos\frac{f_2}{2}\sin\frac{f_1}{2}\right)
$$
$$
= \frac{\sqrt{r_2 r_1}}{\sqrt{a^2(1-e^2)}}\left(\sin\frac{f_2}{2}\cos\frac{f_1}{2} - \cos\frac{f_2}{2}\sin\frac{f_1}{2}\right) \tag{7.95}
$$
$$
= \frac{\sqrt{r_2 r_1}}{\sqrt{ap}}\left(\sin\frac{f_2}{2}\cos\frac{f_1}{2} - \cos\frac{f_2}{2}\sin\frac{f_1}{2}\right)
$$
$$
= \frac{\sqrt{r_2 r_1}}{\sqrt{ap}}\sin\left(\frac{f_2 - f_1}{2}\right),
$$

where we have used Eqs. (3.61) and (3.62), and have used the rule for sin $(A + B)$ twice. Substituting Eq. (7.95) into Eq. (7.94) gives

$$\frac{\sqrt{\mu}\Delta t}{a^{3/2}} = E(t_2) - E(t_1) - \sin(E_2 - E_1) + \frac{2\sqrt{r_2 r_1}}{a}\frac{\sqrt{r_2 r_1}}{\sqrt{ap}}\sin\left(\frac{f_2 - f_1}{2}\right)\cos\left(\frac{f_2 - f_1}{2}\right)$$

$$= E(t_2) - E(t_1) - \sin(E_2 - E_1) + \frac{2r_2 r_1}{a^{3/2}\sqrt{p}}\sin\left(\frac{f_2 - f_1}{2}\right)\cos\left(\frac{f_2 - f_1}{2}\right)$$

$$= E(t_2) - E(t_1) - \sin(E_2 - E_1) + \frac{2r_2 r_1}{2a^{3/2}\sqrt{p}}\sin(f_2 - f_1) \qquad (7.96)$$

$$= E(t_2) - E(t_1) - \sin(E_2 - E_1) + \frac{r_2 r_1}{a^{3/2}\sqrt{p}}\sin(f_2 - f_1),$$

whence

$$\Delta t = \frac{a^{3/2}}{\mu}\left[E(t_2) - E(t_1) - \sin(E_2 - E_1)\right] + \frac{r_2 r_1}{\sqrt{\mu p}}\sin(f_2 - f_1) \qquad (7.97)$$

But from Eqs. (7.70) and (3.62)

$$\frac{1}{Y} = \frac{r_2 r_1 \sin(f_2 - f_1)}{\sqrt{\mu p}\,\Delta t}. \qquad (7.98)$$

Substitution from Eq. (7.97) into Eq. (798) gives

$$\Delta t = \frac{a^{3/2}}{\mu}\left[E(t_2) - E(t_1) - \sin(E_2 - E_1)\right] + \frac{r_2 r_1}{\sqrt{\mu p}}\sin(f_2 - f_1)$$

$$1 = \frac{a^{3/2}}{\mu\,\Delta t}\left[E(t_2) - E(t_1) - \sin(E_2 - E_1)\right] + \frac{r_2 r_1}{\sqrt{\mu p}\,\Delta t}\sin(f_2 - f_1)$$

$$= \frac{a^{3/2}}{\mu\,\Delta t}\left[E(t_2) - E(t_1) - \sin(E_2 - E_1)\right] + \frac{1}{y} \qquad (7.99)$$

$$1 - \frac{1}{Y} = \frac{a^{3/2}}{\mu\,\Delta t}\left[E(t_2) - E(t_1) - \sin(E_2 - E_1)\right].$$

The double angle relationship

$$\sin(f_2 - f_1) = 2\sin\left(\frac{f_2 - f_1}{2}\right)\cos\left(\frac{f_2 - f_1}{2}\right)$$

$$= 2\frac{\sqrt{ap}}{\sqrt{r_2 r_1}}\sin\left(\frac{E_2 - E_1}{2}\right)\cos\left(\frac{f_2 - f_1}{2}\right), \qquad (7.100)$$

in which the last line comes from Eq. (7.95). Substituting this into Eq. (7.98) gives

$$Y = \frac{\sqrt{\mu p}\ \Delta t}{r_2 r_1 \sin\left(f_2 - f_1\right)}$$

$$= \frac{\sqrt{\mu}\ \Delta t}{2\sqrt{a}\ r_2 r_1 \sin\left(\dfrac{E_2 - E_1}{2}\right)\cos\left(\dfrac{f_2 - f_1}{2}\right)}. \tag{7.101}$$

We now eliminate a from Eqs. (7.99) and (7.101). The latter is cubed and multiplied by the former:

$$Y^3\left(1 - \frac{1}{Y}\right) = \left[\frac{\sqrt{\mu}\ \Delta t}{2\sqrt{a}\ \sqrt{r_2 r_1}\sin\left(\dfrac{E_2-E_1}{2}\right)\cos\left(\dfrac{f_2-f_1}{2}\right)}\right]^3 \left[\frac{a^{3/2}}{\sqrt{\mu}\ \Delta t}\left[E\left(t_2\right) - E\left(t_1\right) - \sin\left(E_2 - E_1\right)\right]\right]$$

$$= \frac{a^{3/2}\,\Delta t^3\,\mu^{3/2}}{a^{3/2}\,\Delta t\,\sqrt{\mu}}\left[\frac{1}{2\sqrt{r_2 r_1}\cos\left(\dfrac{f_2-f_1}{2}\right)}\right]^3 \left[\frac{E\left(t_2\right)-E\left(t_1\right)-\sin\left(E_2-E_1\right)}{\sin^3\left(\dfrac{E_2-E_1}{2}\right)}\right] \tag{7.102}$$

$$= \left[\frac{\mu\,\Delta t^2}{\left(2\sqrt{r_2 r_1}\cos\left(\dfrac{f_2-f_1}{2}\right)\right)^3}\right]\left[\frac{E\left(t_2\right)-E\left(t_1\right)-\sin\left(E_2-E_1\right)}{\sin^3\left(\dfrac{E_2-E_1}{2}\right)}\right]$$

$$= m\left[\frac{E\left(t_2\right)-E\left(t_1\right)-\sin\left(E_2-E_1\right)}{\sin^3\left(\dfrac{E_2-E_1}{2}\right)}\right],$$

where the last line of Eq. (7.102) comes from Eq. (7.87).

Let
$$X = \left[\frac{E\left(t_2\right)-E\left(t_1\right)-\sin\left(E_2-E_1\right)}{\sin^3\left(\dfrac{E_2-E_1}{2}\right)}\right]. \tag{7 103}$$

Then we can write that

$$Y^3\left(1-\frac{1}{Y}\right) = Y^3 - Y^2 = Y^2\left(Y-1\right) = mX. \qquad (7.104)$$

Solution of Eqs. (7.89), (7.101) and (7.104)

We will see that these equations will eventually require numerical solution. Having derived these equations, I now give the recipe for solving them. Again, my account largely follows that of Escobal [128].

Combining Eqs. (7.89) and (7.104) gives

$$\left(\frac{m}{l+\xi}\right)(Y-1) = mX$$

$$(Y-1) = mX\left(\frac{l+\xi}{m}\right) \qquad (7.105)$$

$$Y = 1 + X\left(l+\xi\right).$$

We have not yet used that fact, which can be guessed from Fig. 7.19, that the ratio Y is not much greater than one for a real elliptical orbit.

So we make a first guess that

$$Y \oplus 1, \qquad (7.106)$$

And substitute this into Eq. (7.89):

$$\xi = \frac{m}{Y^2} - l. \qquad (7.107)$$

The next iteration might be

$$1 - 2\xi = \cos\left(\frac{E_2 - E_1}{2}\right) \qquad (7.108)$$

from Eq. (7.88).
Now use the fact that

$$0 \le \frac{E_2 - E_1}{2} \le \pi \qquad (7.109)$$

and

$$\cos\left(\frac{E_2-E_1}{2}\right) = \sqrt{1-\sin^2\left(\frac{E_2-E_1}{2}\right)}$$

$$= \sqrt{1-\left(1-2\xi\right)^2} \tag{7.110}$$

$$= \sqrt{1-1-4\xi^2+4\xi}$$

$$= +\sqrt{4\xi\left(1-\xi\right)}.$$

Therefore, the difference in eccentric anomalies $(E_2\text{-}E_1)$ is uniquely determined.

We now plug our known value of $(E_2\text{-}E_1)$ into Eq. (7.103):

$$X = \left[\frac{E\left(t_2\right)-E\left(t_1\right)-\sin\left(E_2-E_1\right)}{\sin^3\left(\dfrac{E_2-E_1}{2}\right)}\right]. \tag{7.103}$$

Then

$$Y = 1+X\left(l+\xi\right). \tag{7.111}$$

This scheme may be iterated from Eqs. (7.107, 7.108, 7.109, 7.110, and 7.111) to desired accuracy.

Now, rewrite Eq. (7.101) in the form

$$\sqrt{a} = \frac{\sqrt{\mu}\ \Delta t}{2\ Y\ r_2 r_1 \sin\left(\dfrac{E_2-E_1}{2}\right)\cos\left(\dfrac{f_2-f_1}{2}\right)} \tag{7.112}$$

Solving Eq. (7.1) with Known Magnitude but Unknown Direction of Sun–Planet Distance

This case occurs in Herget's method if we need to know a value of ρ that satisfies a known magnitude of $\mathbf{r}$.

Equation (7.1) is the first line of Eq. (7.113)

$$\mathbf{r} + \mathbf{R} = \rho$$

$$\rho = \rho\left(\hat{\rho}_x + \hat{\rho}_y + \hat{\rho}_z\right)$$

$$C_1^2 = \left(r^2{}_x + r^2{}_y + r^2{}_z\right)$$

$$\mathbf{R} = \text{known}$$

$$\rho = \rho\left(\hat{\mathbf{x}}\cos\lambda\cos\beta + \hat{\mathbf{y}}\sin\lambda\cos\beta + \hat{\mathbf{z}}\sin\beta\right) \qquad (7.113)$$

$$r_z + R_z = \rho\sin\beta$$

$$r_z = \rho\sin\beta$$

$$r_x + R_x = \rho\cos\lambda\cos\beta$$

$$r_y + R_y = \rho\sin\lambda\cos\beta$$

From this, we have four equations in four unknowns r_x, r_y, r_z and ρ:

$$C_1^2 = \left(r^2{}_x + r^2{}_y + r^2{}_z\right)$$

$$r_z = \rho\sin\beta \qquad (7.114)$$

$$r_x + R_x = \rho\cos\lambda\cos\beta$$

$$r_y + R_y = \rho\sin\lambda\cos\beta$$

$$C_1^2 = \left(r^2{}_x + r^2{}_y + r^2{}_z\right)$$

$$r_z = \rho\sin\beta \qquad (7.115)$$

$$r_x = \rho\cos\lambda\cos\beta - R_x$$

$$r_y = \rho\sin\lambda\cos\beta - R_y$$

$$
\begin{aligned}
C_1^2 &= \left(\left(\rho\cos\lambda\cos\beta - R_x\right)^2 + \left(\rho\sin\lambda\cos\beta - R_y\right)^2 + \rho^2\sin^2\beta\right) \\
&= \left(\begin{array}{l}\rho^2\cos^2\lambda\cos^2\beta + R_x^2 - 2\rho R_x\cos\lambda\cos\beta + \rho^2\sin^2\lambda\cos^2\beta + R_y^2 \\ -2\rho R_y\sin\lambda\cos\beta + \rho^2\sin^2\beta\end{array}\right) \\
&= \left(\rho^2\cos^2\beta + R_x^2 - 2\rho R_x\cos\lambda\cos\beta + R_y^2 - 2\rho R_y\sin\lambda\cos\beta + \rho^2\sin^2\beta\right)
\end{aligned} \qquad (7.116)
$$

$$0 = \rho^2 - 2\rho\left(R_x\cos\lambda\cos\beta + R_y\sin\lambda\cos\beta\right) + R_x^2 + R_y^2 - C_1^2 \qquad (7.117)$$

$$0 = \qquad \rho^2 - 2\rho\cos\beta\left(R_x\cos\lambda + R_y\sin\lambda\right) + R_x^2 + R_y^2 - C_1^2$$

$$\rho = \frac{2\cos\beta\left(R_x\cos\lambda + R_y\sin\lambda\right) \pm \sqrt{4\cos^2\beta\left(R_x\cos\lambda + R_y\sin\lambda\right)^2 - 4\left(R_x^2 + R_y^2 - C_1^2\right)}}{2}$$

$$\rho = \quad \cos\beta\left(R_x\cos\lambda + R_y\sin\lambda\right) \pm \sqrt{\cos^2\beta\left(R_x\cos\lambda + R_y\sin\lambda\right)^2 - \left(R_x^2 + R_y^2 - C_1^2\right)}$$

$$(7.118)$$

Eureka! (Part One)

Yes, we're there. Right at the start of the discussion of Herget's method I said we needed to guess values of $\mathbf{r}_1$ and $\mathbf{r}_n$ and that by implication we would have guessed a value of the difference in the true anomalies of the 1st and nth observations, $(f_n - f_1)$. We know Δt for these observations. From Eq. (7.110), we know $(E_2 - E_1)$. From Eq. (7.111), we know Y. So, we know or have guessed every term in Eq. (7.112).

In Canonical Units, Eq. (3.91) tells us that $\mu = 1$.

Obtaining the Orbital Elements from Two Radius Vectors

There are quite a few ways to do this. I found that they don't all work as the textbooks advertise: for example, if you aren't careful, you end up picking the wrong ($\pm$) root of a square root. I don't know about you, but I want "fit and forget" methods that I can put into a computer program and have them work every time without maintenance, not methods where I have to make manual checks that the correct one of two alternatives was picked.

Escobal [129] cogently argues that the conventional orbital elements aren't a terribly clever choice. The usual definition of elements fails in the cases that the ellipse has zero eccentricity and if it is exactly coplanar with the orbit of the Earth. Nevertheless, people are going to want to compare anything I claim against the published orbital elements, which are usually in the conventional form. So I have stuck with the conventional definition, knowing that the cases where it fails are not very likely to occur with calculated orbits of Mars, be they preliminary or otherwise. At least people will then know what I am claiming.

The usual orbital elements are tabulated below in Table 7.1.

These elements are illustrated in Fig. 7.20.

In my Herget's method program, I calculate the semi-major axis a from Eq. (7.112). Why didn't I use the much simpler Vis Viva Equation (3.107)?

Element	Symbol
Table 7.1 Orbital Elements. See also Fig. 7.20	
Semi-Major Axis	a
Inclination	i
Eccentricity	e
Line of Ascending Node	Ω
Argument of Perihelion	ω
Time of Perihelion Passage (Canonical Units)	T_{Per}

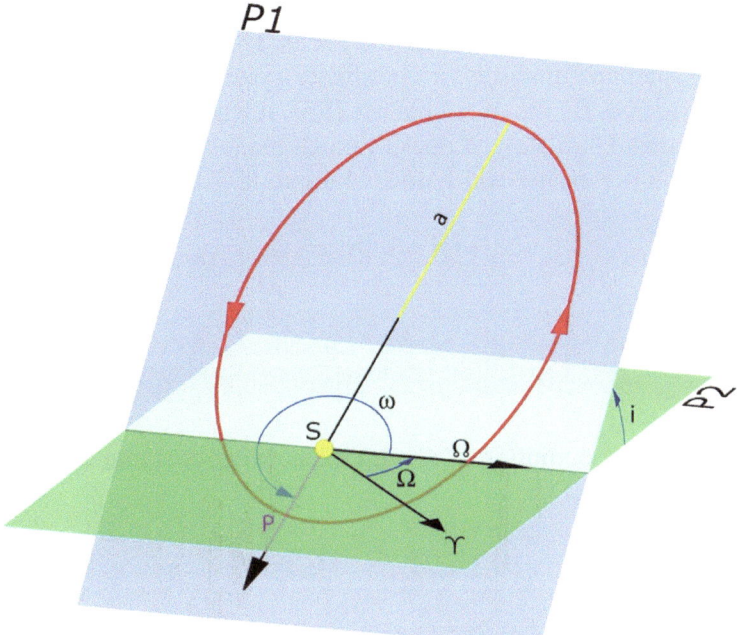

Fig. 7.20 The conventional orbital elements. Plane P2 is the ecliptic, the plane of the Earth's orbit. Plane P2 is that of the orbit of the planet, in our case Mars. The symbol ♈ points to the first point of Aries. (Credit: Brandir~commonswiki, GNU Free Documentation License, Version 1.2)

Was there some frightfully clever reason? Alas, no. I simply forgot about the Vis Viva Equation.

Next, the longitude of the ascending node is calculated. Define a vector normal to the plane of the ecliptic, with its origin at the centre of the Sun, by

$$\mathbf{K} = (0,\ 0,\ 1) \text{ and}$$
$$\mathbf{I} = (1,\ 0,\ 0)\ . \tag{7.119}$$

Then the ascending node lies along the line defined by the cross product of unit vectors

$$\begin{aligned} \mathbf{N} = \; & \mathbf{K} \times \hat{\mathbf{h}} \\ = & \left(-h_y, \; h_x, \; 0\right). \end{aligned} \tag{7.120}$$

The inclination is given by

$$\mathbf{K} \cdot \hat{\mathbf{h}} = \left(0, \; 0, \; 1\right) \cdot \left(\hat{h}_x, \; \hat{h}_y, \; \hat{h}_z\right) = \hat{h}_z = \cos\left(i\right). \tag{7.121}$$
$$\therefore i = \qquad \cos^{-1}\left(\hat{h}_z\right).$$

In a sensible coordinate system, the vector $\mathbf{I}$ would lie along the line to ♈. I chose, perhaps unwisely, a coordinate system in which $\mathbf{I}$ lies in the opposite direction. By the time I realized this, it was way too late to go back and change this. Ultimately, it doesn't much matter anyway, as long as I'm careful. That's my excuse and I am sticking to it. The dot product

$$\mathbf{I} \cdot \mathbf{N} = |\mathbf{I}| \; |\mathbf{N}| \; \cos\Omega \tag{7.122}$$

Therefore

$$\Omega = \cos^{-1} \frac{N_x}{|\mathbf{N}|}. \tag{7.123}$$

To compare my definition of Ω with the conventional one, I simply need to subtract π radians.

$$\Omega_{\text{Conventional}} = \Omega_{\text{Jane}} - \pi. \tag{7.124}$$

Once the velocity $\dot{\mathbf{r}}$ has been calculated, using the Lagrange functions f_{Lag} and g_{Lag}, the angular momentum per unit mass $\mathbf{h}$ comes from Eq. (7.44), with the first observation taken to be the initial condition at time t_0.

So far so good. Calculating the eccentricity e needs care, because it has a tendency to appear in equations in the form e^2. I found the following method to be robust against "wrong" square roots. Eq. (3.62) tells us that

$$\frac{h^2}{\mu} = a \left(1 - e^2\right). \tag{3.62}$$

In Canonical Units, $\mu = 1$. Substituting this value and rearranging gives

$$e = +\sqrt{1 - \frac{h^2}{a}}. \tag{7.125}$$

In this case, it is obvious that the negative square root is the wrong one to take and that the positive square root is the right one to use.

We have two elements left: the argument of perihelion ω and the time T_{Per} since perihelion. These are related, because any uncertainty in one directly translates into uncertainty in the other if we know where Mars is at some time other than the perihelion.

I have seen more than one approach to this issue. The one I like the best is to re-imagine the eccentricity as, not a scalar, but a vector quantity. This is not as crazy as it first seems: there is an orientation to an ellipse as well as an extent of ovality. More precisely, the major axis has an orientation. If we can get a vector describing the ellipse to point towards the perihelion, then we can determine the orientation of the perihelion. Fig. 7.21 illustrates the concept. This idea has an excellent pedigree. It dates back to Sir William Hamilton in 1845 [130].

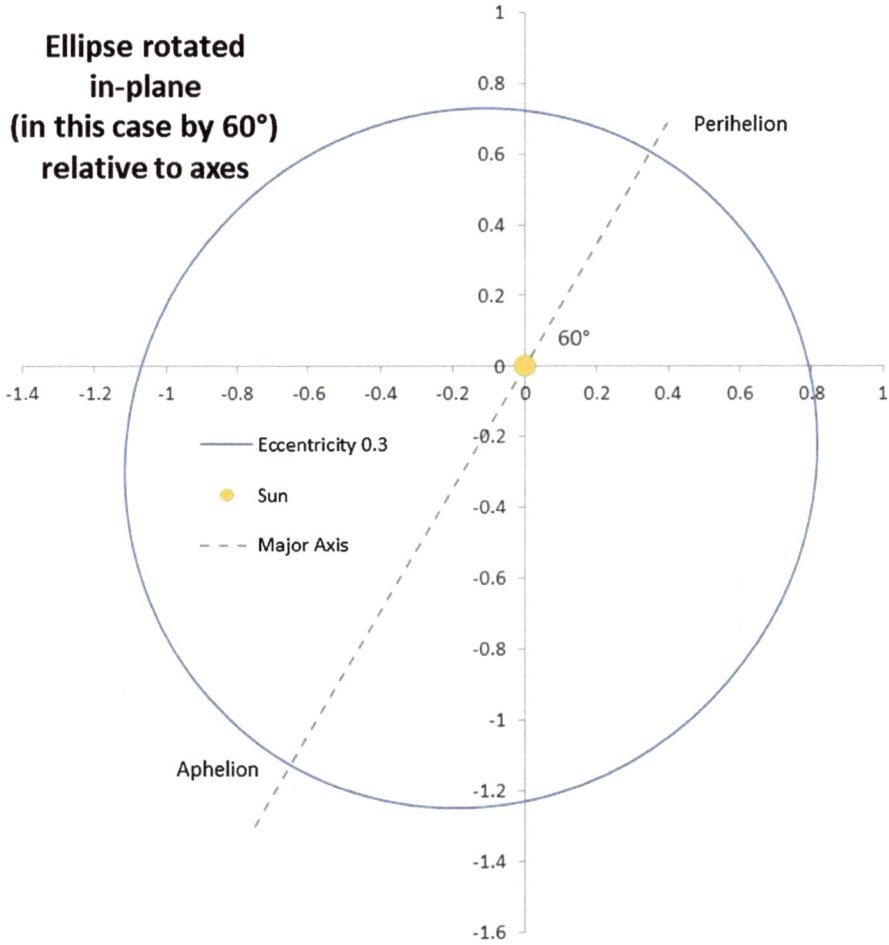

Fig. 7.21 This ellipse has a higher eccentricity than is found in practice, to make the point that the eccentricity has an orientation – that of the major axis – as well as a magnitude

Where to begin? How about with the basic gravitational law, Eq. (3.14)? I'm going to write it in the form

$$\frac{d^2\mathbf{r}}{dt^2} + \frac{\mu}{r^2}\hat{\mathbf{r}} = \ddot{\mathbf{r}} + \frac{\mu}{r^2}\hat{\mathbf{r}} = 0, \tag{7.126}$$

where, as usual, the double dot indicates differentiation twice with respect to time. If I take the cross product of this with $\mathbf{r}$, I get

$$\mathbf{r} \times \ddot{\mathbf{r}} = \frac{-\mu}{r^2}\left(\mathbf{r} \times \hat{\mathbf{r}}\right) = 0 . \tag{7.127}$$

The equation to zero occurs because the vector $\mathbf{r}$ has to be parallel to its own unit vector, and the cross products of parallel vectors are always equal to zero. We have already considered the angular momentum per unit mass and know that it is constant. Using the product rule of differentiation

$$\frac{d}{dt}\left(\mathbf{r} \times \dot{\mathbf{r}}\right) = \left(\mathbf{r} \times \ddot{\mathbf{r}}\right) + \left(\dot{\mathbf{r}} \times \dot{\mathbf{r}}\right)$$
$$= \qquad 0 + 0 \tag{7.128}$$
$$= \qquad 0,$$

by virtue of Eq. (7.127). We have again shown that

$$\mathbf{h} = \mathbf{r} \times \dot{\mathbf{r}} \tag{7.129}$$

is constant.

From Eq. (7.126)
$$\ddot{\mathbf{r}} \times \mathbf{h} = -\frac{\mu}{r^2}\hat{\mathbf{r}} \times \mathbf{h}$$
$$= -\frac{\mu}{r^2}\hat{\mathbf{r}} \times \left(\mathbf{r} \times \dot{\mathbf{r}}\right) \tag{7.130}$$

I now use the relationship, proved in Appendix 2, that

$$\mathbf{A} \times \left(\mathbf{B} \times \mathbf{C}\right) = \left(\mathbf{A} \cdot \mathbf{C}\right)\mathbf{B} - \left(\mathbf{A} \cdot \mathbf{B}\right)\mathbf{C} \tag{7.131}$$

to write

$$\ddot{\mathbf{r}} \times \mathbf{h} - \frac{\mu}{r^2}\left[(\hat{\mathbf{r}} \cdot \dot{\mathbf{r}})\mathbf{r} - (\hat{\mathbf{r}} \cdot \mathbf{r})\dot{\mathbf{r}}\right] = -\frac{\mu}{r^2}\left[\left(\frac{1}{r}\mathbf{r} \cdot \dot{\mathbf{r}}\right)\mathbf{r} - \left(\frac{1}{r}\mathbf{r} \cdot \mathbf{r}\right)\dot{\mathbf{r}}\right]$$
$$= -\frac{\mu}{r^2}\left[\frac{1}{r}(\mathbf{r} \cdot \dot{\mathbf{r}})\mathbf{r} - \frac{1}{r}(\mathbf{r} \cdot \mathbf{r})\dot{\mathbf{r}}\right] \tag{7.132}$$
$$= -\frac{\mu}{r^2}\frac{1}{r}\left[\frac{1}{r}(\mathbf{r} \cdot \dot{\mathbf{r}})\mathbf{r} - r\dot{\mathbf{r}}\right].$$

In Eq. (7.132), I have one slightly awkward term with which I still have to deal, viz. $(\mathbf{r} \cdot \dot{\mathbf{r}})/r$. The slight awkwardness results from a subtlety. Danby [131] catches it best when he writes:

It is very important to realize that $\left|\dfrac{d\mathbf{r}}{dt}\right|$, or $|\dot{\mathbf{r}}|$, is not in general the same as $\dfrac{d|\mathbf{r}|}{dt}$, or $\dfrac{dr}{dt}$, or $\dot{r}$. The former is the speed of [our planet] , while the latter is only the velocity of [our planet] along the radius vector: in general there will also be a *transverse* component of velocity, at right angles to the radius vector.

Bearing this in mind, we can revisit Eq. (7.132) and write

$$
\begin{aligned}
\ddot{\mathbf{r}} \times \mathbf{h} &= -\frac{\mu}{r^2}\frac{1}{r}\left[\frac{r\,\dot{r}}{r}\mathbf{r} - r\dot{\mathbf{r}}\right] \\
&= \frac{\mu}{r^2}\frac{1}{r}\left[r\dot{\mathbf{r}} - \dot{r}\mathbf{r}\right].
\end{aligned}
\tag{7.133}
$$

Our tale has a surprise turn here. Note that, using the quotient rule of differentiation,

$$
\frac{d}{dt}\frac{\mathbf{r}}{r} = \frac{r\dot{\mathbf{r}} - \dot{r}\mathbf{r}}{r^2}.
\tag{7.134}
$$

The plot twists again, just like it did last summer (and the summer before, and so on) For the sheer heck of it, let us differentiate

$$
\begin{aligned}
\frac{d}{dt}\left(\mathbf{r} \times \mathbf{h}\right) &= \left(\ddot{\mathbf{r}} \times \mathbf{h}\right) + \left(\mathbf{r} \times \dot{\mathbf{h}}\right) \\
&= \left(\ddot{\mathbf{r}} \times \mathbf{h}\right) + \left(\mathbf{r} \times \mathbf{0}\right) \\
&= \left(\ddot{\mathbf{r}} \times \mathbf{h}\right).
\end{aligned}
\tag{7.135}
$$

We can use the result of Eqs. (7.135) and (7.134) in Eq. (7.133) to give

$$
\frac{d}{dt}\left(\dot{\mathbf{r}} \times \mathbf{h}\right) = \mu\frac{d}{dt}\left(\frac{\mathbf{r}}{r}\right).
\tag{7.136}
$$

Let us now integrate Eq. (7.136) with respect to time, to give

$$
\dot{\mathbf{r}} \times \mathbf{h} = \mu\left(\frac{\mathbf{r}}{r} + \mathbf{e}\right),
\tag{7.137}
$$

where the vector $\mathbf{e}$ is a constant of integration. I called it $\mathbf{e}$ with mischief aforethought. Let me show you something. I have to use the relationship, proved in Appendix 2, that

$$
(\mathbf{A} \times \mathbf{B}) \cdot \mathbf{C} = (\mathbf{C} \times \mathbf{A}) \cdot \mathbf{B}.
\tag{7.138}
$$

This tells me that, from Eq. (7.137),

$$\left(\dot{\mathbf{r}} \times \mathbf{h}\right) \bullet \mathbf{r} = \quad \left(\mathbf{r} \times \dot{\mathbf{r}}\right) \bullet \mathbf{h} \; ;$$

$$\left(\mathbf{h} \bullet \mathbf{h}\right) = \mu \left(\frac{\mathbf{r} \bullet \mathbf{r}}{r} + \mathbf{e} \bullet \mathbf{r}\right) \qquad (7.139)$$

$$h^2 = \mu \left(r + \mathbf{e} \bullet \mathbf{r}\right) .$$

But

$$\mathbf{e} \bullet \mathbf{r} = e r \cos\phi \; , \qquad (7.140)$$

where ϕ is the included angle between $\mathbf{e}$ and $\mathbf{r}$. Substituting Eq. (7.140) into Eq. (7.139) gives me

$$h^2 = \mu \left(r + e \; r \; \cos\phi\right)$$

$$\frac{1}{r} = \frac{\mu}{h^2} \left(1 + e \; \cos\phi\right) \qquad (7.141)$$

$$r = \frac{h^2}{\mu} \frac{1}{\left(1 + e \; \cos\phi\right)} .$$

Look familiar? It might look even more familiar were I to rewrite Eq. (3.60):

$$r = \frac{h^2}{\mu} \frac{1}{\left(1 + e \; \cos f\right)} . \qquad (3.60)$$

This is nothing but the equation of an ellipse in polar coordinates centred on the attracting focus, with eccentricity e. I have in fact proved that the angle between $\mathbf{e}$ and $\mathbf{r}$ is the true anomaly f:

$$\phi = f. \qquad (7.142)$$

Therefore, the vector eccentricity has the magnitude of the scalar eccentricity and the orientation of the major axis, pointing towards the perihelion.

From Eq. (7.137)

$$\mu\mathbf{e} = \dot{\mathbf{r}} \times \mathbf{h} - \mu \left(\frac{\mathbf{r}}{r}\right)$$

$$\mathbf{e} = \frac{\dot{\mathbf{r}} \times \mathbf{h}}{\mu} - \frac{\mathbf{r}}{r} . \qquad (7.143)$$

Once we have calculated $\mathbf{e}$ as a vector, we can use it to calculate the orbital element ω.

We need to take account of the possibility that ω might be greater than $90°$. We can do this by remembering what the signs of sine and cosine functions in the four quadrants are.

We then calculate

$$\hat{\mathbf{e}} = \frac{\mathbf{e}}{+\sqrt{e_x^2 + e_z^2 + e_z^2}} \tag{7.144}$$

and

$$\hat{\mathbf{N}} = \frac{\mathbf{N}}{+\sqrt{e_x^2 + e_z^2 + e_z^2}} \tag{7.145}$$

Then

$$\cos\omega = \hat{\mathbf{e}} \bullet \hat{\mathbf{N}}, \quad \text{so}$$
$$\therefore \omega = \cos^{-1}\left(\hat{\mathbf{e}} \bullet \hat{\mathbf{N}}\right). \tag{7.146}$$

There's still a problem. Fortran's arc cosine function assumes that ω lies between 0 and π. Most other computer languages will have the same issue, even if the range covered is different: arc cosines are not unique. So we need a trick to tell if $\omega > \pi$.

If indeed $\omega > \pi$, then the z components of the two cross products $\hat{\mathbf{e}} \times \hat{\mathbf{N}}$ and $\mathbf{h} = \mathbf{r} \times \dot{\mathbf{r}}$ will have opposite algebraic signs so long as the inclination i is small. We know that this is true for planets. This gives us a test:

$$\text{if} \left(\hat{\mathbf{e}} \times \hat{\mathbf{N}}\right)_z h_z < 0 \quad \text{then} \quad \omega \leq \pi.$$
$$\text{if} \left(\hat{\mathbf{e}} \times \hat{\mathbf{N}}\right)_z h_z > 0 \quad \text{then} \quad \omega > \pi. \tag{7.147}$$

This test is illustrated in Fig. 7.22.

I will digress for a moment, to point out that in Eq. (7.137), $\mathbf{e}$ was defined as a constant of integration. In other words, it does not change over time. This also means that the orientation of the major axis does not change over time. This point was first noted by Newton in the *Principia*, as Proposition XIV, Book III: "The aphelions are immovable" [79].

Anyway, in the event that $\omega > \pi$, Eq. (7.146) is replaced by

$$\omega = 2\pi - \cos^{-1}\left(\hat{\mathbf{e}} \bullet \hat{\mathbf{N}}\right) \tag{7.148}$$

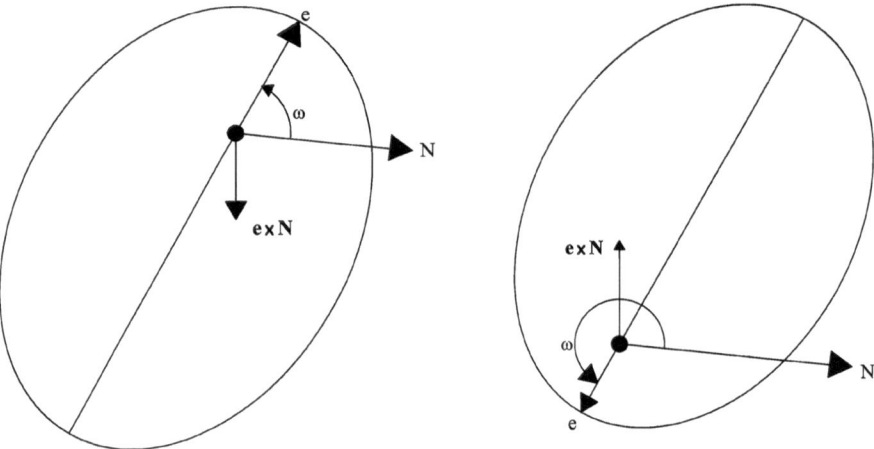

Fig. 7.22 Testing whether ω is greater than π. If not, the vector cross product $\underline{e} \times \underline{N}$ points downwards (left). If so, the vector cross product $\underline{e} \times \underline{N}$ points upwards (right). The angular momentum will point upwards if the planet moves anticlockwise in both scenarios as seen in this figure

Finally, we need to work out the time of perihelion passage. Given that for my ith observation,

$$f_i = \cos^{-1}\left(\hat{\mathbf{e}} \cdot \hat{\mathbf{N}}_i\right) \tag{7.149}$$

from Eq. (3.72) I can work out

$$E_i = 2\tan^{-1}\left(\sqrt{\frac{1-e}{1+e}} \ \tan\frac{f}{2}\right). \tag{7.150}$$

There is some ambiguity about whether I will end up working out the time of the previous or next perihelion passage. You only need one of them and it doesn't matter which.

Finally, I get the mean anomaly from Kepler's Equation (3.79)

$$M_i = E_i - e\sin E_i. \tag{7.151}$$

Finally, the time comes from converting the mean anomaly into a fraction of the period. I'm only working in Canonical Units where the period of the Earth's orbit is 2π and $\mu = 1$. Then the period of any other planet is, from Kepler's third law Eq. (3.88),

$$T_{\text{Planet}} = 2\pi \ a^{3/2} \ . \tag{7.152}$$

The time from my ith observation to perihelion is then

$$t_{PH} - t_i = \frac{M_i}{2\pi} \, 2\pi a^{3/2}$$
$$= M_i a^{3/2} \, . \tag{7.153}$$

The time to an earlier perihelion will obviously be negative.

Software Listing

The main program is quite long. Some tasks were subcontracted to subroutines. These subroutines often have the same names as the subroutines in other programs I list. I sometimes made changes to them depending on the context. Therefore, to be on the safe side I list all the subroutines used in this program.

I make no pretense to be an elegant programmer. My programming is fast and furious. Once it works I don't mess around with it. It's too easy to make something that works no longer work. At least that's my view.

Main Program

```
 1   ! A fortran95 program for G95
 2   ! By JANE CLARK BEGUN AUGUST 30 2020
 3   ! Solves for Preliminary orbit using Herget's method
 4   ! Subroutine LSQFIT written with help from Jeannette M. Fine, Used with Permission
 5   ! Copyright Jane D. Clark & Jeannette M. Fine
 6   ! MANY WRITE STATEMENTS ARE FOR DEBUGGING.  THEY ARE OFTEN COMMENTED OUT.
 7   ! Canonical Units Used
 8   !
 9   program main
10        IMPLICIT DOUBLE PRECISION (A-H,O-Z) ! I'm not a fan of declaring every variable.
11        IMPLICIT INTEGER(I-N)               ! I know that's heresy but I don't care.
12        integer re_i
13        INTEGER YEER(100),MUNTH(100),DAIT(100),HOWR(100),MINIT(100)
14        INTEGER YEER2(100),MUNTH2(100),DAIT2(100),HOWR2(100),MINIT2(100)
15        DOUBLE PRECISION XEARTH(100),YEARTH(100),ZEARTH(100),QTIME(100)
16        DOUBLE PRECISION RLAMBDA(100),RBETA(100),RHO(100),CC1(100),CCN(100),Y1J(100),YJN(100)
17        DOUBLE PRECISION BIG_R(3,100),KC1SQ,C2SQ,RHO_HAT(3,100),SMALL_r(3,100)
18        DOUBLE PRECISION BIG_A(3,100),BIG_D(3,100),P_BIG(100),Q_BIG(100)
19        DOUBLE PRECISION P_DNB(100),Q_DNB(100),DPDRHO1(100),DPDRHON(100),DQDRHO1(100),
                                                                               DQDRHON(100)
20        DOUBLE PRECISION P_PLUSDI(100),P_MINUSDI(100),P_PLUSDF(100),P_MINUSDF(100)
21        DOUBLE PRECISION Q_PLUSDI(100),Q_MINUSDI(100),Q_PLUSDF(100),Q_MINUSDF(100)
22        DOUBLE PRECISION DELTARIP(100),DELTARIM(100),DELTARFP(100),DELTARFM(100)
23        DOUBLE PRECISION R_INIT(3),R_FINAL(3),VEL(3),VELI(3),H(3),HM(3)
24        DOUBLE PRECISION HAY(100,8),HESS(2),EXV(3),EVEC(3)
25        !
26        !Bring the observational data in & check we've read it.
27        !
28        OPEN(UNIT=15,FILE='Earth_Data_09_10.csv')
```

```
29        OPEN(UNIT=25,FILE='op_checks.txt')          ! THIS IS THE DEBUGGING OUTPUT
30        OPEN(UNIT=26,FILE='Outputs.txt')
31        PI=4.0D0*atan(1.0D0)   ! define pi
32        !
33        ! READING & CHECK WRITING
34        !
35        ICOUNT1=0
36        DO I=1,100
37   !    DO I=1,22
38        read(15,*)YEER(I),MUNTH(I),DAIT(I),HOWR(I),MINIT(I),XEARTH(I),YEARTH(I),
                                                ZEARTH(I),RLAMBDA(I),RBETA(I)
39        write(25,*)I,YEER(I),MUNTH(I),DAIT(I),HOWR(I),MINIT(I),XEARTH(I),
                                             YEARTH(I),ZEARTH(I),RLAMBDA(I),RBETA(I)
40        IF (YEER(I).EQ.0)GO TO 400
41        ICOUNT1=ICOUNT1+1
42        END DO
43   400  CONTINUE
44        WRITE(25,*)"ICOUNT1",ICOUNT1
45        !
46        ! GET TIMES INTO CANONICAL UNITS
47        !
48        CALL CONVERT_TIMES(YEER,MUNTH,DAIT,HOWR,
          MINIT,ICOUNT1,QTIME)
49        DO I=1,ICOUNT1
50        WRITE(25,*)"QTIME ",I,QTIME(I)
51        END DO
52   !    STOP 904  ! TEMPORARY FOR DEBUG
53        !
54        ! Calculate components of R and rho-hat
55        !
56        DO I=1,ICOUNT1
57          BIG_R(1,I)=XEARTH(I)
58          BIG_R(2,I)=YEARTH(I)
59          BIG_R(3,I)=ZEARTH(I)
60          RHO_HAT(1,I)=COS(RLAMBDA(I))*COS(RBETA(I))
61          RHO_HAT(2,I)=SIN(RLAMBDA(I))*COS(RBETA(I))
62          RHO_HAT(3,I)=SIN(RBETA(I))
63        END DO
64   !
65        WRITE(25,*)"ICOUNT1: ",ICOUNT1
66        !
67        ! Choose the beginning and end Sun-Planet distances in AU
68        ! THESE ARE SEED VALUES TO INITIALISE THE ANALYSIS
69        ! ALSO CHOOSE OBSERVATION DATA POINTS WITH WHICH TO START AND FINISH
70        !
71        A_INITIAL=1.62
72        A_FINAL=1.61
73        I_INITIAL=15
74        I_FINAL=30
75        !
76        ! BEGIN FORMAL OUTPUT: WRITE HEADERS AND STARTING CONDITIONS
77        !
78        WRITE(26,620)
79   620  FORMAT("HERGET'S METHOD FOR PRELIMINARY ORBIT DETERMINATION",/)
80        WRITE(26,601)YEER(1),YEER(ICOUNT1)
81        WRITE(26,600)A_INITIAL,A_FINAL,I_INITIAL,I_FINAL
82   600  FORMAT("IMPOSED INITIAL & FINAL RADII",1X,F8.4,1X,F8.4," INITIAL & FINAL
                                              OBSERVATIONS",1X,I3,1X,I3)
83   601  FORMAT("YEARS",1X,I5,1X,"-",1X,I5)
84        IF (I_FINAL.LT.I_INITIAL+4)STOP 906 ! TRAP CASES THAT MESS UP THE LOOPING
85        !
86        ! The data on Big R are vectors from the Sun to the Earth.
87        ! It is more conventional to have the vector R go from Earth to Sun.
88        ! Therefore we now reverse the vector R.
89        !
90        DO I=1,ICOUNT1
91          DO J=1,3
```

```
92                 BIG_R(J,I)=-BIG_R(J,I)
93      !          WRITE(25,*)"BIG_R: ",BIG_R(J,I),J,I
94              END DO
95          END DO
96          DO I=I_INITIAL,I_FINAL
97            WRITE(25,*)"BIG R :",BIG_R(1,I),BIG_R(2,I),BIG_R(3,I)
98            WRITE(25,*)"RLAMBDA(I),RBETA(I):",RLAMBDA(I),RBETA(I)
99            WRITE(25,*)"RHO_HAT",RHO_HAT(1,I),RHO_HAT(2,I),RHO_HAT(3,I)
100         END DO
101         !
102         ! Herget's delta
103         !
104         DELTA=0.001D0
105         !
106         ! SET UP SOME ABBREVIATIONS TO CALCULATE RHO AT INITIAL POSITION
107         !
108         C1SQ=A_INITIAL**2    ! This term constrains the magnitude of r to lie on a sphere.
109         CLI=COS(RLAMBDA(I_INITIAL))*COS(RBETA(I_INITIAL))
110         SLI=SIN(RLAMBDA(I_INITIAL))*COS(RBETA(I_INITIAL))
111         SBI=SIN(RBETA(I_INITIAL))
112         BEE_I=(-BIG_R(1,I_INITIAL)*CLI-BIG_R(2,I_INITIAL)*SLI)
113         AY_I=1.0D0
114         CEE_I=+BIG_R(1,I_INITIAL)**2+BIG_R(2,I_INITIAL)**2-C1SQ
115         !
116         ! CALCULATE RHO INITIAL
117         ! Trap negative square roots
118         !
119         IF ((BEE_I**2-AY_I*CEE_I).GE.0) THEN
120         RHO(I_INITIAL)=(-BEE_I+SQRT(BEE_I**2-AY_I*CEE_I))/AY_I
121          write(25,*)'Rho initial ',RHO(I_INITIAL)
122         else
123          stop 901
124         endif
125         !
126         ! SET UP SOME ABBREVIATIONS TO CALCULATE RHO AT FINAL POSITION
127         !
128         C2SQ=A_FINAL**2    ! This term constrains the magnitude of r to lie on a sphere.
129         CLF=COS(RLAMBDA(I_FINAL))*COS(RBETA(I_FINAL))
130         SLF=SIN(RLAMBDA(I_FINAL))*COS(RBETA(I_FINAL))
131         SBF=SIN(RBETA(I_FINAL))
132         BEE_F=(-BIG_R(1,I_FINAL)*CLF-BIG_R(2,I_FINAL)*SLF)
133         AY_F=1.0D0
134         CEE_F=BIG_R(1,I_FINAL)**2+BIG_R(2,I_FINAL)**2-C2SQ
135         !
136         ! CALCULATE RHO FINAL
137         ! Trap negative square roots
138         !
139         IF ((BEE_F**2-AY_F*CEE_F).GE.0) THEN
140          RHO(I_FINAL)=(-BEE_F+SQRT(BEE_F**2-AY_F*CEE_F))/AY_F
141          DRHONP=RHO(I_FINAL)+DELTA
142          DRHONM=RHO(I_FINAL)-DELTA
143          write(25,*)'Rho final ',RHO(I_FINAL)
144         else
145          stop 902
146         endif
147         DRHO_INIT=0
148         DRHO_FINAL=0
149         DO ITERATE=1,5
150            !
151            ! Calculate components of r(initial) and r(final)
152            !
153            DO I=1,3
154               SMALL_r(I,I_INITIAL)=RHO_HAT(I,I_INITIAL)*(RHO(I_INITIAL)+DRHO_INIT)
                                                          -BIG_R(I,I_INITIAL)
155               SMALL_r(I,I_FINAL)=RHO_HAT(I,I_FINAL)*(RHO(I_FINAL)+DRHO_FINAL)-BIG_R(I,I_FINAL)
156            enddo
157            !
```

```
158          WRITE(25,*)"SMALL_r INIT: ",
SMALL_r(1,I_INITIAL),SMALL_r(2,I_INITIAL),SMALL_r(3,I_INITIAL)
159          WRITE(25,*)"SMALL_r FINAL: ",SMALL_r(1,I_FINAL),SMALL_r(2,I_FINAL),SMALL_r(3,I_FINAL)
160          !
161          !CHECK
162          !
163          XLITTLE_R_INIT=SQRT(SMALL_r(1,I_INITIAL)**2+SMALL_r(2,I_INITIAL)**2
                                                          +SMALL_r(3,I_INITIAL)**2)
164          XLITTLE_R_FINAL=SQRT(SMALL_r(1,I_FINAL)**2+SMALL_r(2,I_FINAL)**2
                                                          +SMALL_r(3,I_FINAL)**2)
165          WRITE(25,*)"XLITTLE_R_INIT,XLITTLE_R_FINAL: ",XLITTLE_R_INIT,XLITTLE_R_FINAL
166          !
167          ! Set up A & D per Danby Equation 7.4.2
168          !
169          DO I=I_INITIAL,I_FINAL
170            BIG_A(1,I)=-SIN(RLAMBDA(I))
171            BIG_A(2,I)=COS(RLAMBDA(I))
172            BIG_A(3,I)=0.0D0
173            BIG_D(1,I)=-SIN(RBETA(I))*COS(RLAMBDA(I))
174            BIG_D(2,I)=-SIN(RBETA(I))*SIN(RLAMBDA(I))
175            BIG_D(3,I)=COS(RBETA(I))
176            WRITE(25,*)"A: ",BIG_A(1,I),BIG_A(2,I),BIG_A(3,I)
177            WRITE(25,*)"D: ",BIG_D(1,I),BIG_D(2,I),BIG_D(3,I)
178          END DO
179          !
180          DO I=1,3
181            r_INIT(I)=SMALL_r(I,I_INITIAL)
182            r_FINAL(I)=SMALL_r(I,I_FINAL)
183          END DO
184          DELTAT=QTIME(I_FINAL)-QTIME(I_INITIAL)
185          WRITE(25,*)"DELTAT: ",DELTAT
186          !
187          ! USE ESCOBAL'S METHODS FROM HIS §6.2 TO CALCULATE SOME USEFUL QUANTITIES
188          ! INCLUDING SEMIMAJOR AXIS A
189          !
190          CALL ESCOBAL_V(r_INIT,r_FINAL,DELTAT,VEL,H,A,Y_RATIO,H_MAG)
191          HINCLINATION=ACOS(H(3))*180.0D0/PI
192          VELMAG=SQRT((VEL(1)**2)+(VEL(2)**2)+(VEL(3)**2))
193          DO J=1,3
194            HM(J)=H(J)*H_MAG
195          END DO
196          WRITE(25,*)"VEL: ",VEL(1),VEL(2),VEL(3)
197          WRITE(25,*)"VEL MAG:",VELMAG
198          WRITE(25,*)"h  : ",H(1),H(2),H(3)
199          WRITE(25,*)"a  :",A
200          WRITE(25,*)"i  :",HINCLINATION,ACOS(H(3))
201          WRITE(25,*)"Y:",Y_RATIO
202          !
203          ! BEGIN TO SET UP DANBY EQ 7.4.1
204          ! USE HERGET EQ 4
205          !
206          ! Derivatives
207          ! FIRST WORK OUT RHO+DELTA & RHO-DELTA
208          ! I DON'T THINK I USED THIS
209          !
210          DO J=1,3
211            DELTARIP=r_INIT(J)+RHO_HAT(J,I_INITIAL)*DELTA
212            DELTARFP=r_FINAL(J)+RHO_HAT(J,I_FINAL)*DELTA
213            DELTARIM=r_INIT(J)-RHO_HAT(J,I_INITIAL)*DELTA
214            DELTARIM=r_FINAL(J)-RHO_HAT(J,I_FINAL)*DELTA
215          END DO
216          WRITE(25,*)"REACHED LINE 189" ! FOR DEBUGGING
217          DO I=I_INITIAL+1,I_FINAL-1
218            QTJ=QTIME(I)-QTIME(I_INITIAL)
219            TN=QTJ/DELTAT
220            TM=1.0D0-TN
221            Y1J(I)=1.0D0+((TN**2)*(Y_RATIO-1))
```

```
222            YJN(I)=1.0D0+((TM**2)*(Y_RATIO-1))
223            CC1(I)=TM*Y_RATIO/YJN(I)
224            CCN(I)=TN*Y_RATIO/Y1J(I)
225            WRITE(25,*)"I,QTJ,TN,TM,Y1J,YJN,CC1,CCN:",I,QTJ,TN,TM,Y1J(I),YJN(I),
                                                                        CC1(I),CCN(I)
226            DO J=1,3
227              SMALL_r(J,I)=(CC1(I)*SMALL_r(J,i_INITIAL))+(CCN(I)*SMALL_r(J,i_FINAL))
228            END DO
229            !
230            ! DANBY'S P AND Q DOT PRODUCTS
231            !
232            P_DNB(I)=0
233            Q_DNB(I)=0
234            DO J=1,3
235              P_DNB(I)=P_DNB(I)+((SMALL_r(J,I)+BIG_R(J,I))*BIG_A(J,I))
236              Q_DNB(I)=Q_DNB(I)+((SMALL_r(J,I)+BIG_R(J,I))*BIG_D(J,I))
237            END DO
238            WRITE(25,*)"P,Q,SMALL_r:",P_DNB(I),Q_DNB(I),SMALL_r(1,I),SMALL_r(2,I),
                                                                        SMALL_r(3,I)
239            !
240            ! NOW CALCULATE THE FOUR PARTIAL DERIVATIVES
241            !
242            DPDRHO1(I)=0
243            DPDRHON(I)=0
244            DQDRHO1(I)=0
245            DQDRHON(I)=0
246            WRITE(25,*)"REACHED LINE 218"
247            DO J=1,3
248              DPDRHO1(I)=DPDRHO1(I)+(RHO_HAT(J,I_INITIAL)*BIG_A(J,I))
249              DPDRHON(I)=DPDRHON(I)+(RHO_HAT(J,I_FINAL)*BIG_A(J,I))
250              DQDRHO1(I)=DQDRHO1(I)+(RHO_HAT(J,I_INITIAL)*BIG_D(J,I))
251              DQDRHON(I)=DQDRHON(I)+(RHO_HAT(J,I_FINAL)*BIG_D(J,I))
252            END DO
253            WRITE(25,*)"DERIVATIVES",DPDRHO1(I),DPDRHON(I),DQDRHO1(I),DQDRHON(I)
254          END DO
255          !
256          ! SET UP DANBY EQUATION 7.4.8.
257          ! I MAY NOT USE DANBY'S NOTATION
258          !
259          INT1=I_INITIAL+1
260          INT2=I_FINAL-1
261          INT3=INT2-INT1+1
262          INT4=2*INT3
263          INT5=I_INITIAL
264            WRITE(25,*)"REACHED LINE 236: INT 1-5",I_INITIAL,I_FINAL,INT1,INT2,INT3,INT4,INT5
265          DO II=1,INT4,2
266            INT5=INT5+1
267            HAY(II,1)=DPDRHO1(INT5)
268            HAY(II,2)=DPDRHON(INT5)
269            HAY(II,3)=-P_DNB(INT5)     ! Plus?
270            HAY(II+1,1)=DQDRHO1(INT5)
271            HAY(II+1,2)=DQDRHON(INT5)
272            HAY(II+1,3)=-Q_DNB(INT5)    ! Plus?
273            WRITE(25,*)"INT5:",INT5
274
275          END DO
276          MMM=2
277          !
278          ! THIS SUBROUTINE IS TO DO THE LEAST SQUARES SOLUTION
279          ! OF AN OVERDETERMINED SYSTEM OF EQUATIONS
280          !
281          CALL LSQFIT(HAY,INT4,MMM,HESS)
282          !
283          ! THE OUTPUT HESS IS THE ARRAY OF CHANGES FOR RHO
284          !
285          DO I=1,MMM
286            WRITE(25,*)"HESS(",I,") = ",HESS(I)
```

```
287         END DO
288         DRHO_INIT=HESS(1)
289         DRHO_FINAL=HESS(2)
290       END DO
291       DO I=1,3
292         SMALL_r(I,I_INITIAL)=RHO_HAT(I,I_INITIAL)*(RHO(I_INITIAL)+DRHO_INIT)
                                                          -BIG_R(I,I_INITIAL)
293         SMALL_r(I,I_FINAL)=RHO_HAT(I,I_FINAL)*(RHO(I_FINAL)+DRHO_FINAL)
                                                          -BIG_R(I,I_FINAL)
294       enddo
295       !
296       WRITE(25,*)"SMALL_r INIT:
     ",SMALL_r(1,I_INITIAL),SMALL_r(2,I_INITIAL),SMALL_r(3,I_INITIAL)
297       WRITE(25,*)"SMALL_r FINAL: ",SMALL_r(1,I_FINAL),
     SMALL_r(2,I_FINAL),SMALL_r(3,I_FINAL)
298       !
299       !CHECK
300       !
301       XLITTLE_R_INIT=SQRT(SMALL_r(1,I_INITIAL)**2+SMALL_r(2,I_INITIAL)**2
                                                          +SMALL_r(3,I_INITIAL)**2)
302       XLITTLE_R_FINAL=SQRT(SMALL_r(1,I_FINAL)**2+SMALL_r(2,I_FINAL)**2
                                                          +SMALL_r(3,I_FINAL)**2)
303       WRITE(25,*)"XLITTLE_R_INIT,XLITTLE_R_FINAL: ",XLITTLE_R_INIT,
                           XLITTLE_R_FINAL, "RATIO:", XLITTLE_R_FINAL/XLITTLE_R_INIT
304       !
305       ! CALCULATE PERIOD
306       !
307       T_PERIOD=2.0D0*PI*(A**(1.5))
308       T_PERIOD_YRS=A**(1.5)
309       !
310       ! Calculate Eccentricity Using my Method (EQ. 6.118)
311       !
312       rx=SMALL_r(1,I_INITIAL)
313       ry=SMALL_r(2,I_INITIAL)
314       rz=SMALL_r(3,I_INITIAL)
315       !
316       EXS=SQRT(1.0D0-((H_MAG**2)/A))
317       !
318       ! THE ASCENDING NODE CALCULATION IS LONG ENOUGH TO JUSTIFY A SUBROUTINE
319       !
320       CALL ASCNODE(HM,OMEGA)    ! ASCENDING NODE
321       WRITE(25,*)"OMEGA:",OMEGA
322       !
323       ! LONGITUDE OF PERIHELION
324       ! FIRST GET TRUE ANOMALY
325       ! THIS LOOKS LIKE OLD BITS OF CODE I DON'T NOW USE
326       !
327       XKOZ_F=((A*(1-(EXS**2))-XLITTLE_R_INIT)/(EXS*XLITTLE_R_INIT))
328       XF_INIT=ACOS(XKOZ_F)
329       XF_INIT_DEG=180.0D0*XF_INIT/PI
330       WRITE(25,*)"XKOZ_F,XF_INIT:",XKOZ_F,XF_INIT,XF_INIT_DEG
331       !
332       ! THIS IS WHERE WE WORK OUT THE LONGITUDE OF PERIHELION SMALL OMEGA
333       ! AND THE TIME OF PERIHELION PASSAGE.
334       !
335       CALL PERIHLONG(rx,ry,rz,VEL,A,EVEC,XLITL_OMEGA,T_PHP)
336       WRITE(25,*)"XLITL_OMEGA:",XLITL_OMEGA,XLITL_OMEGA*180.0D0/PI
337       WRITE(25,*)"EVEC      :",EVEC(1),EVEC(2),EVEC(3)
338       !
339       ! WORK OUT TIME OF PERIHELION PASSAGE IN DAYS
340       !
341       T_PHP_D=ABS(T_PHP)*365.25/(2.0D0*PI)
342       !
343       ! Major outputs go to File 26, the formal output
344       !
345       WRITE(26,603)SMALL_r(1,I_INITIAL),SMALL_r(2,I_INITIAL),SMALL_r(3,I_INITIAL)
346   603 FORMAT(//,"r INITIAL",1P,3(1X,E14.6))
```

```
347          WRITE(26,604)SMALL_r(1,I_FINAL),SMALL_r(2,I_FINAL),SMALL_r(3,I_FINAL)
348    604   FORMAT("r FINAL  ",1P,3(1X,E14.6))
349          WRITE(26,605)XLITTLE_R_INIT,XLITTLE_R_FINAL
350    605   FORMAT("MAGNITUDES: ",3(1X,F8.4))
351          WRITE(26,606)VEL(1),VEL(2),VEL(3),VELMAG
352    606   FORMAT(//,"VELOCITY: X,Y,Z,MAGNITUDE",1P,3(1X,E14.6),1X,0P,"(",F8.4,")")
353          WRITE(26,607)H(1),H(2),H(3)
354    607   FORMAT(/,"ANGULAR MOMENTUM h: ",3(1X,F8.4))
355          WRITE(26,608)A,T_PERIOD,T_PERIOD_YRS
356    608   FORMAT(/,"SEMI-MAJOR AXIS:",9X,F8.4,10X,"(PERIOD:",1X,F8.4," = ",F8.4," YEARS)")
357          WRITE(26,609)HINCLINATION*PI/180.0D0,HINCLINATION
358    609   FORMAT(/,"INCLINATION:",1P,13X,E14.6," rad = ",0P,F8.4,1X,"DEGREES")
359          WRITE(26,610)EXS
360    610   FORMAT(/,"ECCENTRICITY:",12X,F8.4)
361          WRITE(26,611)OMEGA,180.0D0*OMEGA/PI
362    611   FORMAT(/,"ASCENDING NODE OMEGA:",1P,4X,E14.6," rad = ",0P,F8.4,1X,"DEGREES")
363          WRITE(26,612)XLITL_OMEGA,180.0*XLITL_OMEGA/PI
364    612   FORMAT(/,"PERIHELION LONGITUDE:",1P,4X,E14.6," rad = ",0P,F8.4,1X,"DEGREES")
365          WRITE(26,613)T_PHP,T_PHP_D
366    613   FORMAT(/,"PERIHELION PASSAGE TIME:",1P,1X,E14.6," rad = ",0P,F8.4,1X,"DAYS")
367          STOP 900
368          end PROGRAM MAIN
```

Calculate Ascending Node

```
369    !
370          SUBROUTINE ASCNODE(HM,OMEGA)
371          ! SUBROUTINE TO CALCULATE THE ASCENDING NODE
372          ! USE BOULET'S METHOD
373          IMPLICIT DOUBLE PRECISION (A-H,O-Z)
374          IMPLICIT INTEGER(I-N)
375          DOUBLE PRECISION HM(3),XN(3)
376          !
377          PI=4.0D0*atan(1.0D0)    ! define pi
378          XN(1)=-HM(2)
379          XN(2)=HM(1)
380          XN(3)=0.0D0
381          XN_MAG=SQRT((XN(1)**2)+(XN(2)**2)+(XN(3)**2))
382          COSOMEGA=XN(1)/XN_MAG
383          WRITE(25,*)"XN:",XN(1),XN(2),XN(3)
384          WRITE(25,*)"COSOMEGA:",COSOMEGA
385          OMEGA=ACOS(COSOMEGA)
386          IF(XN(2).LT.0.0D0)OMEGA=(2*PI)-OMEGA
387          RETURN
388          END SUBROUTINE ASCNODE
```

Convert Observation Times to Canonical Units

```
389    !
390          SUBROUTINE CONVERT_TIMES(YEER,MUNTH,DAIT,HOWR,MINIT,ICOUNT1,QTIME)
391          !
392          ! Subroutine to get common calendar dates into Canonical time units
393          !
394          IMPLICIT DOUBLE PRECISION (A-H,O-Z)
395          IMPLICIT INTEGER(I-N)
396          INTEGER YEER(100),MUNTH(100),DAIT(100),HOWR(100),MINIT(100),
                                                      DAYSINMONTH(12),DAYS
397          DOUBLE PRECISION QTIME(100),FRACDAY(100)
398          PI=4.0D0*ATAN(1.0D0)
399          DAYSPERYR=365.25
400          !
```

```
401       ! EARTH'S PERIHELION
402       !
403       PERIHELION=DAYSPERYR+3+(7.0D0/24.0D0)+(12.0D0/(24.0D0*60.0D0))
          ! 3 JAN 2010 @ 07:12
404       !
405       DAYSINMONTH(1)=31
406       DAYSINMONTH(2)=28
407       DAYSINMONTH(3)=31
408       DAYSINMONTH(4)=30
409       DAYSINMONTH(5)=31
410       DAYSINMONTH(6)=30
411       DAYSINMONTH(7)=31
412       DAYSINMONTH(8)=31
413       DAYSINMONTH(9)=30
414       DAYSINMONTH(10)=31
415       DAYSINMONTH(11)=30
416       DAYSINMONTH(12)=31
417       WRITE(25,*)"PI,DAYSPERYR,PERIHELION,ICOUNT1",PI,DAYSPERYR,PERIHELION
                                                                   ,ICOUNT1
418       DO K =1,ICOUNT1
419         !WRITE(25,*)"K",K
420         !
421         ! COUNT YEARS
422         !
423         YRS=-1.0D0
424         IF(YEER(K).EQ.2010)YRS=DAYSPERYR
425         IF(YEER(K).EQ.2009)YRS=0.0D0
426         !WRITE(25,*)"YRS",YRS
427         IF(YRS.LT.0.0D0) STOP 903  ! TRAP TO
            CATCH WRONG YEAR ERRORS
428         DAYS=0
429         IF (MUNTH(K).GT.2) THEN
430           DO J=1,MUNTH(K)-1
431             DAYS=DAYS+DAYSINMONTH(J)
432           END DO
433         ELSE IF (MUNTH(K).EQ.2) THEN
434           DAYS=DAYSINMONTH(1)
435         ELSE
436           DAYS=0
437         END IF
438         !WRITE(25,*)"DAYS",K,DAYS
439         DAYS=DAYS+DAIT(K)
440         RDAYS=DAYS
441         !WRITE(25,*)"RDAYS",K,RDAYS
442         HR=HOWR(K)
443         XMIN=MINIT(K)
444         FRACDAY(K)=(HR/24.0D0)+(XMIN/(24.0D0*60.0D0))
445         !WRITE(25,*)"HR,XMIN,FRACDAY",HR,XMIN,FRACDAY(K)
446         QTIME(K)=YRS+RDAYS+FRACDAYS        ! RELATIVE TO 2009 01 01 00:00
447         !WRITE(25,*)"QTIME(K)",QTIME(K)
448         QTIME(K)=QTIME(K)-PERIHELION      ! RELATIVE TO PERIHELION
449         !WRITE(25,*)"QTIME(K)",QTIME(K)
450         QTIME(K)=QTIME(K)*2*PI/DAYSPERYR  ! CONVERT TO CANONICAL UNITS
451         !WRITE(25,*)"QTIME(K)",QTIME(K)
452       END DO
453       RETURN
454       END SUBROUTINE CONVERT_TIMES
455     !
```

Use Escobal's Methods to Calculate Various Quantities

```
456       SUBROUTINE ESCOBAL_V(RADI,RADJ,DELTAT,VELI,H,A,Y_RATIO,H_MAG)
457     ! ESCOBAL'S METHOD FOR OBTAINING A VELOCITY, SEMIMAJOR AXIS AND A
                                                               NGULAR MOMENTUM
458     ! MANY WRITE STATEMETS ADDED FOR DEBUGGING, NOW COMMENTED
```

```
459        IMPLICIT DOUBLE PRECISION (A-H,O-Z)
460        IMPLICIT INTEGER(I-N)
461        DOUBLE PRECISION RADI(3),RADJ(3),VELI(3),H(3)
462        PI=4.0D0*ATAN(1.0D0)
463        !
464        ! WORK OUT MAGNITUDES OF ri & rj
465        !
466        RI_DOT_RJ=0.0D0
467        XMAG_SQ_I=0
468        XMAG_SQ_J=0
469        DO K=1,3
470          RI_DOT_RJ=RI_DOT_RJ+RADI(K)*RADJ(K)
471          XMAG_SQ_I=XMAG_SQ_I+RADI(K)*RADI(K)
472          XMAG_SQ_J=XMAG_SQ_J+RADJ(K)*RADJ(K)
473        END DO
474        r_I=SQRT(XMAG_SQ_I)
475        r_J=SQRT(XMAG_SQ_J)
476        ROOTRIRJ=SQRT(r_I*r_J)
477        WRITE(25,*)"ROOTRIRJ:",ROOTRIRJ
478        XMU=1.0D0 ! Canonical Units
479        COZ_DELTAF=(RI_DOT_RJ/(r_I*r_J))
480        SIN_DELTAF=SQRT(1.0D0-(COZ_DELTAF**2))
481        TAN_DELTAF=SIN_DELTAF/COZ_DELTAF
482    !    CHECK_1=COZ_DELTAF**2+SIN_DELTAF**2
483    !    WRITE(25,*)"CHECK_1: ",CHECK_1
484        DELTAF=ATAN(TAN_DELTAF)    ! N.B. FOR ANGLES FROM 0 TO PI, ACOS IS UNIQUE.  NOT SO FOR ASIN.
485        DEG_DELTAF=180.0D0*DELTAF/PI
486        WRITE(25,*)"DELTAF: ",DELTAF,DEG_DELTAF
487        ! Initially set Y=1
488        !
489        Y_RATIO=1.0D0
490        Y_RATIO_OLD=Y_RATIO-1.0D0
491        DO WHILE (ABS((Y_RATIO_OLD-Y_RATIO)/ABS(Y_RATIO)).GT.1.0D-12)
492          !
493          ! Escobal's constants l (L.C. L) and m
494          !
495          XM=(DELTAT**2)/((2.0D0*ROOTRIRJ*COS(DELTAF/2.0D0))**3)
496          XL=((r_I+r_J)/(4.0D0*ROOTRIRJ*COS(DELTAF/2.0D0)))-0.5D0
497          X_ESCOBAL=(XM/(Y_RATIO**2))-XL
498          WRITE(25,*)"X_ESCOBAL:",X_ESCOBAL
499          COZ_DELTAEO2=1.0D0-(2.0D0*X_ESCOBAL)
500          SIN_DELTAEO2=SQRT(4.0D0*X_ESCOBAL*(1.0D0-X_ESCOBAL))
501          TAN_DELTAEO2=SIN_DELTAEO2/COZ_DELTAEO2
502          DELTAE=2.0D0*ATAN(TAN_DELTAEO2)
503          write(25,*)"TAN_DELTAEO2,DELTAE:",TAN_DELTAEO2,DELTAE
504          BIGX_ESCOBAL=((DELTAE-SIN(DELTAE))/(SIN(DELTAE/2)**3))
505          Y_RATIO_OLD=Y_RATIO
506          Y_RATIO=1.0D0+(BIGX_ESCOBAL*(XL+X_ESCOBAL))
507    !      WRITE(25,*)"Y_RATIO: ",Y_RATIO
508        END DO
509    !    WRITE(25,*)"ESCOBAL'S L & M: ",XL,XM
510    !    write(25,*)"x, BIGX_ESCOBAL: ",X_ESCOBAL, BIGX_ESCOBAL
511        ROOTA=(DELTAT*SQRT(XMU))/(2.0D0*Y_RATIO*ROOTRIRJ*COS(DELTAF/2.0D0)
                                                         *SIN(DELTAE/2.0D0))
512        A=ROOTA*ROOTA
513        F_LAG=1.0D0-((A/r_I)*(1.0D0-COS(DELTAE)))
514        G_LAG=DELTAT-((ROOTA**3/SQRT(XMU))*(DELTAE-SIN(DELTAE)))
515        WRITE(25,*)"ROOTRIRJ, ROOTA, a: ",ROOTRIRJ,ROOTA, A
516        WRITE(25,*)"COZ_DELTAE: ",COZ_DELTAEO2
517        WRITE(25,*)"DELTAE, F_LAG,G_LAG: ",DELTAE,F_LAG,G_LAG
518        !
519        ! WORK OUT VELOCITY
520        !
521        DO K=1,3
522          VELI(K)=(RADJ(K)-F_LAG*RADI(K))/G_LAG
523        END DO
524        !
525        ! Work out vector h  = r x v as a vector cross product & make it a unit vector
526        !
```

```
527        H(1)=RADI(2)*VELI(3)-RADI(3)*VELI(2)
528        H(2)=-RADI(1)*VELI(3)+RADI(3)*VELI(1)
529        H(3)=RADI(1)*VELI(2)-RADI(2)*VELI(1)
530        H_MAG=SQRT(H(1)**2+H(2)**2+H(3)**2)
531        H(1)=H(1)/H_MAG
532        H(2)=H(2)/H_MAG
533        H(3)=H(3)/H_MAG
534        WRITE(25,*)"H_MAG:",H_MAG
535        RETURN
536        END SUBROUTINE ESCOBAL_V
537        !
```

Use Methods Given Above to Calculate the Longitude and Passage Time of Perihelion

```
538        SUBROUTINE PERIHLONG(rx,ry,rz,VEL,A,EVEC,XLITL_OMEGA,TPH)
539        !
540        ! TO CALCULATE THE ECCENTRICITY VECTOR OF HAMILTON,
541        ! AND TIME OF PERIHELION PASSAGE
542        ! USED MY EQUATIONS (6.118)-(6.152)
543        !
544        IMPLICIT DOUBLE PRECISION (A-H,O-Z)
545        IMPLICIT INTEGER(I-N)
546        DOUBLE PRECISION
XLITL_R(3),H(3),VEL(3),EVEC(3),VXH(3),r_HAT(3),XN(3),XN_HAT(3),E_HAT(3),EXN(3)
547        !
548        PI=4.4D0*ATAN(1.0D0)
549        !
550        ! GATHER UP COMPONENTS rx, ry & rz INTO A VECTOR
551        !
552        XLITL_R(1)=rx
553        XLITL_R(2)=ry
554        XLITL_R(3)=rz
555        !
556        ! GET H
557        !
558        H(1)=(XLITL_R(2)*VEL(3))-(XLITL_R(3)*VEL(2))
559        H(2)=(XLITL_R(3)*VEL(1))-(XLITL_R(1)*VEL(3))
560        H(3)=(XLITL_R(1)*VEL(2))-(XLITL_R(2)*VEL(1))
561        !
562        ! MY EQ.(6.136)
563        ! GET V CROSS H
564        !
565        VXH(1)=(VEL(2)*H(3))-(VEL(3)*H(2))
566        VXH(2)=(VEL(3)*H(1))-(VEL(1)*H(3))
567        VXH(3)=(VEL(1)*H(2))-(VEL(2)*H(1))
568        !
569        !NOW GET UNIT RADIUS VECTOR
570        !
571        RMAG=0.0D0
572        DO I=1,3
573          RMAG=RMAG+(XLITL_R(I)*XLITL_R(I))
574        END DO
575        DO I=1,3
576          r_HAT(I)=XLITL_R(I)/SQRT(RMAG)
577        END DO
578        WRITE(25,*)"r_HAT:",r_HAT(1),r_HAT(2),r_HAT(3)
579        WRITE(25,*)"VEL   :",VEL(1),VEL(2),VEL(3)
580        WRITE(25,*)"H     :",H(1),H(2),H(3)
581        WRITE(25,*)"VXH   :",VXH(1),VXH(2),VXH(3)
582        !
583        ! COMBINE TERMS
584        !
```

```
585          DO I=1,3
586            EVEC(I)=VXH(I)-r_HAT(I)
587          END DO
588          !
589          ! NOW WORK OUT E_HAT
590          !
591          EMAG=0.0D0
592           DO I=1,3
593             EMAG=EMAG+(EVEC(I)*EVEC(I))
594          END DO
595          EVALUE=SQRT(EMAG)
596          WRITE(25,*)"EVEC :",EVEC(1),EVEC(2),EVEC(3)
597          WRITE(25,*)"EMAG :",EMAG
598          WRITE(25,*)"EVALUE:",EVALUE
599          DO I=1,3
600            E_HAT(I)=EVEC(I)/SQRT(EMAG)
601          END DO
602          WRITE(25,*)"E_HAT:",E_HAT(1),E_HAT(2),E_HAT(3)
603          !
604          ! NOW WORK OUT N HAT FROM MY EQS (6.119) & (6.144)
605          !
606          XN(1)=-H(2)
607          XN(2)=H(1)
608          XN(3)=0.0D0
609          XNMAG=0.0D0
610          DO I=1,3
611            XNMAG=XNMAG+(XN(I)*XN(I))
612          END DO
613          DO I=1,3
614            XN_HAT(I)=XN(I)/SQRT(XNMAG)
615          END DO
616          WRITE(25,*)"XN   :",XN(1),XN(2),XN(3)
617          WRITE(25,*)"XNMAG:",XNMAG
618          WRITE(25,*)"XN_HT:",XN_HAT(1),XN_HAT(2),XN_HAT(3)
619          !
620          ! DOT PRODUCT E_HAT DOT N_HAT
621          !
622          COSOMEGA=0.0D0
623          DO I=I,3
624            COSOMEGA=COSOMEGA+(E_HAT(I)*XN_HAT(I))
625          END DO
626          !
627          ! TEST FOR WHETHER XLITL_OMEGA IS GREATER THAN PI
628          ! CROSS PRODUCT E_HAT CROSS N_HAT
629          !
630          EXN(1)=(E_HAT(2)*XN_HAT(3))-(E_HAT(3)*XN_HAT(2))
631          EXN(2)=(E_HAT(3)*XN_HAT(1))-(E_HAT(1)*XN_HAT(3))
632          EXN(3)=(E_HAT(1)*XN_HAT(2))-(E_HAT(2)*XN_HAT(1))
633          !
634          IF(EXN(3)*H(3).LT.0.0D0)THEN
635            XLITL_OMEGA=ACOS(COSOMEGA)
636          ELSE
637            XLITL_OMEGA=(2.0D0*PI)-ACOS(COSOMEGA)
638          END IF
639          !
640          ! MOVE ON TO TIME OF PERIHELION PASSAGE
641          !
642          ! TRUE ANOMALY
643          !
644          EDOTR=0.0D0
645          DO I=1,3
646            EDOTR=EDOTR+(E_HAT(I)*r_HAT(I))
647          END DO
648          EFF=ACOS(EDOTR)
649          WRITE(25,*)"EDOTR:",EDOTR
650          WRITE(25,*)"EFF  :",EFF
651          !
```

```
652       ! ECCENTRIC ANOMALY
653       !
654       EAN=2.0D0*ATAN((SQRT((1.0D0-EVALUE)/(1.0D0+EVALUE)))*TAN(EFF/2.0D0))
655       WRITE(25,*)"EAN   :",EAN
656       !
657       ! MEAN ANOMALY
658       !
659       XMA=EAN-(EVALUE*SIN(EAN))
660       WRITE(25,*)"XMA   :",XMA
661       TPH=-XMA*(A**1.5)
662       WRITE(25,*)"TPH   :",TPH
663       RETURN
664       END SUBROUTINE PERIHLONG
665       !
```

This Is Basically the Subroutine Described in Chap. 4

```
666   !
667         SUBROUTINE LSQFIT (A,N,M,S)
668   !-------------------------------------------------------------------------
669   !    LSQFIT:
670   !    solution of an overdetermined system of linear equations
671   !    A(i,1)*s(1)+...A(i,m)*s(m) - A(i,m+1) = 0    (i=1,..,n)
672   !    according to the method of least squares using Givens rotations
673   !    A: matrix of coefficients
674   !    N: number of equations   (rows of A)
675   !    M: number of unknowns    (M+1=columns of A, M=elements of S)
676   !    S: solution vector
677   !    Partial Credit: Jeannette M. Fine
678   !-------------------------------------------------------------------------
679         INTEGER I,J,K,M,N
680         double precision A(100,8),P,Q,H,H2,H3,S(M),EPS
681
682   EPS = 1.0D-10
683     DO J=1,M ! loop over columns 1...M of A
684       ! eliminate matrix elements A(i,j) with i>j from column j
685       DO I=J+1,N
686         IF (A(I,J).NE.0.0D0)THEN
687             ! calculate p, q and new A(j,j); set A(i,j)=0
688             IF (ABS(A(J,J)).LT.EPS*ABS(A(I,J))) THEN
689                 P=0.0D0
690                 Q=1.0D0
691                 A(J,J)=-A(I,J)
692                 A(I,J)=0.0D0
693             ELSE
694                 H=SQRT(A(J,J)*A(J,J)+A(I,J)*A(I,J))
695                 IF (A(J,J).LT.0.0D0)  H=-H
696                 P=A(J,J)/H
697                 Q=-A(I,J)/H
698                 A(J,J)=H
699                 A(I,J)=0.0D0
700             END IF
701             ! calculate rest of the line
702             DO K=J+1,M+1
703                 H2    = P*A(J,K) - Q*A(I,K)
704                 A(I,K) = Q*A(J,K) + P*A(I,K)
705                 A(J,K) = H2
706             END DO
707         END IF
708       END DO
709     END DO
710     ! backsubstitution
711
712     DO I = M,1,-1
713       H3=A(I,M+1)
```

```
714        DO K=I+1,M
715           H3=H3-A(I,K)*S(K)
716        END DO
717        S(I) = H3/A(I,I)
718     END DO
719  !
720     RETURN
721     END SUBROUTINE LSQFIT
```

Eureka! (Part Two)

The program listed above has to be used judiciously. You can't just put in any old input and expect it to work. Danby [89] makes no bones about this. He encourages people to play around with the method because it won't always converge to a sensible answer. I found it to be really quite ticklish.

You have to guess two initial radii for the Sun–Mars distance at given times, and let the program guide you to better answers. Basically, I tried lots of values until I found that the guesses I input and the values it output were very similar.

Once that had happened, I claimed this as my preliminary orbit. Here is the output for the initial values in the listing above (Table 7.2).

The input data were those of Table 6.4.

The perihelion time translates to 6 March 2009 at 03:26 UT.

These results are pretty much what the label says: the elements of a preliminary orbit. The next stage is to refine them.

Table 7.2 Results output from my program for Herget's method

HERGET'S METHOD FOR PRELIMINARY ORBIT DETERMINATION
YEARS 2009–2010
IMPOSED INITIAL & FINAL RADII 1.6200 1.6100 INITIAL & FINAL
   OBSERVATIONS 15 30
r INITIAL 7.033714E-01 -1.469551E+00 4.683596E-02
r FINAL 1.454420E+00 -7.428639E-01 4.819375E-02
MAGNITUDES: 1.6299 1.6339
VELOCITY: X,Y,Z,MAGNITUDE 6.916611E-01 3.078046E-01 8.908687E-03
   (0.7571)
ANGULAR MOMENTUM h: −0.0223 0.0212 0.9995
SEMI-MAJOR AXIS: 1.5294 (PERIOD: 11.8839 = 1.8914 YEARS)
INCLINATION: 3.076228E-02 rad = 1.7625 DEGREES
ECCENTRICITY: 0.0714
ASCENDING NODE OMEGA: 3.952702E+00 rad = 226.4731 DEGREES
PERIHELION LONGITUDE: 5.340708E+00 rad = 306.0000 DEGREES
PERIHELION PASSAGE TIME: −5.128313E+00 rad = 298.1157 DAYS

Chapter 8

Refining the Preliminary Orbit

I have now told you my worst mathematical horror stories. This chapter will be much less mathematically burdensome than the previous one.

Least-Squares Curve Fitting

The game is to find a way to get a least-squares fit to my observations, having seeded them with the results in Table 7.2. What was output becomes input.

The actual least-squares fit will be done using the method of Chap. 4, due to Givens, and ultimately deriving from the ideas of Gauss.

First, however, I need to work out exactly what I am applying the technique to.

I know from Chap. 7 that knowing the position and velocity of a planet relative to the Sun at any one time tells me enough to deduce its entire orbit. (If you did not read that chapter, please just take my word for it.) So we refine the values of the initial radius and velocity reported in Table 7.2. We don't *directly* refine the elements deduced. Rather, I got the best values I could of $\mathbf{r}$ and $\dot{\mathbf{r}}$ and calculated the orbital elements afresh.

I have a set of measurements of λ and β, the celestial longitudes and latitudes, at my observation times. These are listed in Table 6.4.

My tried and trusted references Danby [89], Escobal [128] and Boulet [113] all offer much the same way to skin this particular cat. In what follows, I am largely following the equations given in Boulet [113], §§12.1–3, although I did not use or adapt his computer program. To be honest, I found it easier to write my own than try to understand his. No doubt there are some similarities, because, ultimately, we are solving the same equations.

The method is to vary $\mathbf{r}$ and $\dot{\mathbf{r}}$ for one observation, calculate a new set of orbital elements and predict the observations I would have made, had the orbit actually had this set of elements.

So, at the time of my chosen ith observation, I have

$$\mathbf{r}_i = (x_i, y_i, z_i), \text{ and}$$
$$\dot{\mathbf{r}}_i = (\dot{x}_i, \dot{y}_i, \dot{z}_i) . \tag{8.1}$$

If I were to vary these values and re-calculate the orbital elements, at every jth observation, my *prediction* of the celestial latitude and longitude will depend on $\mathbf{r}$ and $\dot{\mathbf{r}}$:

$$\lambda_j = \lambda_j \left(x_i, y_i, z_i, \dot{x}_i, \dot{y}_i, \dot{z}_i \right) ,$$
$$\beta_j = \beta_j \left(x_i, y_i, z_i, \dot{x}_i, \dot{y}_i, \dot{z}_i \right) . \tag{8.2}$$

In practice, my aim is to hope that a finite change in $\mathbf{r}$ and $\dot{\mathbf{r}}$ will be small enough for Eq. (8.3) to be valid for all i and j.

$$\lambda_j \simeq \frac{\partial \lambda_j}{\partial x_i} \ x_i + \frac{\partial \lambda_j}{\partial y_i} \ y_i + \frac{\partial \lambda_j}{\partial z_i} \ z_i + \frac{\partial \lambda_j}{\partial \dot{x}_i} \ \dot{x}_i + \frac{\partial \lambda_j}{\partial \dot{y}_i} \ \dot{y}_i + \frac{\partial \lambda_j}{\partial \dot{z}_i} \ \dot{z}_i \ ;$$
$$\beta_j \simeq \frac{\partial \beta_j}{\partial x_i} \ x_i + \frac{\partial \beta_j}{\partial y_i} \ y_i + \frac{\partial \beta_j}{\partial z_i} \ z_i + \frac{\partial \beta_j}{\partial \dot{x}_i} \ \dot{x}_i + \frac{\partial \beta_j}{\partial \dot{y}_i} \ \dot{y}_i + \frac{\partial \beta_j}{\partial \dot{z}_i} \ \dot{z}_i \ . \tag{8.3}$$

Equation (8.3) is in fact a set of overdetermined equations, in which I have only one value of i but as many values of j as I have observations. I have more than six equations to solve for six variables x_i, y_i, z_i, $\dot{x}_i$, $\dot{y}_i$, and $\dot{z}_i$. By now you may have gotten the idea that I will do this using Givens' method set out in Chap. 4.

Before doing that, however, I need to work out how to deal with the derivatives in Eq. (8.3). I will show you how I do that with one example, the derivative of λ_j with respect to z_i:

$$\frac{\partial \lambda_j}{\partial z_i} = \frac{\lambda_j \left(x_i, y_i, z_i + \Delta z_i, \dot{x}_i, \dot{y}_i, \dot{z}_i \right) - \lambda_j \left(x_i, y_i, z_i, \dot{x}_i, \dot{y}_i, \dot{z}_i \right)}{\Delta z_i} . \tag{8.4}$$

So I have to make my computer program work out the consequences for every case like Eq. (8.4). I need to work out six perturbed cases of each of λ_j and β_j and do this for every value of j. I also need to work out a seventh case: the unperturbed one.

There's one thing I need to tell you about numerical differentiation: it is a dangerous game. Small fluctuations in the quantities you are differentiating can lead to enormous fluctuations in their derivatives. It is what mathematicians warn their teenage children about before they go out at night. Not Bigfoot. Not ghosts. Not spiders. No: it's numerical differentiation.

Consider the simple example shown in Fig. 8.1. The point that numerical noise realty influences the values of derivatives is succinctly made.

In my case, I found that I had to tune the values of my x_i, y_i, z_i, $\dot{x}_i$, $\dot{y}_i$, $\dot{z}_i$ very carefully. On the one hand, I want them to be as small as possible because they are supposed to be emulating infinitesimal quantities. On the other hand, the bigger they are, the more they smooth out random noise and the more chance I have of getting a halfway decent

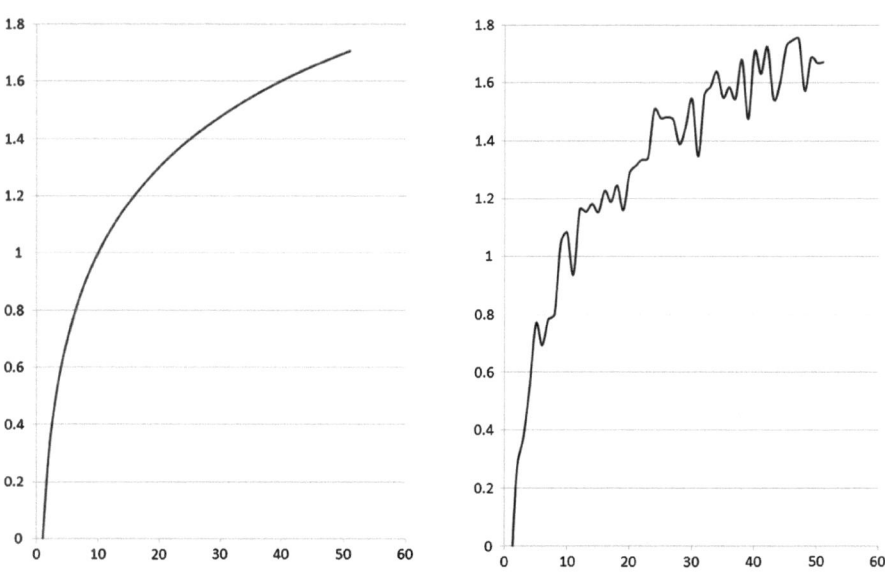

Fig. 8.1 Consider the two curves shown. They are very similar, except that random noise has been added to the one on the right. Were you to integrate the curves to get the area under them, there would not be much difference between the values you obtain. (I tried it: the difference is well under 1%.) But if you were to differentiate them, the one on the left would give predictable, well-behaved values. The one on the right would give wildly fluctuating values with no rhyme or reason to them

Table 8.1 Results of a least-squares fit from my orbit improvement program

MARS ORBIT REFINEMENT 2009 2010
Orbital Elements
$a = 1.52386172$
$i = 1.81461914$
$e = 0.12735710$
Omega $= 50.29953495$
w_PERIH $= 306.00000000$
t_PERIH $= -3.56790853$

result. For derivatives with respect to x_i, y_i, $\dot{x}_i$, $\dot{y}_i$ I settled on a Δ value of 9×10^{-3}. For derivatives with respect to z_i and $\dot{z}_i$ I settled on a Δ value of 9×10^{-4}.

Once I have this information, I can assemble Eq. (8.3) and feed them to my least-squares fitting routine. That's what my computer program does.

My results after programming this method up in Fortran are given in Table 8.1.

These results are not very convincing. It is not even worth putting them on a graph to compare with the 2009–10 observations, because the predicted epicycle is not at all close to the observed one.

Therefore, I am also not going to bother listing the Fortran program. It won't help you.

Another heroic failure. This method was supposed to work, but it doesn't. Now what?

Gambling on Monte Carlo

The essence of Monte Carlo methods is simple. You randomly vary the variables for which you are trying to solve and run a very large number of trial cases on a computer. You work out the square of the error, fitted versus predicted, values of λ and β. The quantity I actually minimized was

$$S_{\text{umsq}} = \sum_{\text{last}}^{i=\text{first}} \left(\lambda_i^2 + \beta_i^2 \right). \tag{8.5}$$

I have done this in the past using Microsoft Excel. That is OK for a few thousand runs, but if you want to run millions of cases, you need something faster.

The "space" we need to explore is six-dimensional (6-D), with a dimension for each of the six orbital elements a, e, i, Ω, ω and T_{Per}. To explore a large region of this 6-D "space" would take a huge number of runs, billions, if not trillions. I went as far as

$$2^{27} = 134,217,728 \tag{8.6}$$

trials per set of parameters before I ran out of patience and gave up. It was convenient to scale the number of runs in powers of two while trying things out, but there is no reason why you have to do this. Overall, 2^{27} cases took about five hours on my home PC. I could do better if I were to use a computer with a few dozen processors. You can buy them for a few thousand dollars, but that's really outside the reach of what it's reasonable to expect an amateur astronomer to do.

Monte Carlo methods are slow to converge. There comes a point at which you can do better by manual trial and error after the Monte Carlo method has given you some clues.

You can also hold some of the orbital elements constant and explore the rest of the elements.

At various times I did all of these things. I could have done this via a data input file, but the program compiles and links in the blink of an eye, so I did it in the Fortran code.

Monte Carlo Source Code

```
 1 Program main
 2 !********************************************************************************
 3 ! Initiated 2020 12 19 by Jane Clark (c)
 4 ! Uses Monte Carlo method to propose refinement of a Keplerian elliptical orbit
 5 ! Given a Preliminary Orbit, which has been deduced elsewhere
 6 ! Start with 2**20 = 1,048,576 trials
 7 ! Version 1.0 Origination
 8 !********************************************************************************
 9 !
10 ! TIME, EARTH ORBIT DATA AND OBSERVATIONS
11 ! FILE25 IS FOR DEBUGGING, FILE 26 IS THE DESIRED OUTPUT
12 !
13 IMPLICIT DOUBLE PRECISION (A-H,O-Z)
14 IMPLICIT INTEGER(I-N)
15 INTEGER YEER(100),MUNTH(100),DAIT(100),HOWR(100),MINIT(100)
16 INTEGER YEER2(100),MUNTH2(100),DAIT2(100),HOWR2(100),MINIT2(100)
17 DOUBLE PRECISION XEARTH(100),YEARTH(100),ZEARTH(100),QTIME(100)
18 DOUBLE PRECISION RLAMBDA(100),RBETA(100),RHO(100),CC1(100),CCN(100),
Y1J(100),YJN(100)
19 DOUBLE PRECISION RLAMBDA2(100),RBETA2(100),RDSQ(100),RHOX(100),
RHOY(100),RHOZ(100)
20 DOUBLE PRECISION XLAMBDA(11),E(11),TRU_ANOM(100),T_OBS(100),M_ANOM_OBS(100)
21 DOUBLE PRECISION RR(100),XX(100),YY(100),ZZ(100),COL1(3),COLRAD(3),COL2(3)
22 DOUBLE PRECISION ROT_W_PERIH(3,3),ROT_OMEGA(3,3),ROT_XI(3,3),
```

```
XMPROD1(3,3),XMPROD2(3,3)
 23
 24 !
 25 ! TIME, EARTH ORBIT DATA AND OBSERVATIONS
 26 ! FILE25 IS FOR DEBUGGING, FILE 26 IS THE DESIRED OUTPUT
 27 !
 28 OPEN(UNIT=15,FILE='Earth_Data_09_10_11_12_20.csv')
 29 OPEN(UNIT=25,FILE='op_checks.txt')
 30 OPEN(UNIT=26,FILE='Outputs.txt')
 31 !
 32 PI=4.0D0*atan(1.0D0) ! define pi
 33 VSMALL=1.0D-9 ! tiny number for later use in coordinate transforms
 34 !
 35 ICOUNT1=0
 36 DO I=1,100
 37 read(15,*)YEER(I),MUNTH(I),DAIT(I),HOWR(I),MINIT(I),XEARTH(I),
YEARTH(I),ZEARTH(I),RLAMBDA(I),RBETA(I)
 38 write(25,*)I,YEER(I),MUNTH(I),DAIT(I),HOWR(I),MINIT(I),XEARTH(I),
YEARTH(I),ZEARTH(I),RLAMBDA(I),RBETA(I)
 39 IF (YEER(I).EQ.0)GO TO 400
 40 ICOUNT1=ICOUNT1+1
 41 END DO
 42 400 CONTINUE
 43 !
 44 ! Fix an Error: reverse XEARTH & YEARTH data
 45 !
 46 DO I=1,ICOUNT
 47 XEARTH(I)=-XEARTH(I)
 48 YEARTH(I)=-YEARTH(I)
 49 ZEARTH(I)=-ZEARTH(I)
 50 END DO
 51 WRITE(25,*)"ICOUNT1",ICOUNT1
 52 CALL CONVERT_TIMES(YEER,MUNTH,DAIT,HOWR,MINIT,ICOUNT1,QTIME)
 53 DO I=1,ICOUNT1
 54 WRITE(25,*)"QTIME ",I,QTIME(I)
 55 END DO
 56 !
 57 ! Elements from Preliminary orbit
 58 ! FINAL ZERO IN NAME IMPLIES THAT THESE ARE STARTING VALUES
 59 !
 60 A0=1.524125D0 ! SEMI-MAJOR AXIS
 61 XI0=3.3835075548E-002 ! INCLINATION, RADIANS
 62 ECC0=0.09458634D0 ! ECCENTRICITY
 63 OMEGA0=228.60068224D0*PI/180.0D0 ! ASCENDING NOTE WRT COODINATE SYSTEM
 64 W_PERIH0=300.28319723*PI/180.0D0 ! ARGUMENT OF PERIHELION
 65 !
 66 ! FINAL ARGUMENT, TIME OF PERIHELION, T_PERIHEL, MUST BE CALCULATED,
AS IT DEPENDS ON A
 67 !
 68 PERIOD=2*PI*A0**1.5
 69 !
 70 ! TIME OF APHELION PASSAGE IS ROUGHLY KNOWN FOR MARS, 2010
 71 !
 72 T_PERIHEL0=-5.61542634D+00
 73 !
 74 WRITE(25,*)"A0,XI0,ECC0,OMEGA0,W_PERIH0,T_PERIHEL0:",A0,XI0,ECC0,
OMEGA0,W_PERIH0,T_PERIHEL0
 75 ! PREPARE MONTE CARLO LOOPING
 76 !
 77 !
 78 ! Stored values at minimum least squares
 79 !
 80 A_S=A0 ! SEMI-MAJOR AXIS
 81 XI_S=XI0 ! INCLINATION, RADIANS
 82 ECC_S=ECC0 ! ECCENTRICITY
 83 OMEGA_S=OMEGA0 ! ASCENDING NOTE WRT COODINATE SYSTEM
 84 W_PERIH_S=W_PERIH0 ! ARGUMENT OF PERIHELION
```

```
85 T_PERIHEL_S=T_PERIHEL0 ! TIME OF PERIHELION PASSAGE
86 !
87 SUMSQ=1.0D9 ! INTITIAL VALUE
88 !
89 ! MONTE CARLO LOOPING
90 !
91 !
92 JTOT=2**27 ! THE NUMBER OF INSTANCES EXPLORED. I PLAYED AROUND
WITH THIS NUMBER
93 !
94 DO JLOOP=1,JTOT
95 !
96 ! RANDOMISE I SPENT A LOT OF TIME PLAYING AROUND WITH THESE VARIABLES.
97 !
98 A=A0*(1.0D0+(0.04D0*RAND())-0.02D0) ! 2% VARIATION
99 !A=A0 ! 0% VARIATION
100 XI=XI0*(1.0D0+(0.2D0*RAND())-0.1D0) ! 10% VARIATION
101 !XI=XI0 ! 0% VARIATION
102 !ECC=ECC0*(1.0D0+(0.4D0*RAND())-0.2D0) ! 20% VARIATION
103 ECC=ECC0 ! 0% VARIATION
104 OMEGA=OMEGA0-0.1D0+(0.20*RAND()) ! Â±0.1rad
105 !OMEGA=OMEGA0 ! 0% VARIATION
106 W_PERIH=W_PERIH0-0.1D0+(0.20*RAND()) ! Â±0.1rad
107 T_PERIHEL=T_PERIHEL0*(1.0D0+(0.40D0*RAND())-0.20D0) ! 20% VARIATION
108 !T_PERIHEL=T_PERIHEL0 ! 0% VARIATION
109 !
110 !WRITE(25,*)"A,XI,ECC,OMEGA,W_PERIH,T_PERIHEL:",A,XI,ECC,OMEGA,W_PERIH,T_PERIHEL
111 !
112 ! GET OBSERVATION TIMES & MEAN ANOMALIES IN TERMS OF PERIHELION PASSAGE TIME
113 !
114 DO K=1,ICOUNT1
115 T_OBS(K)=QTIME(K)-T_PERIHEL
116 M_ANOM_OBS(K)=T_OBS(K)*(2*PI)/PERIOD
117 !WRITE(25,*)"QTIME(K),T_OBS(K),M_ANOM_OBS(K)",K,QTIME(K),T_OBS(K),M_ANOM_OBS(K)
118 END DO
119 !
120 ! SOLVE KEPLER'S EQUATION AT OBSERVATION TIMES
121 !
122 SUMSQ2=0.0D0
123 DO K=1,ICOUNT1
124 XLAMBDA(1)=M_ANOM_OBS(K)
125 DO WHILE (M_ANOM_OBS(K).GT.2.0D0*PI)
126 M_ANOM_OBS(K)=M_ANOM_OBS(K)-2.0D0*PI
127 END DO
128 E(1)=M_ANOM_OBS(K)
129 !WRITE(25,*)"E(1):",E(1)
130 DO L=2,6
131 E(L)=E(L-1)-((E(L-1)-(ECC*SIN(E(L-1))-M_ANOM_OBS(K)))/
(1.0D0-(ECC*COS(E(L-1)))))
132 !WRITE(25,*)"E(L-1),:",L,E(L-1)
133 END DO
134 TRU_ANOM(K)=2*ATAN(SQRT((1.0D0+ECC)/(1.0D0-ECC))*TAN(E(6)/2))
135 IF(TRU_ANOM(K).LT.0.0D0)TRU_ANOM(K)=TRU_ANOM(K)+(2.0D0*PI)
136 !WRITE(25,*)"TRU_ANOM(K),ECC,:",K,TRU_ANOM(K),ECC
137 !
138 ! RADIUS
139 !
140 RR(K)=(A*(1-(ECC*ECC)))/(1.0D0+(ECC*COS(TRU_ANOM(K))))
141 !WRITE(25,*)"RR(K):",K,RR(K)
142 !
143 ! CARTESIAN COORDINATES
144 !
145 XX(K)=RR(K)*COS(TRU_ANOM(K))
146 YY(K)=RR(K)*SIN(TRU_ANOM(K))
147 ZZ(K)=0
148 !
149 COL1(1)=XX(K)
```

```
150 COL1(2)=YY(K)
151 COL1(3)=ZZ(K)
152 !
153 !WRITE(25,*)"COL1:",COL1(1),COL1(2),COL1(3)
154 !
155 !
156 ! Set up the rotation matrices ready to put into ecliptic coordinates
157 ! First initialise the elements to zero
158 !
159 DO I=1,3
160 DO J=1,3
161 ROT_W_PERIH(I,J)=0.0D0
162 ROT_OMEGA(I,J)=0.0D0
163 ROT_XI (I,J)=0.0D0
164 END DO
165 END DO
166 !WRITE(25,*)" "
167 ROT_W_PERIH(1,1)=COS(W_PERIH)
168 ROT_W_PERIH(1,2)=-SIN(W_PERIH)
169 ROT_W_PERIH(2,1)=SIN(W_PERIH)
170 ROT_W_PERIH(2,2)=COS(W_PERIH)
171 ROT_W_PERIH(3,3)=1.0D0
172 !DO I=1,3
173 ! WRITE(25,*)"ROT_W_PERIH:",ROT_W_PERIH(I,1),
ROT_W_PERIH(I,2),ROT_W_PERIH(I,3)
174 !END DO
175 !WRITE(25,*)" "
176 !
177 ROT_OMEGA(1,1)=COS(OMEGA)
178 ROT_OMEGA(1,2)=-SIN(OMEGA)
179 ROT_OMEGA(2,1)=SIN(OMEGA)
180 ROT_OMEGA(2,2)=COS(OMEGA)
181 ROT_OMEGA(3,3)=1.0D0
182 !DO I=1,3
183 ! WRITE(25,*)"ROT_OMEGA :",ROT_OMEGA(I,1),ROT_OMEGA(I,2),ROT_OMEGA(I,3)
184 !END DO
185 !WRITE(25,*)" "
186 !
187 ROT_XI(1,1)=1.0D0
188 ROT_XI(2,2)=COS(XI)
189 ROT_XI(2,3)=-SIN(XI)
190 ROT_XI(3,2)=SIN(XI)
191 ROT_XI(3,3)=COS(XI)
192 !DO I=1,3
193 ! WRITE(25,*)"ROT_XI :",ROT_XI(I,1),ROT_XI(I,2),ROT_XI(I,3)
194 !END DO
195 !WRITE(25,*)" "
196 !
197 ! MULTIPLY THESE MATRICES TO TRANSFORM RADIUS VECTOR TO
EARTH'S ECLIPTIC PLANE
198 !
199 CALL MMULT_3X3(ROT_XI,ROT_W_PERIH,XMPROD1)
200 CALL MMULT_3X3(ROT_OMEGA,XMPROD1,XMPROD2)
201 CALL M3X3XCOLVEC(XMPROD2,COL1,COLRAD)
202 !WRITE(25,*)"COLRAD:",COLRAD(1),COLRAD(2),COLRAD(3)
203 !WRITE(25,*)"VSMALL:",VSMALL
204 !
205 ! WORK OUT THE PREDICTED RHO(I)
206 !
207 RHOX(K)=COLRAD(1)+XEARTH(K)
208 RHOY(K)=COLRAD(2)+YEARTH(K)
209 RHOZ(K)=COLRAD(3)+ZEARTH(K)
210 !
211 COL2(1)=RHOX(K)
212 COL2(2)=RHOY(K)
213 COL2(3)=RHOZ(K)
214 !
```

```
215 !WRITE(25,*)"COL2:",COL2(1),COL2(2),COL2(3)
216
217 !
218 ! Cases for azimuth in Cartesian to Spherical Polar transformation:
219 ! x>0 & y>0: standard.
220 ! x<0 & y>0: Make x >0 then subtract the angle from PI
221 ! x<0 & y<0: Make both > 0 than add PI to the angle
222 ! x>0 & y<0: standard but add 2*PI
223 ! y=0 & x>0: Angle =0
224 ! y=0 & x<0: Angle= PI
225 ! x=0 & y>0: Angle = PI/2
226 ! x=0 & y<0: Angle = 3*PI/2
227 ! x=y=0: Should not happen
228 !
229 IF((COL2(1).GE.VSMALL).AND.(COL2(2).GE.VSMALL))
 RLAMBDA2(K)=ATAN(COL2(2)/COL2(1))
230 IF((COL2(1).LE.VSMALL).AND.(COL2(2).GE.VSMALL))
 RLAMBDA2(K)=PI+(ATAN(COL2(2)/COL2(1)))
231 IF((COL2(1).LE.VSMALL).AND.(COL2(2).LE.VSMALL))
 RLAMBDA2(K)=PI+(ATAN(ABS(COL2(2)/COL2(1))))
232 IF((COL2(1).GE.VSMALL).AND.(COL2(2).LE.VSMALL))
 RLAMBDA2(K)=(2.0D0*PI)+(ATAN(COL2(2)/COL2(1)))
233 IF((COL2(1).GE.VSMALL).AND.(ABS(COL2(2)).LT.VSMALL)) RLAMBDA2(K)=0.0D0
234 IF((COL2(1).LE.VSMALL).AND.(ABS(COL2(2)).LT.VSMALL)) RLAMBDA2(K)=PI
235 IF((ABS(COL2(1)).LT.VSMALL).AND.(COL2(2).GE.VSMALL))
 RLAMBDA2(K)=PI/2.0D0
236 IF((ABS(COL2(1)).LT.VSMALL).AND.(COL2(2).LE.VSMALL))
 RLAMBDA2(K)=3.0D0*PI/2.0D0
237 IF((ABS(COL2(1)).LT.VSMALL).AND.(ABS(COL2(2)).LT.VSMALL)) STOP 901
! Error trap - this case should not occur
238 !
239 IF(ABS(COL2(3)).LT.VSMALL)THEN
240 RBETA2(K)=0.0D0
241 ELSE
242 RBETA2(K)=ASIN(COL2(3))
243 ! WRITE(25,*)"SQRT((COL2(1)**2)+(COL2(2)**2))/COL2(3):",
SQRT((COL2(1)**2)+(COL2(2)**2))/COL2(3)
244 END IF
245 !WRITE(25,*)"RLAMBDA2,RBETA2",K,RLAMBDA2(K),RBETA2(K)
246 !
247 !BUILD UP SQUARE OF DIFFERENCES BETWEEN PREDICTED AND OBSERVED ANGLES
248 !
249 RLSQ=(RLAMBDA2(K)-RLAMBDA(K))**2
250 RBSQ=(RBETA2(K)-RBETA(K))**2
251 RDSQ(K)=RKLS+RBSQ
252 SUMSQ2=SUMSQ2+RDSQ(K)
253 END DO
254 !
255 !WRITE(25,*)"SUMSQ2:",SUMSQ2
256 !
257 IF(SUMSQ2.LE.SUMSQ)THEN
258 SUMSQ=SUMSQ2
259 A_S=A ! SEMI-MAJOR AXIS
260 XI_S=XI ! INCLINATION, RADIANS
261 ECC_S=ECC ! ECCENTRICITY
262 OMEGA_S=OMEGA ! ASCENDING NOTE WRT COODINATE SYSTEM
263 W_PERIH_S=W_PERIH ! ARGUMENT OF PERIHELION
264 T_PERIHEL_S=T_PERIHEL ! TIME OF PERIHELION
265 END IF
266 END DO
267 XLIT_OME=OMEGA_S-PI
268 POMEGA=XLIT_OME+W_PERIH_S
269 !
270 WRITE(25,*)"SUMSQ",SUMSQ
271 WRITE(25,*)"A,XI,ECC,OMEGA,W_PERIH,T_PERIHEL:",A_S,XI_S,ECC_S,OMEGA_S,W_PERIH_S,T_
PERIHEL_S
272 write(26,602)YEER(1),YEER(ICOUNT1-1)
```

```
273 602 FORMAT("MARS ORBIT REFINEMENT",1X,I4,1X,I4)
274 WRITE(26,600)A_S,XI_S*180.0D0/PI,ECC_S
275 600 FORMAT("Orbital Elements",/,"a = ",F12.8,/,"i = ",
F12.8,/,"e = ",F12.8)
276 WRITE(26,601)OMEGA_S*180.0D0/PI,XLIT_OME*180.0D0/PI,
W_PERIH_S*180.0D0/PI,POMEGA*180.0D0/PI,T_PERIHEL_S
277 601 FORMAT("Omega = ",F13.8,2X,F13.8,/,"w_PERIH = ",F13.8,2X,F13.8,/
,"t_PERIH = ",F13.8,2X," w.r.t 3 JAN 2010 @ 07:12")
278 !
279 STOP 900 ! Normal Stop
280 End Program main
281 !
282 SUBROUTINE MMULT_3X3(MATX1,MATX2,MATX3)
283 !
284 ! 3X3 MATRIX MULTIPLICATION
285 !
286 IMPLICIT DOUBLE PRECISION (A-H,O-Z)
287 IMPLICIT INTEGER(I-N)
288 DOUBLE PRECISION MATX1(3,3),MATX2(3,3),MATX3(3,3)
289 !
290 DO I=1,3
291 DO J=1,3
292 SUM=0.0D0
293 DO K=1,3
294 SUM=SUM+(MATX1(I,K)*MATX2(K,J))
295 END DO
296 MATX3(I,J)=SUM
297 END DO
298 END DO
299 !
300 RETURN
301 END SUBROUTINE MMULT_3X3
302 !
303 SUBROUTINE M3X3XCOLVEC(MATX,COLVEC1,COLVEC2)
304 !
305 ! 3X3 MATRIX MULT BY 3X1 COL VECTOR
306 !
307 IMPLICIT DOUBLE PRECISION (A-H,O-Z)
308 IMPLICIT INTEGER(I-N)
309 DOUBLE PRECISION MATX(3,3),COLVEC1(3),COLVEC2(3)
310 !
311 DO I=1,3
312 SUM=0.0D0
313 DO K=1,3
314 SUM=SUM+(MATX(I,K)*COLVEC1(K))
315 END DO
316 COLVEC2(I)=SUM
317 END DO
318 !
319 RETURN
320 END SUBROUTINE M3X3XCOLVEC
321 SUBROUTINE CONVERT_TIMES(YEER,MUNTH,DAIT,HOWR,MINIT,ICOUNT1,QTIME)
322 ! CONVERT TIMES TO CANONICAL UNITS
323 IMPLICIT DOUBLE PRECISION (A-H,O-Z)
324 IMPLICIT INTEGER(I-N)
325 INTEGER YEER(100),MUNTH(100),DAIT(100),HOWR(100),MINIT(100),DAYSINMONTH(12),DAYS
326 DOUBLE PRECISION QTIME(100),FRACDAY(100)
327 PI=4.0D0*ATAN(1.0D0)
328 DAYSPERYR=365.25
329 PERIHELION=DAYSPERYR+3+(7.0D0/24.0D0)+(12.0D0/(24.0D0*60.0D0)) ! 3 JAN 2010 @ 07:12
330 DAYSINMONTH(1)=31
331 DAYSINMONTH(2)=28
332 DAYSINMONTH(3)=31
333 DAYSINMONTH(4)=30
334 DAYSINMONTH(5)=31
335 DAYSINMONTH(6)=30
336 DAYSINMONTH(7)=31
```

```
337 DAYSINMONTH(8)=31
338 DAYSINMONTH(9)=30
339 DAYSINMONTH(10)=31
340 DAYSINMONTH(11)=30
341 DAYSINMONTH(12)=31
342 WRITE(25,*)"PI,DAYSPERYR,PERIHELION,ICOUNT1",PI,DAYSPERYR,PERIHELION,ICOUNT1
343 DO K =1,ICOUNT1
344 !WRITE(25,*)"K",K
345 !
346 ! COUNT YEARS
347 !
348 YRS=-1.0D0
349 IF(YEER(K).EQ.2020)YRS=11*DAYSPERYR+1
350 IF(YEER(K).EQ.2019)YRS=10*DAYSPERYR
351 IF(YEER(K).EQ.2018)YRS=9*DAYSPERYR
352 IF(YEER(K).EQ.2017)YRS=8*DAYSPERYR
353 IF(YEER(K).EQ.2016)YRS=7*DAYSPERYR+1
354 IF(YEER(K).EQ.2015)YRS=6*DAYSPERYR
355 IF(YEER(K).EQ.2014)YRS=5*DAYSPERYR
356 IF(YEER(K).EQ.2013)YRS=4*DAYSPERYR
357 IF(YEER(K).EQ.2012)YRS=3*DAYSPERYR+1
358 IF(YEER(K).EQ.2011)YRS=2*DAYSPERYR
359 IF(YEER(K).EQ.2010)YRS=DAYSPERYR
360 IF(YEER(K).EQ.2009)YRS=0.0D0
361 !WRITE(25,*)"YRS",YRS
362 IF(YRS.LT.0.0D0) STOP 903 ! TRAP TO CATCH WRONG YEAR ERRORS
363 DAYS=0
364 IF (MUNTH(K).GT.2) THEN
365 DO J=1,MUNTH(K)-1
366 DAYS=DAYS+DAYSINMONTH(J)
367 END DO
368 ELSE IF (MUNTH(K).EQ.2) THEN
369 DAYS=DAYSINMONTH(1)
370 ELSE
371 DAYS=0
372 END IF
373 !WRITE(25,*)"DAYS",K,DAYS
374 DAYS=DAYS+DAIT(K)
375 RDAYS=DAYS
376 !WRITE(25,*)"RDAYS",K,RDAYS
377 HR=HOWR(K)
378 XMIN=MINIT(K)
379 FRACDAY(K)=(HR/24.0D0)+(XMIN/(24.0D0*60.0D0))
380 !WRITE(25,*)"HR,XMIN,FRACDAY",HR,XMIN,FRACDAY(K)
381 QTIME(K)=YRS+RDAYS+FRACDAYS ! RELATIVE TO 2009 01 01 00:00
382 !WRITE(25,*)"QTIME(K)",QTIME(K)
383 QTIME(K)=QTIME(K)-PERIHELION ! RELATIVE TO PERIHELION
384 !WRITE(25,*)"QTIME(K)",QTIME(K)
385 QTIME(K)=QTIME(K)*2*PI/DAYSPERYR ! CONVERT TO CANONICAL UNITS
386 !WRITE(25,*)"QTIME(K)",QTIME(K)
387 END DO
388 RETURN
389 END SUBROUTINE CONVERT_TIMES
390
391
```

Manual Adjustment of Orbits

This, of course, is a fancy name for "messing around until the orbit looks believable".

As advertised, the Monte Carlo results were not always spot on. In particular, the times to perihelion were, frankly, all over the place. This particular orbital element is not sensitive to trying to minimize the least squares. It is possible that that's why the least-squares method I took from Boulet's [132] equations performed so poorly.

Anyway, here are a three of datasets I used and the results I obtained. Each of these took about five hours to run.

If you were to plot the predicted right ascensions and declensions from the sets of elements in Table 8.2 or 8.3, the results would appal you, I promise.

But they did give me a basis on which to start trying to make manual fits. I did this in Microsoft Excel™, because it was easier and quicker to get good graphs out without having to debug my own graphics routine. I don't think I can compete with the small army of programmers at Microsoft. No. Bad idea.

(If you don't have Microsoft Office, you can download the free LibreOffice suite of very similar programs. They are much the same as the suite in Open Office, except that Open Office is no longer being developed. LibreOffice is the one being developed.)

Table 8.2 Monte Carlo results obtained from $2^{27} = 134{,}217{,}728$ instances of the randomized trial. The Semi-major axis a was held constant

MARS ORBIT REFINEMENT 2009 2020
Orbital Elements
$a = 1.52412500$
$i = 1.93860703$
$e = 0.09458634$
Omega = 228.60068224   48.60068224
w_PERIH = 300.28319723   348.88387948
t_PERIH = −5.61542634   w.r.t 3 JAN 2010 @ 07:12

Table 8.3 Monte Carlo results obtained from $2^{27} = 134{,}217{,}728$ instances of the randomized trial. The eccentricity e was held constant

MARS ORBIT REFINEMENT 2009 2020
Orbital Elements
a = 1.55397996
i = 2.13194944
e = 0.09458634
Omega = 228.97056183   48.97056183
w_PERIH = 294.56344152   343.53400335
t_PERIH = −5.38629767   w.r.t 3 JAN 2010 @ 07:12

What did I do? I solved Kepler's Eq. (3.79) for the trial set of elements for every day from 1 January 2009 to 31 December 2020 for Mars in one worksheet. I worked out the values of the actor **r** at daily intervals.

I did the same for the Earth in another worksheet and worked out the values of **R** at daily intervals, measured from my calculated Earth perihelion, from Eq. (5.54).

Then in a third, linked worksheet, I worked out my values of **ρ** and hence calculated the right ascension α and declination δ that an observer at the Earth's centre would see.

I then copied these into a separate spreadsheet, where I plotted these predicted values and the observed values, to enable me to compare the results.

The results are given below for two sets of data.

Figures 8.2 and 8.3 show a comparison of data generated from this set of elements with observational data from the Mars apparitions of 2009–10, 2011–12 and 2020.

Figures 8.4 and 8.5 show a comparison of data generated from this set of elements with observational data from the Mars apparitions of 2009–10, 2011–12 and 2020.

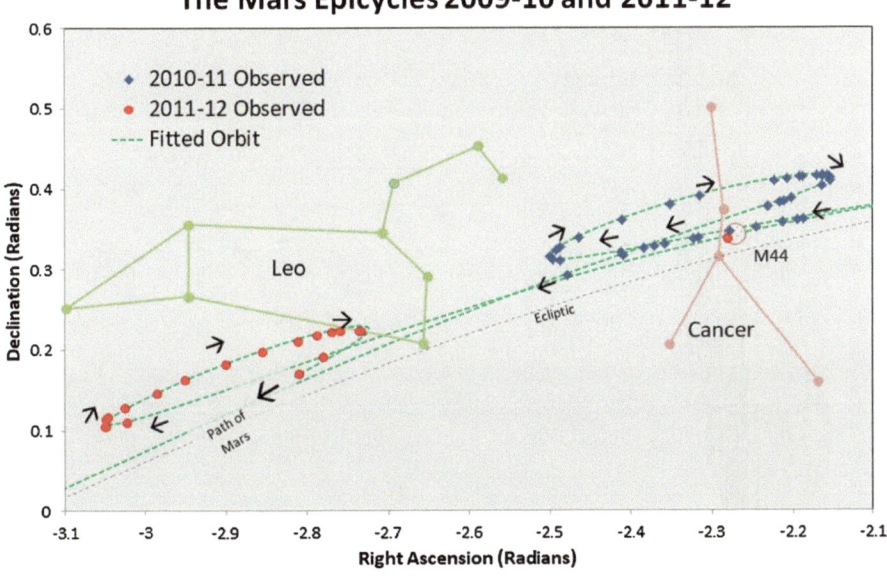

Fig. 8.2 Fitted orbits for 2009–10 from Table 8.4

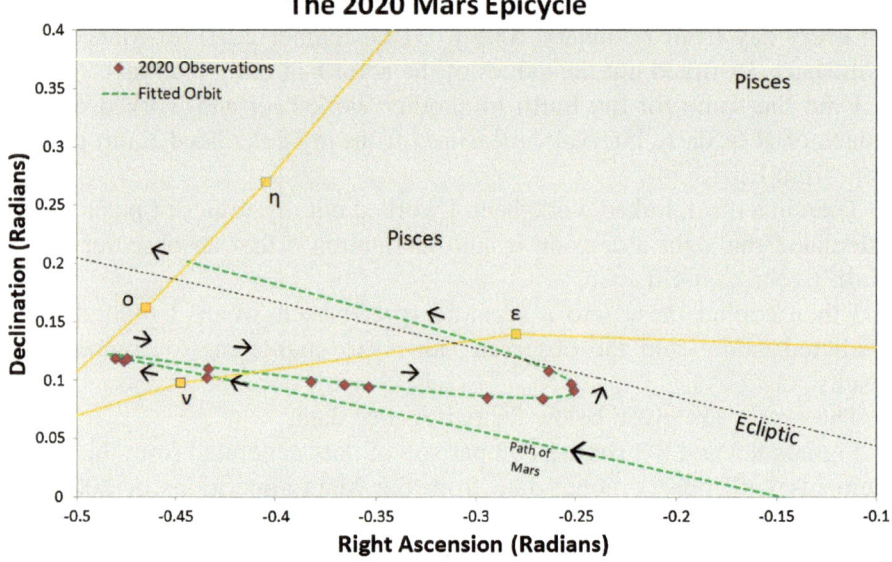

Fig. 8.3 Fitted orbits for 2020 from Table 8.4

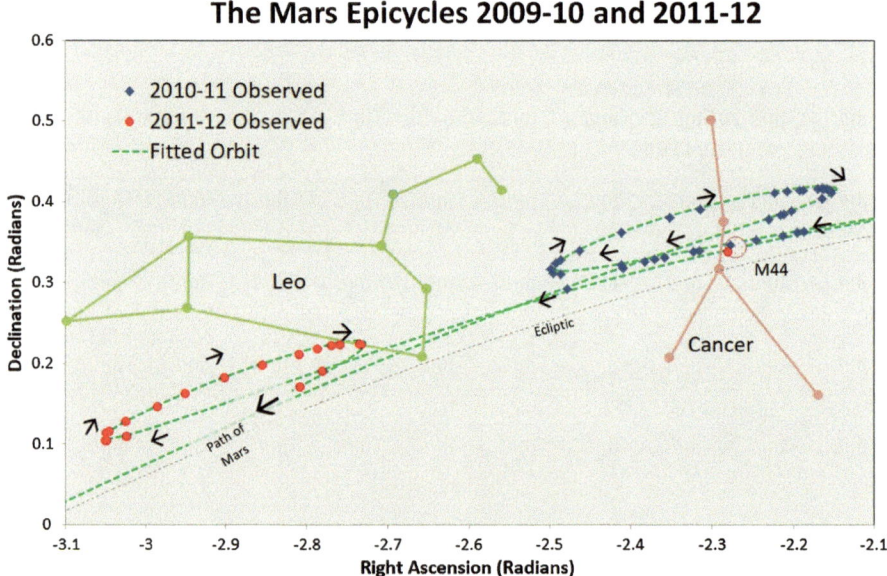

Fig. 8.4 Fitted orbits for 2009–10 from Table 8.5

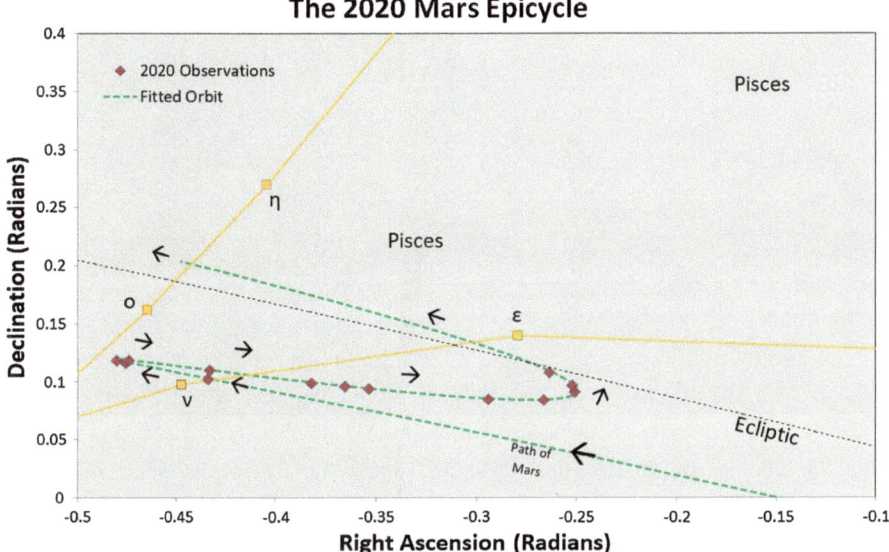

Fig. 8.5 Fitted orbits for 2020 from Table 8.5

Using Knowledge of the 2020 Opposition

At this point I realized that I can constrain my data further by trying to measure and predict the opposition time.

Of course, I knew perfectly well when the 2020 opposition was going to be. The astronomy clubs were abuzz with excitement, because, in British skies, Mars was going to rise higher than the magic 30° altitude, above which good quality photography is possible. I also noticed that my fitted Mars orbits woefully failed to get the date right.

In any event, oppositions are hard to miss. The planet is due south at the local midnight, solar time, and it rises at sunset and sets at sunrise. The local midnight I found from a website that allowed me to enter my home coordinates [133]. I would have preferred to use the US Naval observatory's web site [97], a more respectable source, but at the time it was down. I checked against another sunrise/sunset website and got pretty much the same answer. Anyway, in mid-October 2020, the solar midday at my location was two minutes earlier than the UT midday. The solar midnight would therefore also be two minutes before midnight Universal Time.

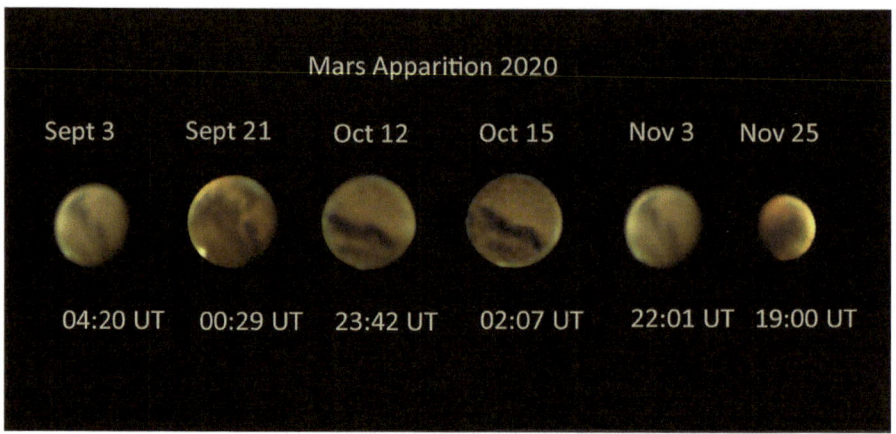

Fig. 8.6 Photographs of Mars during the 2020 apparition. The planet is full, suggesting opposition in mid-October 2020. Before opposition, the planet is around in the early hours. After opposition, it is an evening object. This can be seen in the times the photographs were taken
Taken with an 11″ Celestron SCT, 2× (nominal) Barlow, ZWO ASI290MC camera, postprocessed in Autostakkert 3, Registax, Photoshop Elements, then Topaz Sharpen AI. (Images: Author)

Another clue about opposition time comes from photographs of the planet. It is full at opposition and gibbous away from opposition (Fig. 8.6).

From Table 6.6, I measured the right ascensions in radians of Mars to be −0.365 rad on 2020/10/12 at 22:54 and −0.353 rad on 2020/10/15 at 01:41. I used Cartes du Ciel to look up the right ascensions of stars on the meridian line at 11:58 UT on those two days. I found these longitudes to be −0.357 rad on 2020/10/12 at 23:58 UT and −0.322 rad on 2020/10/14 at 23:58 UT. I am therefore tempted to say that since −0.365 rad ≈ −0.357 rad, I have measured the opposition to be at about 2020/10/12 at 23:58 UT. If I allow for the difference of 0.002 rad, this corresponds to about 0.116 days or 2 h 47 min (ignoring the difference between sidereal and mean time). That would make my measured opposition time 2020/10/12 at 21:11 UT.

At opposition, the angle between the Sun–Earth line and the Sun–Mars line will be zero when looking down onto the Solar System from above the Sun's North Pole. I searched for the time when this was true in my models, in other words when the angle between **r** and −**R** in the plane of the ecliptic in Eq. (7.1) and Fig. 7.2 is zero.

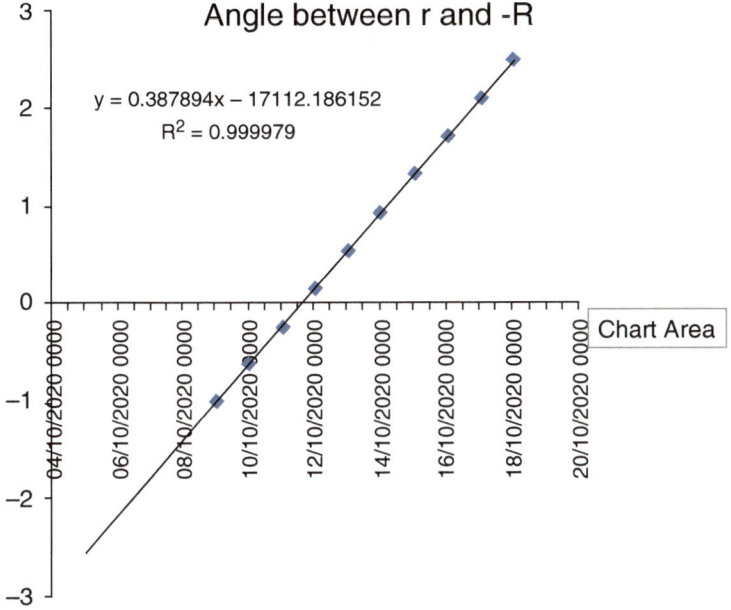

Fig. 8.7 From this chart I calculate the model prediction of opposition time to be 2020/10/11 at 15:45 UT
I did this by solving the fitted equation and using Excel to convert the number ~17,000 into its date format. (Images: Author)

I manually adjusted my fits to get this to be as nearly the case as possible. The predicted opposition time is 2020/10/11 at 15:45 UT (Fig. 8.7). I could not get closer than this and still predict the path of Mars. I suspect that to do better, I would have to improve my knowledge of the orbit of the Earth.

Be that as it may, I made several attempts manually to fit the data and nearly always came up with something very close to the following parameters.

How do the fits look? They are shown in Figs. 8.8 and 8.9. To the unaided eye, they look like good fits.

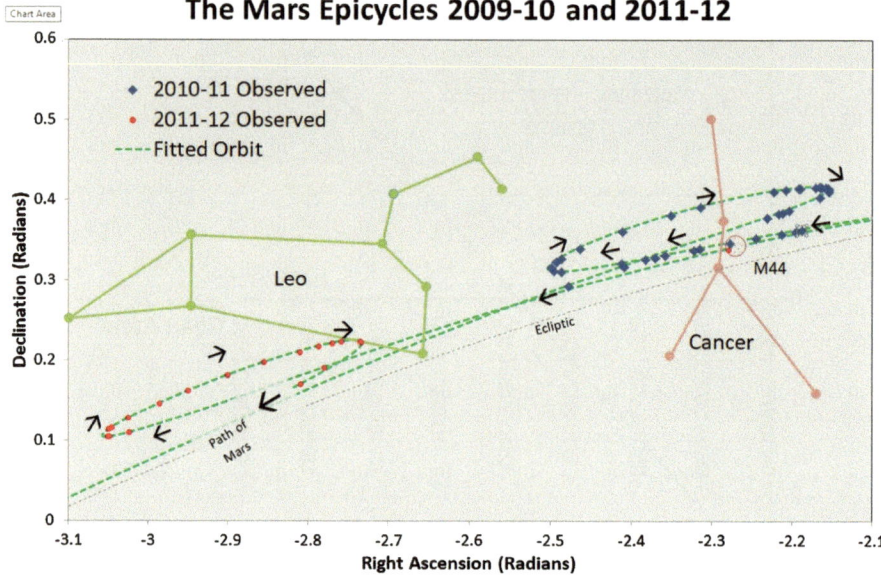

Fig. 8.8 Fitted orbits for 2009–12 from Table 8.6

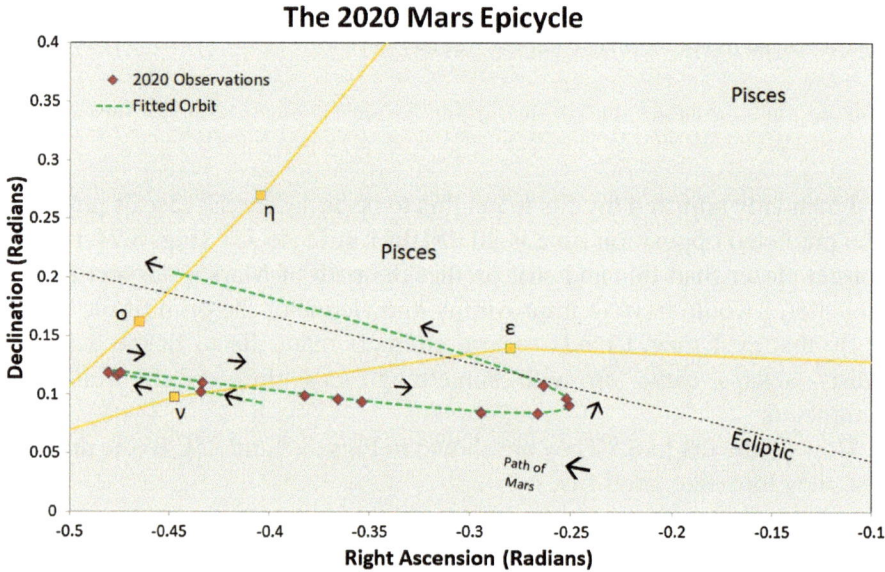

Fig. 8.9 Fitted orbits for 2020 from Table 8.6

Conclusions Drawn

It is difficult to distinguish between the quality of the orbital elements in Tables 8.4, 8.5, and 8.6 by eyeballing Figs. 8.2, 8.3, 8.4, 8.5, 8.6, 8.7, 8.8, and 8.9. So which is better? The least-squares fit suggests the data from

Table 8.4 My first set of manually obtained elements

MARS ORBIT 2009 2020
Orbital Elements
$a = 1.524125$
$i = 1.85°$
$e = 0.094$
$\Omega = 229.558°$ (Conventional Value 49.558°)
$\omega = 283.66241169°$
T_Per $= -4.574554559$ rad w.r.t 3 JAN 2010 @ 07:12
$= 12$ April 2009 @ 09:00 approx.
Sum of squares of errors $= 0.0230$ radians^2

Table 8.5 My Second set of manually obtained elements

MARS ORBIT 2009 2020
Orbital Elements
$a = 1.5235$
$i = 1.85°$
$e = 0.094$
$\Omega = 229.6°$ (Conventional Value 49.6°)
$\omega = 287.25°$
T_Per $= -4.425466$ rad w.r.t 3 JAN 2010 @ 07:12
$= 21$ April 2009 @ 01:00 approx.
Sum of squares of errors $= 0.0303$ radians^2

Table 8.6 My third set of manually obtained elements

MARS ORBIT REFINEMENT 2009 2020
Orbital Elements
$a = 1.52372500$
$i = 1.85000000$
$e = 0.09400000$
Omega $= 229.50000000$ 49.50000000
w_PERIH $= 286.90000000$ 336.40000000
t_PERIH $= -258.99999998$ Days w.r.t 3 JAN 2010 @ 07:12
$= 19$ April 2009 @ 07:12 approx.
Opposition
$= 11$ October 2020 @ 15:45
Sum of squares of errors $= 0.0290$ radians^2

Table 8.4 is marginally the better fit, but the prediction of opposition is poor. By trying to predict the opposition, I have constrained the otherwise rather uncertain and variable values of the argument of perihelion ω and the time of perihelion passage T_{Per}. I therefore choose my third set of orbital elements as the best.

The longitude of the perihelion and the time at which it is passed do not vary independently. You have to vary the one to compensate for changes in the other.

I conclude that the data in Eq. (8.7) gives a best fit to my observations of Mars from 2009 to 2020, assuming the orbit of Mars to be Keplerian and assuming the Earth to be a point at its centre.

$$a = 1.523725 \text{ AU}$$
$$i = 1.85°$$
$$e = 0.094$$
$$' = 49.5 \pm 0.1° \text{ w.r.t. } \Upsilon \tag{8.7}$$
$$\omega = 286.9°$$
$$T_{Per} = 19 \text{ April } 2009 @ 07:12$$
$$T_{opp} = 13 \text{ October } 2020 @ 15:45$$

Da Capo al Fine: Can I Fit Tycho's Data?

In Figs. 1.22 and 1.23, I showed Tycho's RA and declination data from 1582 to 1589.

I would not expect to fit these data from my orbital elements for Mars as they stand. For a start, the precession of the equinoxes over 430 or so years will have been significant. I estimated from Fig. 8.6 that in Tycho's day Polaris would have been about 3° from the Celestial North Pole. Indeed, Tycho himself mentions this on page 381 of Vol X of his Complete Works [39], where he measures the maximum and minimum altitudes of Polaris:

$$
\begin{array}{lcl}
\text{Alt. Max. Polaris} & = & 58°49'40 \\
\text{Alt. Min. Polaris} & = & 52°59'40 \\
\textit{Difference} & = & 5°50'0
\end{array}
\tag{8.8}
$$

These measurements were made on the morning and evening of 7 January 1589 (Julian).

Interactions with other planets would also be significant over 430 years. Boulet [113], pp. 307–310, suggests that for a single orbit, differences of up to 1 arcminute arise when taking into account planet–planet gravitational interaction.

Fig. 8.10 Showing the Position of the North Celestial Pole as the Earth's axis precesses. From this diagram, I estimate that the distance of Polaris from the North Celestial Pole in Tycho's day would have been about 3°. Original Diagram: Wikimedia Commons, Author: Tau'olunga. Creative Commons Attribution-Share Alike 2.5 Generic license. Red lines and text added by the Author

I can correct for the precession of the equinoxes using a formula from Bretagnon & Simon [134] (Fig. 8.10).

The formulae in question are, with M equal to the number of millennia between 2000 CE and the year in question, as follows. The precession of longitude

$$p_A = 50290.966'' \ M + 111.191'' \ M^2 + 0.07732'' M^3 - 0.235316'' \ M^4 + \ldots \qquad (8.9)$$

The precession of the obliquity of the Earth's axis

$$\varepsilon_A = 23°26'21.448'' - 4468.093'' \ M - 0.0155'' \ M^2 + 1.99925'' \ M^3 - 0.005138'' \ M^4 + \ldots \qquad (8.10)$$

I took M to be 0.4 millennia, that is, the time from 2000 to 1600 CE.

Consequently, I am not going to hold my breath about whether I can even almost match Tycho's data with my calculated orbit. But let's see. Results are shown in Figs. 8.11 and 8.12.

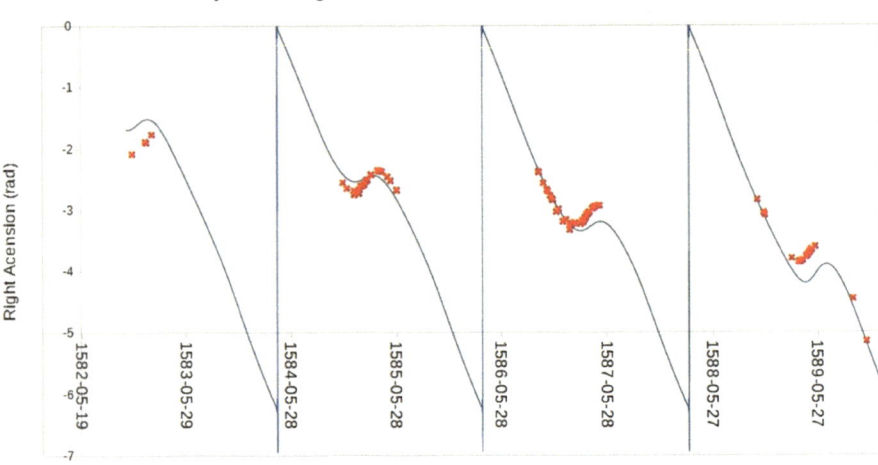

Fig. 8.11 Tycho's RA data compared against my orbit calculation with slightly modified elements. (Image: Author)

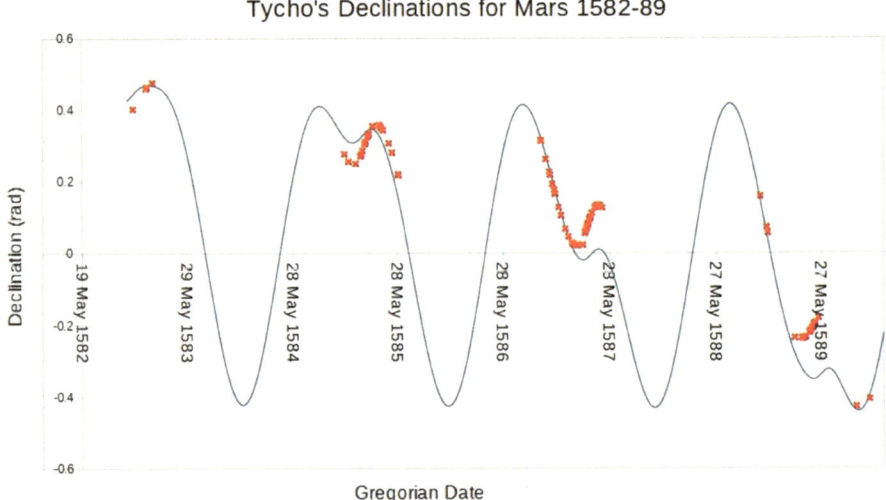

Fig. 8.12 Tycho's declination data compared against my orbit calculation with slightly modified elements. (Image: Author)

The agreement is not brilliant, but not awful either. The time-dependent phenomena like oppositions and epicycles are modelled at approximately the right times. But the extents of the back-and-forth motion over the epicycles are poorly modelled.

The orbital elements I used are the same as before.

$$a = 1.523725 \text{ AU}$$
$$i = 1.85°$$
$$e = 0.094$$
$$' = 49.5° \text{ w.r.t. } Υ \tag{8.11}$$
$$ω = 286.9°$$
$$T_{\text{Per}} = 19 \text{ April } 2009 @ 07:12$$

I'm confident enough in the interpretation of my own twenty-first-century data to conclude that something has happened to the orbit of Mars in 430 years. It has changed. I don't yet know how. But there's something there to investigate further. It must be rememebred that the Earth's orbit may also have changed, so I would need a way to correct for this.

How Well Did I Do?

What of the "official" answer? These elements are given by Simon et al. [135], except for the perihelion date, which I found on Wikipedia for 2020, and back-calculated from there assuming an orbital period of 686.98 days calculated from Kepler's third law (Eq. 3.88) from the value of a in Eq. (8.11), assuming that the Earth is 1 AU from the Sun with a year of 365.25 days.

So, officially

$$a = 1.523679 \text{ AU}$$
$$i = 1.84973°$$
$$e = 0.093401$$
$$' = 49.5581° \text{ w.r.t. } Υ \tag{8.12}$$
$$ω = 286.50°$$
$$T_{\text{Per}} = 21 \text{ April } 2009$$
$$T_{\text{opp}} = 13 \text{ November } 2020 @ 23:00$$

These values are within 1% of those in Eq. (8.7), except that my time of perihelion passage is about two days early. My modelled opposition time is also about two days early compared to the widely quoted one of 2020/10/13 at 23:00 UT from, for example, https://earthsky.org/tonight/mars-to-reach-opposition-october-13, accessed on 14 April 2021.

The coincidence of this two-day error makes me suspect that the two errors are related and are systematic not random. As mentioned, the prime suspect is the crudity of my Earth orbit.

There are other sources of error: planet–planet interactions and the fact that I have used geocentric not topocentric coordinates, that is, those based on my position on the Earth's surface. And of course, the Earth and the Moon orbit around a common centre of mass. This also could usefully be allowed for.

Nevertheless, I have succeeded in my quest to understand the orbit of Mars – at least the orbit of Mars in the present epoch.

By looking at Tycho's data, I have discovered that there are long-term changes in the orbit. I am keen to investigate these further. Indeed, I have already begun to.

Finally, thank you, dear reader, for persevering this far. I wish you clear skies and good luck with your astronomical endeavours.

Appendix 1: Parabolas and Hyperbolas

Deriving the Equation of a Hyperbola in Polar

Let us start from the equation of a hyperbola in Cartesian coordinates,

$$\frac{x^2}{a^2} - \frac{y^2}{b^2} = 1$$

$$\frac{x^2}{a^2} - \frac{y^2}{c^2 - a^2} = 1 \tag{A1.1}$$

where

$$b = +\sqrt{a^2 - c^2}, \tag{A1.2}$$

and

$$e = \frac{c}{a} = +\sqrt{1 + \frac{b^2}{a^2}}. \tag{A1.3}$$

We can show that this equation is consistent with the older definition of a hyperbola

J. Clark, *Calculate the Orbit of Mars!*,
https://doi.org/10.1007/978-3-030-78267-2,
© Springer Nature Switzerland AG 2021

$$
\begin{aligned}
x^2\left(c^2-a^2\right)-a^2 y^2 &= a^2\left(c^2-a^2\right) \\
-a^2 y^2 &= +a^2\left(c^2-a^2\right)-x^2\left(c^2-a^2\right) \\
+a^2 y^2 &= -a^2\left(c^2-a^2\right)+x^2\left(c^2-a^2\right) \\
a^2 x^2+a^2 c^2-2a^2 xc+a^2 y^2 &= a^4+x^2 c^2-2a^2 xc \\
a^2\left((x-c)^2+y^2\right) &= a^4+x^2 c^2-2a^2 xc \\
a\sqrt{(x-c)^2+y^2} &= a^2-xc \\
-4a\sqrt{(x-c)^2+y^2} &= +4xc-4a^2 \\
-2xc+4a^2-4a\sqrt{(x-c)^2+y^2} &= +2xc \\
x^2+c^2-2xc+y^2+4a^2-4a\sqrt{(x-c)^2+y^2} &= x^2+c^2+2xc+y^2 \\
(x-c)^2+y^2+4a^2-4a\sqrt{(x-c)^2+y^2} &= (x+c)^2+y^2 \\
\left(\sqrt{(x-c)^2+y^2}-2a\right)^2 &= (x+c)^2+y^2 \\
\sqrt{(x-c)^2+y^2}-2a &= \sqrt{(x+c)^2+y^2} \\
\sqrt{(x-c)^2+y^2}-\sqrt{(x+c)^2+y^2} &= 2a.
\end{aligned}
$$

(A1.4)

In other words, the difference in the two distances from the foci is constant. (In an ellipse, the *sum* of the distances from the two foci is constant.)

Let

$$
\sqrt{(x-c)^2+y^2}=r_1
$$
$$
\sqrt{(x+c)^2+y^2}=r_2
$$

(A1.5)

From Eqs. (A1.1), (A1.2) and (A1.3),

$$
\begin{aligned}
\frac{x^2}{a^2}-1 &- \frac{y^2}{b^2} \\
\frac{b^2 x^2}{a^2}-b^2 &= y^2 \\
y^2 &= \left(e^2-1\right)x^2-a^2\left(e^2-1\right) \\
y^2 &= \left(e^2-1\right)\left(x^2-a^2\right).
\end{aligned}
$$

(A1.6)

Then

$$
\begin{aligned}
r_1^2 &= (x+ae)^2 + y^2 \\
&= x^2 + a^2 e^2 + 2aex + y^2 \\
&= x^2 + a^2 e^2 + 2aex + (e^2 - 1)(x^2 - a^2) \\
&= x^2 + a^2 e^2 + 2aex + e^2 x^2 - e^2 a^2 - x^2 + a^2 \\
&= 2aex + e^2 x^2 + a^2 \\
&= (ex + a)^2
\end{aligned}
\tag{A1.7}
$$

Similarly

$$
r_2^2 = (ex - a)^2
\tag{A1.8}
$$

Hence

$$
\begin{aligned}
r_1 &= -ex - a; \\
r_2 &= -ex + a.
\end{aligned}
\tag{A1.9}
$$

In polar coordinates, the left-hand branch of the hyperbola takes the form

$$
\begin{aligned}
y &= r_2 \sin\varphi \\
x &= -ae + r_2 \cos\varphi \\
r_2 &= -ex - a \\
&= -e(-ae + r_2 \cos\varphi) - a \\
r_2(1 + e\cos\varphi) &= a(e^2 - 1) \\
r_2 &= \frac{a(e^2 - 1)}{(1 + e\cos\varphi)}.
\end{aligned}
\tag{A1.10}
$$

This is our equation of a hyperbola in polar coordinates about a focus.

Deriving the Equation of a Parabola in Polar Coordinates

$$y^2 = 4ax$$

$$y^2 = a^2 + x^2 - 2ax - \left(a^2 + x^2 - 2ax\right)$$

$$(x-a)^2 + y^2 = (a-x)^2$$

$$\sqrt{(x-a)^2 + y^2} = a - x$$

$$\sqrt{(x-a)^2 + y^2} + x - a = 2a$$

$$\sqrt{(x-a)^2 + y^2}\left(1 + \frac{x-a}{\sqrt{(x-a)^2 + y^2}}\right) = 2a \tag{A1.11}$$

$$\sqrt{(x-a)^2 + y^2} = \frac{2a}{1 + \dfrac{x-a}{\sqrt{(x-a)^2 + y^2}}}$$

$$i.e. \quad r = \frac{2a}{1 + \cos\varphi}$$

$$where \quad \tan\varphi = \frac{y}{x-a}$$

$$and \quad r^2 = \sqrt{(x-a)^2 + y^2}.$$

Hyperbola with Origin at Focus

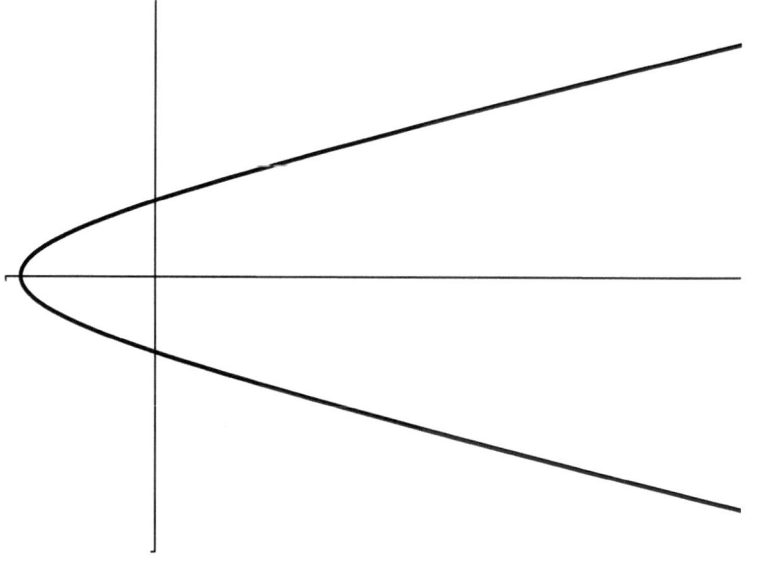

Fig. A1.1 A hyperbola, showing axes whose origin is at one focus

Paraobola, with origin at focus

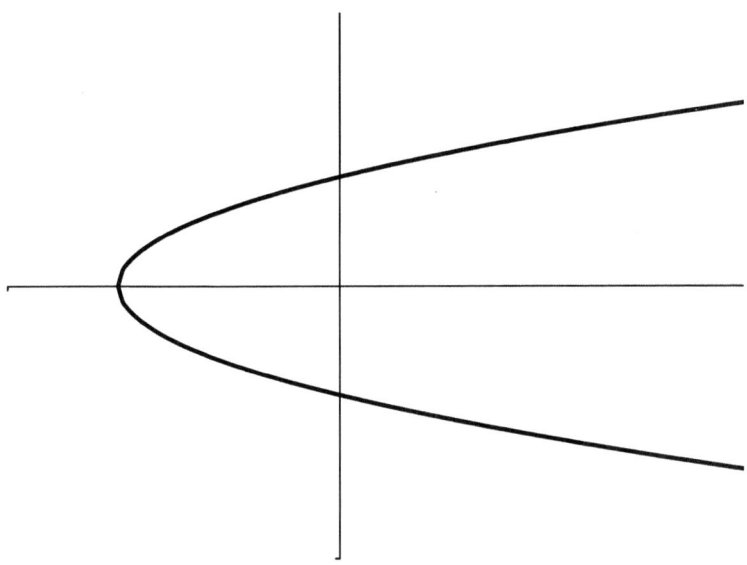

Fig. A1.2 A parabola, showing axes whose origin is at its focus

Appendix 2: Multiplying Three Vectors

In this Appendix, three theorems will be proved. I include them because I did not learn them when I was a student, so I am guessing that most readers did not either.

First comes the case where we multiply a dot and a cross product. Using the language of determinants

$$\mathbf{A} \bullet (\mathbf{B} \times \mathbf{C}) = \mathbf{A} \bullet \begin{vmatrix} \hat{\mathbf{x}} & \hat{\mathbf{y}} & \hat{\mathbf{z}} \\ B_x & B_y & B_z \\ C_x & C_y & C_z \end{vmatrix}$$

$$= \begin{vmatrix} A_x & A_y & A_z \\ B_x & B_y & B_z \\ C_x & C_y & C_z \end{vmatrix} . \tag{A2.1}$$

Next, I prove that

$$\mathbf{A} \bullet (\mathbf{B} \times \mathbf{C}) = \mathbf{B} \bullet (\mathbf{C} \times \mathbf{A}) = \mathbf{C} \bullet (\mathbf{A} \times \mathbf{B}) . \tag{A2.2}$$

These three expressions are written out as determinants, and I use the rule that interchanging two rows of determinants changes the algebraic sign of the determinant. This in itself is easily proved by multiplying out determinants. First

J. Clark, *Calculate the Orbit of Mars!*,
https://doi.org/10.1007/978-3-030-78267-2,
© Springer Nature Switzerland AG 2021

$$\mathbf{A} \bullet (\mathbf{B} \times \mathbf{C}) = \begin{vmatrix} A_x & A_y & A_z \\ B_x & B_y & B_z \\ C_x & C_y & C_z \end{vmatrix}$$

$$= - \begin{vmatrix} B_x & B_y & B_z \\ A_x & A_y & A_z \\ C_x & C_y & C_z \end{vmatrix}$$

(A2.3)

$$= \begin{vmatrix} B_x & B_y & B_z \\ C_x & C_y & C_z \\ A_x & A_y & A_z \end{vmatrix}$$

$$= \mathbf{B} \bullet (\mathbf{C} \times \mathbf{A}) \ .$$

Next

$$\mathbf{A} \bullet (\mathbf{B} \times \mathbf{C}) = \begin{vmatrix} A_x & A_y & A_z \\ B_x & B_y & B_z \\ C_x & C_y & C_z \end{vmatrix}$$

$$= - \begin{vmatrix} C_x & C_y & C_z \\ B_x & B_y & B_z \\ A_x & A_y & A_z \end{vmatrix}$$

(A2.4)

$$= \begin{vmatrix} C_x & C_y & C_z \\ A_x & A_y & A_z \\ B_x & B_y & B_z \end{vmatrix}$$

$$= \mathbf{C} \bullet (\mathbf{A} \times \mathbf{B}) \ .$$

Finally, we look at the triple vector product to prove that

$$\mathbf{A} \times (\mathbf{B} \times \mathbf{C}) = \mathbf{B} (\mathbf{A} \bullet \mathbf{C}) - \mathbf{C} (\mathbf{A} \bullet \mathbf{B}) = \mathbf{B} (\mathbf{A} \bullet \mathbf{C}) - \mathbf{A} (\mathbf{B} \bullet \mathbf{C}) \ . \quad (A2.5)$$

$$\mathbf{A} \times (\mathbf{B} \times \mathbf{C}) = \mathbf{A} \times \begin{vmatrix} \hat{\mathbf{x}} & \hat{\mathbf{y}} & \hat{\mathbf{z}} \\ A_x & A_y & A_z \\ B_2 C_3 - B_3 C_2 & B_3 C_1 - B_1 C_3 & B_1 C_2 - B_2 C_1 \end{vmatrix}$$

$$= \hat{\mathbf{x}} \left(A_y B_x C_y - A_y B_y C_x - A_z B_z C_x + A_z B_x C_z \right)$$

$$+ \hat{\mathbf{y}} \left(A_z B_y C_z - A_z B_z C_y - A_x B_x C_y + A_x B_y C_x \right)$$

$$+ \hat{\mathbf{z}} \left(A_x B_z C_x - A_x B_x C_z - A_y B_y C_z + A_y B_z C_y \right)$$

$$= \left(B_x \hat{\mathbf{x}} + B_y \hat{\mathbf{y}} + B_z \hat{\mathbf{z}} \right) \left(A_x C_x + A_y C_y + A_z C_z \right) - \left(C_x \hat{\mathbf{x}} + C_y \hat{\mathbf{y}} + C_z \hat{\mathbf{z}} \right) \left(A_x B_x + A_y B_y + A_z B_z \right)$$

$$= \mathbf{B}(\mathbf{A} \cdot \mathbf{C}) - \mathbf{C}(\mathbf{A} \cdot \mathbf{B}) \,. \tag{A2.6}$$

Similarly

$$\mathbf{A} \times (\mathbf{B} \times \mathbf{C}) = \mathbf{A} \times \begin{vmatrix} \hat{\mathbf{x}} & \hat{\mathbf{y}} & \hat{\mathbf{z}} \\ A_x & A_y & A_z \\ B_2 C_3 - B_3 C_2 & B_3 C_1 - B_1 C_3 & B_1 C_2 - B_2 C_1 \end{vmatrix}$$

$$= \hat{\mathbf{x}} \left(A_y B_x C_y - A_y B_y C_x - A_z B_z C_x + A_z B_x C_z \right)$$

$$+ \hat{\mathbf{y}} \left(A_z B_y C_z - A_z B_z C_y - A_x B_x C_y + A_x B_y C_x \right)$$

$$+ \hat{\mathbf{z}} \left(A_x B_z C_x - A_x B_x C_z - A_y B_y C_z + A_y B_z C_y \right)$$

$$= \left(B_x \hat{\mathbf{x}} + B_y \hat{\mathbf{y}} + B_z \hat{\mathbf{z}} \right) \left(A_x C_x + A_y C_y + A_z C_z \right) - \left(A_x \hat{\mathbf{x}} + A_y \hat{\mathbf{y}} + A_z \hat{\mathbf{z}} \right) \left(B_x C_x + B_y C_y + B_z C_z \right)$$

$$= \mathbf{B}(\mathbf{A} \cdot \mathbf{C}) - \mathbf{A}(\mathbf{B} \cdot \mathbf{C}) \,. \tag{A2.7}$$

Appendix 3: Proof of Trigonometric Addition Formulae for All Angles

Use can be made of Euler's well-known formulae that

$$\sin x = \frac{e^{ix} - e^{-ix}}{2i} \tag{A3.1}$$

and

$$\cos x = \frac{e^{ix} + e^{-ix}}{2} \tag{A3.2}$$

where i is the square root of minus one.

The proofs are most obvious when done backwards, by assuming the results and demonstrating equality to the original quantity.

$$
\begin{aligned}
\sin A \cos B + \cos A \sin B &= \frac{\left(e^{iA} - e^{-iA}\right)\left(e^{iB} + e^{-iB}\right)}{2i} \cdot \frac{1}{2} + \frac{\left(e^{iA} + e^{-iA}\right)\left(e^{iB} - e^{-iB}\right)}{2} \cdot \frac{1}{2i} \\
&= \frac{1}{4i}\left[\left(e^{iA} - e^{-iA}\right)\left(e^{iB} + e^{-iB}\right) + \left(e^{iA} + e^{-iA}\right)\left(e^{iB} - e^{-iB}\right)\right] \\
&= \frac{1}{4i}\left[e^{iA}e^{iB} + e^{iA}e^{-iB} - e^{-iA}e^{iB} - e^{-iA}e^{-iB} + e^{iA}e^{iB} - e^{iA}e^{-iB} + e^{-iA}e^{iB} - e^{-iA}e^{-iB}\right] \\
&= \frac{1}{4i}\left[2e^{iA}e^{iB} - 2e^{-iA}e^{-iB}\right] \\
&= \frac{1}{2i}\left[e^{iA}e^{iB} - e^{-iA}e^{-iB}\right] \\
&= \frac{e^{i(A+B)} - e^{-i(A+B)}}{2i} \\
&= \sin(A+B).
\end{aligned} \tag{A3.3}
$$

Finally

J. Clark, *Calculate the Orbit of Mars!*,
https://doi.org/10.1007/978-3-030-78267-2,
© Springer Nature Switzerland AG 2021

$$\cos A \cos B - \sin A \sin B = \frac{\left(e^{iA}+e^{-iA}\right)\left(e^{iB}+e^{-iB}\right)}{2} \cdot \frac{}{2} - \frac{\left(e^{iA}-e^{-iA}\right)\left(e^{iB}-e^{-iB}\right)}{2i} \cdot \frac{}{2i}$$

$$= \frac{1}{4}\left[\left(e^{iA}+e^{-iA}\right)\left(e^{iB}+e^{-iB}\right)+\left(e^{iA}-e^{-iA}\right)\left(e^{iB}-e^{-iB}\right)\right]$$

$$= \frac{1}{4}\left[e^{iA}e^{iB}+e^{iA}e^{-iB}+e^{-iA}e^{iB}+e^{-iA}e^{-iB}+e^{iA}e^{iB}-e^{iA}e^{-iB}-e^{-iA}e^{iB}+e^{-iA}e^{-iB}\right] \qquad \text{(A3.4)}$$

$$= \frac{1}{4}\left[2e^{iA}e^{iB}+2e^{-iA}e^{-iB}\right]$$

$$= \frac{1}{2}\left[e^{iA}e^{iB}+e^{-iA}e^{-iB}\right]$$

$$= \frac{e^{i(A+B)}+e^{-i(A+B)}}{2}$$

$$= \cos(A+B).$$

Equations (A3.3) and (A3.4) are valid for all angles A and B.

References

1. Clark, J.D., 2009, *Measure Solar System Objects and their Movements for Yourself!*, Springer, ISBN 978-0-387-89561-1. (The book was written before I underwent gender reassignment and changed my name.)
2. Scott, C. P., 1921, *A Hundred Years*, Op-Ed piece in the Manchester Guardian, May 5th 1921.
3. Koestler, A., 1959, *The Sleepwalkers: a History of Man's Changing Vision of the Universe*, Penguin Classics, ISBN-13 : 978-0141394534
4. Rømer, O., 1676, *Démonstration touchant le mouvement de la lumière trouvé par M. Römer de l'Académie Royale des Sciences*, Journal des Sçavans: 233–36, 1676. (in French)
5. Orchiston, W., 1996, *Australian Aboriginal , Polynesian and Maori Astronomy* in *Astronomy Before the Telescope*, Ed. By C. Walker, BCA in association with the British Museum, ISBN 9780714117461.
6. https://en.wikipedia.org/wiki/Stadion_(unit), Retrieved 2021/07/03
7. Cleomedes, dates unknown, *Caeliestia*.
8. Aristarchos of Samos, 301-230 BCE, *On the sizes and distances of the Sun and Moon*, Translated by Sir Thomas Heath in *Aristarchus of Samos*, Oxford, Clarendon Press, 1913.
9. Heath, Sir Thomas, 1913, *Aristarchus of Samos*, Work cited, p. 302.
10. Dreyer, J. L. E., 1953, *A History of Astronomy from Thales to Kepler*, 2nd Ed., Dover, ISBN 0486-60079-3, p. 138.
11. Dreyer, J. L. E., 1953, Work Cited, p. 108.
12. Ptolemy, Claudius, c. 150 CE, *The Almagest*, Translated as *Ptolemy's Almagest* by G. J. Toomer, 1998, 2nd Ed., Princeton University Press, ISBN 978-0691002606.
13. Fitzpatrick, R., 2010, *A Modern Almagest*, self-published, downloadable from http://farside.ph.utexas.edu/Books/Syntaxis/Almagest.pdf or orderable in print from https://www.lulu.com/content/9011549?page=1&pageSize=4
14. Gingerich, O., 1993, *The Eye of Heaven: Ptolemy, Copernicus, Kepler*, American Institute of Physics, ISBN 978-0-88318-863-7.
15. Dreyer, J. L. E., 1953, Work cited, pp 149-170.
16. Dreyer, J. L. E., 1953, Work cited, p. 324.
17. Misner, C., Thorne, K and Wheeler, J. A., 1973, *Gravitation*, W. H. Freeman, ISBN 0716703440.

J. Clark, *Calculate the Orbit of Mars!*,
https://doi.org/10.1007/978-3-030-78267-2,
© Springer Nature Switzerland AG 2021

18. Thoren, V. E., 1990, *The Lord of Uraniborg: a Biography of Tycho Brahe*, Cambridge University Press, ISBN 978-0-521-03307-1

19. https://www.bbc.co.uk/news/science-environment-20344201

20. Swerdlow, N. M., 1996, *Astronomy in the Renaissance*, in *Astronomy Before the Telescope*, Ed. By C. Walker, BCA in association with the British Museum, ISBN 9780714117461.

21. Dreyer, J. L. E, 1890, *Tycho Brahe: a Picture of Scientific Life and Work in the Sixteenth Century*, Cambridge University Press, ISBN 978-1-108-06871-0.

22. Mosley, A., 2007, *Bearing the heavens: Tycho Brahe and the Astronomical Community of the Later Sixteenth Century*, Cambridge University Press, ISBN 978-1-107-40365-9.

23. King, D. A., 1996, *Islamic Astronomy*, in *Astronomy Before the Telescope*, Ed. By C. Walker, BCA in association with the British Museum, ISBN 9780714117461.

24. Dreyer, J. L. E., 1916, *On Tycho Brahe's Manual of Trigonometry*, The Observatory 39, 127–131.

25. Thoren, V. E., 1988, *Prosthaphaeresis Revisited*, Historia Mathematica 15, 32–39.

26. Dreyer, J. L. E, 1890, *Tycho Brahe: a Picture of Scientific Life and Work in the Sixteenth Century*, work cited, Ch. III.

27. Brahe, T., 1598, *Astronomiæ Instauratæ Mecanica*. Manuscript now in the Royal Library of Copenhagen.

28. Thoren, V. E., 1988, Work Cited, Ch. 8

29. Dreyer, J. L. E, 1890, Ch. VII.

30. Brahe, Tycho, 1588, *De Mundi Aetherei Recentioribus Phaenomenis*.

31. Thoren, V. E., 1973, *New Light on Tycho's Instruments*, Journal for the History of Astronomy, 4, 25– 45.

32. Blaeu, Willem & Joan, c. 1660, *Novus Atlas*.

33. Krehbiel, D. G., 1990, *Jacob's Staff*, The Ontario Land Surveyor, Spring 1990, accessed from http://www.surveyhistory.org/jacob's_staff1.htm

34. http://www.covingtoninnovations.com/michael/blog/1712/index.html#x171219A retrieved 2021/03/20.

35. Amery, H. and Cartwright, S., 2004, *The Usborne First Thousand Words in Arabic*, Usborne, ISBN 9780746046517

36. Chevalier, Patrick et al., 2017, https://www.ap-i.net/skychart//en/start

37. https://en.wikipedia.org/wiki/Bellatrix

38. https://en.wikipedia.org/wiki/Gregorian_calendar

39. Brahe, T., c. 1575-1601, *Tychonis Brahe Dani Opera Omni* (Complete Works of Tycho Brahe), in Latin, Edited by J. L. E. Dreyer, Volumes X and XI, accessed from https://archive.org/details/operaomniaedidit10brahuoft/page/174/mode/2up and https://archive.org/details/operaomniaedidit11brahuoft/page/176/mode/2up.

40. Drake, S., 1989, *A History of Free Fall*, published as an appendix to his translation of Galileo's Two New Sciences, 2000, Wall & Emeron, Inc., ISBN 0-921332-50-5.

41. Heilbron, J. L., 2010, *Galileo*, Oxford University Press, ISBN 978-0-19-965598-4.

42. https://en.wikipedia.org/wiki/Giordano_Bruno

43. Cooper, L., 1935, *Aristotle, Galileo and the Tower of Pisa*, Cornell University Press

44. http://ircamera.as.arizona.edu/NatSci102/NatSci/text/extgalleg.htm

45. Charles River Editors, 2019, *The Unification of Italy: The History of the Risorgimento and the Conflicts that Unified the Italian Nation*, Independently Published, ISBN-13 : 978-1079529043.

46. Sobel, Dava, 1999, *Galileo's Daughter*, Fourth Est ate, ISBN 978-1-85702-861-9.

47. https://en.wikipedia.org/wiki/Sector_(instrument)
48. Drake, S., 1778, *Galileo at Work*, Re-published by Dover Publications Inc., as ISBN 0-486-28631-2. 49 Truesdell, C. 1968, Essays in The History of Mechanics, Springer-Verlag, New York, 1968, ISBN 9783642866494, p. 30
49. Grant, E., 1974, *Sourcebook in Medieval Science 1974*, Harvard University Press, ISBN 0674823605, p. 252.
50. http://galileo.rice.edu/sci/instruments/pendulum.html
51. Galileo, 1638, *Two New Sciences*, Translated by Stillman Drake, 1989, 2nd Edition, Wall & Emerson Inc., ISBN 0-921332-50-5.
52. Galileo, 1613, *Sunspot Letters*
53. Galileo, 1610, *Siderius Nuncius* (*Starry Messenger*)
54. Galileo, 1623, *Il Saggiatore* (*The Assayer*)
55. Willings, D., 2000, Private Communication to the author.
56. Willings, D., 1980, The Creatively Gifted, Woodhead-Faulkner, ISBN 0-85941-120-6
57. Caspar, M., 1948, *Kepler*, Tr. C. D. Hellmann, republished by Dover, 2012.
58. Kepler, J. 1596, *Mysterium Cosmographicum* (*The Cosmographic Mystery*)
59. Kepler, J., 1604, *Astronomiae Pars Optica* (*The Optical Part of Astronomy*).
60. Kepler, J, 1609, *Astronomia Nova*, translated into English by William H. Donahue, 2015, Green Lion Press, ISBN 9781888009477
61. Gingerich, O., 1972, *Johannes Kepler and the New Astronomy*, Q. Jl. R. Astr. Soc., 13, 346-373
62. Wilson, C., 1968, *Kepler's Derivation of the Elliptical Path*, Isis, 59(1), pp. 4–25
63. https://en.wikipedia.org/wiki/John_Napier
64. Kepler, J. 1619, *Harmonices Mundi* (*The Harmonies of the World*)
65. Bakich, M., 2000, The Cambridge Planetary Handbook, ISBN 978-0-521-63280-3
66. Murray, C.D. and Dermott, S.F., 1999, *Solar System Dynamics*, Cambridge University Press, ISBN 00521575974, Appendix A, pp.526–530
67. Kepler, J. (1618-21) *Epitome of Copernican Astronomy*, Parts I-V, published in three parts.
68. MacCulloch, D., 2003, *Reformation: Europe's House Divided 1490-1700*, Penguin, ISBN 0-140-28534-2
69. Kepler, J., 1609, *Astronomia Nova*, Work cited.
70. Thomas, George B., Jr., *Calculus and Analytic Geometry*, 3rd Edition, Addison-Wesley, 1960, p.867 Recent editions of this excellent book are available on Amazon and Abebooks
71. Harper, W. L., 2011, *Isaac Newton's Scientific Method*, Oxford University Press, ISBN 978-0-19-870942-8, Chapter 5.
72. French, A. P., 1971, *Newtonian Mechanics*, Nelson, ISBN 17-771074-8, p. 308.
73. Jardine, L, 2002, *On a Grander Scale: The Outstanding Career of Sir Christopher Wren*, Harper Collins, ISBN 0-00-710776-5
74. Bennett, J. A., 1982, *The Mathematical Science of Christopher Wren*, Cambrsdge University Press, ISBN 0-521-24608-3
75. Jardine, L., 2003, *The Curious Life of Robert Hooke, The Man who Measured London*, Harper Collins, ISBN 978-0007151752
76. Sellers, D., 2006, *A letter from William Gascoigne to Sir Kenelm Digby*, Journal for the History of Astronomy 37(4), pp. 405 – 416
77. Westfall, R. S., 1983, Never at Rest: A Biography of Isaac Newton, Cambridge University Press, ISBN 978-0521274357

78. Jardine, L., 2003, *The Curious Life of Robert Hooke, the Man who Measured London*, Work cited.

79. Newton, I., 1687, *Philosophiæ Naturalis Principia Mathematica* (*Mathematical Principles of Natural Philosophy*), later Editions 1713 and 1726, Translated by I. Bernard Cohen and Anne Whitman, preceded by A Guide to Newton's Principia by I. Bernard Cohen, University of California Press, 1999, ISBN 978-0-520-08817-7

80. Fara, P., 2002, republished 2011, *The Making of Genius* , Pan Macmillan, ISBN 978-1-447-20453-4

81. Harper, W. L., 2014, *Isaac Newton's Scientific Method*, Work cited

82. Chandrasekhar, S., 1995, republished 2012, *Newton's Principia for The Common Reader*, Oxford University Press, ISBN: 978-0198088271

83. Murray, C.D. and Dermott, S.F., 1999, *Solar System Dynamics*, Work Cited

84. These formulae are proved in many undergraduate-level mathematics books. Thomas, George B., Jr., Calculus and Analytic Geometry, work cited.

85. Colwell, Peter, 1993, *Solving Kepler's Equation over Three Centuries*, Willmann-Bell, ISBN 0943396409.

86. https://en.wikipedia.org/wiki/Newton%27s_method

87. Danby, J.M.A., 1988, *Fundamentals of Celestial Mechanics*, Willmann-Bell, ISBN 9780943396200, Chapters 12 and 14.

88. Brewster, D., 1855, *Memoirs of the Life, Writings and Discoveries of Sir Isaac Newton*, Vol. 2, Chapter XVIII

89. Danby, J.M.A., 1988, Work Cited, p.146.

90. Crida, A. (2009). *Solar System Formation, Reviews in Modern Astronomy*. 21. pp. 215–227. arXiv:0903.3008. Bibcode:2009RvMA...21..215C

91. Milgrom, M., 1983, *A Modification of the Newtonian Dynamics as a Possible Alternative to the Hidden Mass Hypothesis*. Astrophys. J. 270, 365–370.

92. Peebles, P. J. E., 2017, *Growth of the nonbaryonic dark matter theory*, Nature Astronomy, 1, 0057.

93. Bond, H. E. et al., 2017, *The Sirius System and Its Astrophysical Puzzles: Hubble Space Telescope and Ground-based Astrometry*, The Astrophysical Journal, 840:70 (17pp)

94. Adibekyan, V. et al., 2018, *The AMBRE project: searching for the closest solar siblings*, Astronomy & Astrophysics, 619, Article Number A130, 19 pages

95. Gauss, C. F, 1809, *Theoria Motus Corporum Coelestium in sectionibus conicis solem ambientium*, translated by C. H. Davis as *Theory of the Motion of the Heavenly Bodies Moving About the Sun in Conic Sections, a Translation of Gauss' "Theoria Motus" With an Appendix*, 1857, Franklin Classics, ISBN 9780342834815.

96. Givens, W., 1958, *Computation of Plain Unitary Rotations Transforming a General Matrix to Triangular Form*, Journal of the Society for Industrial and Applied Mathematics, 6(1), 26–50.

97. Danby, J.M.A., 1988, *Fundamentals of Celestial Mechanics*, Work Cited

98. Curtis, H. D., 2013, *Orbital Mechanics for Engineering Students*, 3rd Ed., Butterworth-Heinemann, ISBN 9780080977478.

99. https://aa.usno.navy.mil/data/docs/RS_OneYear.php

100. https://en.wikipedia.org/wiki/Habash_al-Hasib_al-Marwazi

101. Danby, J.M.A., 1988, Work cited, §6.14.

102. Wilson, E. B., 1901, *Vector Analysis*, downloadable from https://ia800200. us.archive.org/18/items/cu31924001246341/cu31924001246341.pdf.

103. Todhunter, I. *Spherical Trigonometry*, 1871, 3rd Ed., https://archive.org/details/ in.ernet.dli.2015.511900

104. Euclid, c. 300 BCE, translated by I. Todhunter, 1869, https://archive.org/details/ elementsofeuclid00todhuoft.

105. Milne, R. (1921). 593. *Note on the Equation of Time*, The Mathematical Gazette, 10(155), 372–375. doi:https://doi.org/10.1017/S0025557200232944

106. https://aa.usno.navy.mil/faq/docs/RST_defs.php

107. Clark, J. D., 2015, *Viewing and Imaging the Solar System: a Guide for Amateur Astronomers*, Springer, ISBN 9781461451785.

108. https://www.ap-i.net/skychart/en/start

109. https://www.ap-i.net/skychart/en/download

110. Montenbruck, O. & Pfleger, T., 1994, *Astronomy on the Personal Computer*, 2nd Edition,

111. Montenbruck, O. & Pfleger, T., 1999, *Astronomy on the Personal Computer*, 4th Edition, translated by Storm Dunlop, Springer-Verlag, ISBN3540672214.

112. https://www.pgroup.com/products/community.htm

113. http://www.codeblocks.org

114. https://visualstudio.microsoft.com/vs/older-downloads

115. Boulet, D., 1991, *Methods of Orbit Determination for the Microcomputer*, Willmann-Bell, ISBN 0943396344.

116. Gauss, C. F, 1809, Work cited.

117. Bauschinger, J., 1906, *Die Bahnbestimmung der Himmelskörper; mit 84 Figuren im Text*, Wilhelm Engelmann, available online, in German, at https://diglib.uibk. ac.at/ulbtirol/content/pageview/360713.

118. Karimi, R. R. & Mortari, D.,2010, *Orbit Determination Using Prescribed Orbits*, AAS 10-236, AAS/AIAA Space Flight Mechanics Meeting Conference, San Diego, CA, February 14-18, 2010.

119. Karimi, R.R. and Mortari, D., 2009, *On Preliminary Orbit Determination: A New Approach*, Paper AAS 09-106 of the 2009 AAS/AIAA Space Flight Mechanics Meeting Conference, Savannah, GA, February 9-12, 2009.

120. Karimi, R.R. and Mortari, D., 2010, *Initial Orbit Determination Using Multiple Observations*, Celestial Mechanics and Dynamical Astronomy, Vol. 109, No. 2, 2010, pp.167–180.

121. Karimi, R.R. and Mortari, D., 2012 *Orbit Determination Based on Variation of Orbital Error*, Paper AAS 12-201 of the 2012 AAS/AIAA Space Flight Mechanics Meeting Conference, Charleston, SC, Jan 29 –Feb 2, 2012

122. Karimi, R.R. and Mortari, D., 2011, *On Laplace's Orbit Determination Method: Some Modifications*, Paper AAS 11-121 of the 2011 AAS/AIAA Space Flight Mechanics Meeting Conference, New Orleans, LA, February 13-17, 2011.

123. Karimi, R.R. and Mortari, D., 2013, *A Performance Based Comparison of Angle-only Initial Orbit Determination Methods*, AAS 13-823, 2013 AAS/AIAA Astrodynamics Specialist Conference, Hilton Head, SC, August 11-15, 2013.

124. Herget, P., 1965, *Computation of Preliminary Orbits*, Astronomical Journal, 70,1–3

125. https://en.wikipedia.org/wiki/International_Celestial_Reference_Frame

126. Clark, J. D., 2009, Work Cited, Chapter 4 shows the observational evidence for this phenomenon.

127. Herget, P., 1948, *The Computation of Orbits*, Privately Published.

128. Escobal, P. R., 1976, *Methods of Orbit Determination*, Robert E. Krieger Publishing Co., Malabar, FL.
129. Escobal, P. R., Work Cited, §3.7.1, pp. 96–98
130. Hamilton, W. R., 1845, Proc. R. Irish Acad., III, Appendix p. 36.
131. Danby, J. M. A., Work Cited, p. 28
132. Boulet, D., Work Cited, §§12.2-3.
133. https://www.sunrisesunset.com/calendar.asp?comb_city_info=Home;3.08;51.6;0; 2&ianatz=Europe/London&month=10&year=2020&back=&time_type=0&wsom =0&txsz=M&supr=924172
134. Bretagnon, P. & Simon, J-L, 1986, *Planetary Programs and Tables from -4000 to +2800*, Willmann-Bell, ISBN 0-943396-08-5, Equations (14) & (15)
135. Simon, J.L., Bretagnon, P., Chapront, J., Chapront-Touzé, M., Francou, G. & Lasker, J., 1994, *Numerical Expressions for the Precession Formulae and mean elements for the Moon and the Planets*, Astron. Astrophys., 282, 663–683.

Index

A

Acceleration, 35
Accuracy, 16, 21, 27
Alidade, 21–23
Almagest, 4
Angular momentum, 71–73, 96, 223, 225, 248
Anomaly, 78–86, 88, 91
Aphelion, 206, 253
Apparition, 281, 284
Apsis, 206
Archimedes, 2
Area of a triangle, 71, 72
Area of an ellipse, 58
Argument of perihelion, 249
Aristotelian, 6, 16, 38
Aristotle, 3, 32, 37, 38, 41
Armillary, 18, 19
Astronomia Nova, 47–52
Astronomy, 1

B

Bauschinger's method, 202
Big science, 15
Boulet, D., 269, 280, 288
Bruno, G., 32, 33
Business, 34

C

Calculus, 62, 98
Camera sensor, 169, 170
Canonical units, 93, 94, 254
Cartes du Ciel, 26–28, 160–163, 194

Cartesian coordinate system, 171, 207, 293–295
Catadioptric lens, 151, 152, 194
Celestial Equator, 139, 145
Celestial North Pole, 139, 288
Celestial sphere, 171
Celestron CGX mount, 190
Central America, 1
Centre of mass (CoM), 64, 99
Centrifugal force, 59
Centripetal force, 59
Ceres, 107
Circles, 14
Circular motion, 3
Civilizations, 1
Comet, 16
Computer-aided drafting (CAD) software, 157–161, 163, 164
Conic sections, 73–78
Conservation of linear momentum, 59
Conventional orbital elements, 247
Copernican Revolution, 6, 7
Copernicus, 6–8, 31
Coplanar, 201, 202, 204, 206, 246
Crescent, 42
Cross-hairs, 60

D

Danby, 269
Dark frame, 155
Declination, 17, 19, 24, 26–30, 171
Deep Sky Stacker, 155, 195, 196
Deferent, 5, 6

J. Clark, *Calculate the Orbit of Mars!*,
https://doi.org/10.1007/978-3-030-78267-2,
© Springer Nature Switzerland AG 2021

Density variation, 71
Didymium filter, 153, 195, 196
Digital single-lens reflex (DSLR) camera, 151,
        157, 168, 192, 194–196
du Châtelet, É., 98
Dusk, 119

E
Earth, 2–6, 16, 18, 38, 40, 43, 291
    celestial equator, 145
    elliptical, 121, 128
    ellipticity, 146
    Equation of Time, 146, 147
    Excel, 146
    King's Lynn, Norfolk, England, 113
    major axis, 207
    Mars, 150
    mean anomaly, 145
    Monte-Carlo simulation, 146
    obliquity, 137–144, 146
    orbital mechanics, 145
    orbital parameters, 147
    parameters, 146
    Right Ascension, 121
    sidereal day, 122, 123
    solar midday, 145
    sunrise and sunsets (*see* Sunrise and
        sunsets)
    the Sun, 113, 149, 150
    tilt, 121, 139
Eccentric anomaly, 79–81, 83, 128, 130, 136,
        227, 228, 244
Eccentricity, 49, 56, 87–89, 91, 215, 234, 246,
        248, 249, 252
Ecliptic, 5, 18, 121, 124, 136, 139, 144, 145,
        150, 186, 187, 190, 203
    latitude, 190, 192
    longitude, 190, 191
Elements, 269, 270, 273, 281, 287, 288,
        290, 291
Ellipse, 84, 105, 201, 203–206, 223, 224, 227,
        246, 249, 252
    eccentricity, 56
    equation, polar coordinates,
        53, 54
    focuses/foci, 55, 56
    hyperbolas, 56
    Kepler, J., 53
    locus of points, 55
    mathematical methods, 52
    parabolas, 56
    Pythagoras' theorem, 53

stretched circle, 54
stretching, 58
true anomaly, 58
Tycho Brahe Museum, 52
Elliptical, 121, 128
Elliptical orbit with eccentricity, 87–91
Ellipticity, 146
Energy, 94–99, 103
Epicycle, 5, 6, 16, 186, 189, 291
Equation of motion, 101
Equation of the Centre, 135
Equation of Time, 146, 147
Equatorial armillary, 17, 18, 20
Eratosthenes Aristarchus, 2
Escobal's method, 221
Euclid, 2
Excel, 85, 161

F
Fitted orbits, 281–283, 286
Five regular spheres, 46
Fixed stars, 207
Florence, 34, 42, 43
Focus, 47, 55
Force law, 98–105
Fortran, 167, 174, 272, 273
Fortran's arc cosine function, 253
Free fall, 31–35, 37

G
Galileo, G., 3, 6, 31, 32, 34–38, 40–43
Gascoigne, W., 61
Gauss, C.F., 107, 269
Gauss' method, 201, 202
Geocentric, 3, 41
Geometric astrometry, 168, 169
Geometry, 10, 223
Gibbous, 42, 286
Gibbs, J.W., 202
Givens, W., 108
Givens' method, 269, 270
GNU Fortran, 168
GPS-based satellite navigation devices, 115
Gravesande, W., 98
Gravity, 6, 35, 37, 67
Gravity due to point masses and spheres, 67,
        69, 71
Greenwich Mean Time, 122
Gregorian, 25–28, 31
Grey area, 72, 73
Guidemaster system, 191

H
Halley, E., 62, 93
Halley's Comet, 87, 91
Hamilton, W., 249
Harmonices Mundi, 50
Heavenly spheres, 16
Heliocentric, 6, 8
Herget's method, 203, 244
   derivatives, 221
   ellipses, 223
   geometry, 223
   hypergeometric functions, 221
   least-squares fit solutions, 221
   observation times, 219
   preliminary orbit, 218, 219
   results, 267
   semi-major axis, 246
   spherical polar coordinates, 220
   true anomalies, 246
   unit vector, 219
   variables, 220
   vectors, 220
Hipparchus, 6
Home coordinates, 283
Hooke, R., 60, 62
Huyghens, C., 59, 60
Hyperbola, 56, 293–295
Hypotheses non fingo, 67

I
Inclination, 204, 205, 216, 217, 219, 248, 253
Inclined plane, 34–36
Indirect methods, 115
Inquisition, 32, 42, 43
International Celestial Reference Frame,
   139, 207
Italy, 33, 35

J
Jacob's Ladder, 24
JPEG, 155, 157
Julian, 24, 25, 27, 28, 31
Jupiter's moons, 39, 40

K
Karimi, R.R., 202
Kepler, J.
   action at a distance, 47
   Astronomia Nova, 47–49, 52
   eccentricity, 49
   ellipse, 47

   equation, 85, 129, 228, 239–242, 254
   five regular spheres, 46
   focus, 47
   Harmonices Mundi, 50, 51
   Mars data, 47
   mathematical prodigy, 45
   orbit of Mars, 50
   orbit, oval, 48, 50, 51
   planetary laws, 45
   range-finding, 48, 49
   solar system, 52
   telescope, 47
   third law, 51–52
Kepler's laws
   planetary motion, 73, 87, 91
Kinetic energy, 95, 101
King's Lynn, Norfolk, England, 113, 114, 122,
   124–126

L
Lagrange's f and g functions, 223–231
Layers, 161
Least squares
   equation, 109
   Gauss, C.F., 107
   Givens, W., 108
   numerical algorithm, 110
   subroutine, 110
Least-squares curve fitting
   curves, 271
   Fortran program, 272
   Givens' method, 270
   Givens, W., 269
   infinitesimal quantities, 271
   initial radius and velocity, 269
   Monte Carlo methods, 272, 273
   numerical differentiation, 271
   position and velocity, 269
   prediction, 270
   random noise, 271
   set of elements, 270
   set of measurements, 269
Leibniz, G.W., 98
Lens hood, 152, 153
LibreOffice, 280
Live view shooting, 151
Logarithm, 10, 14
Longitude of the ascending node, 205, 247

M
Mach, 6
Magnetism, 47

Manual adjustment, 279, 280, 283, 287
Mars, 4, 5, 24, 26–30, 281, 283–285, 288,
       291, 292
    angle unit changing subroutines,
       180–181
    C++, 167, 168
    celestial sphere, 169–171
    declination, 171
    distance unit changing subroutines, 183
    DLSR, 154
    equipment, 151–154
    errors, 172, 183
    Fortran, 167
    GNU Fortran, 168
    guiding, 154
    Least Squares Fitting Subroutine
       LSQFIT, 181
    measurements, 157–166
    night's observation, 156–158
    observatory
       2009–2010 apparition, 186–188, 190
       2011–2012 apparition, 190, 191, 193,
          194, 197
       2020 apparition, 190, 191, 193,
          194, 199
    off-axis sensor, 168
    opposition, 155
    orientation, sensor, 168
    Pascal, 167
    PGI Fortran, 168
    photographic technique, 155, 156
    reference stars, 170
    right ascension, 171
    seeing, 156
    triangulating planet positions, 167
Mathematical model, 4
*Mathematical Principles of Natural
       Philosophy*, 62
Mathematical prodigy, 45
Mathematics, 10–12, 14, 15
Mean and solar noon, 137
Mean anomaly, 83, 91, 135, 136, 145,
       239, 254
Meridian, 120, 137, 139, 284
Merton rule, 35
Microsoft Excel, 272, 280, 285
Modified Newtonian dynamics (MOND), 99
Monte Carlo methods, 272, 273
Moon, 2, 16, 18, 38, 40, 41, 92, 93
Mortari, 202, 203
Motion, 3, 4, 6, 34, 35, 37
Moving average, 123, 127, 128
Musician, 31, 35

N
Newton, I., 3, 4, 6, 37, 61–63
Newton's laws
    gravity, 63, 99
    motion, 63, 91
    telescopes, 62
Newton-Raphson technique, 85, 86
Non-conservative gravitational potential, 101
Non-coplanar orbits, 203, 204
North Celestial Pole, 289

O
Obliquity, 137–144, 146
Observational astronomer, 8
Observatory, 156, 160
Off-axis sensor, 168
Opposition, 283–285, 288, 291, 292
Orbit, 269, 270, 272, 279–281, 283, 286, 288,
       290–292
    elements, 247
    mechanics, 54, 145
Orientation, 168, 169

P
Padua, 32–34, 37
Parabolas, 56, 297
Parallax, 16
Pascal, 167
Patent, 34, 38
Pendulum, 36, 37
Perihelion, 206, 207, 249, 252, 255
PGI Fortran, 168
Photoshop, 158, 159
    elements, 284
Piggyback, 155, 168, 192
Pisa, 32, 33
Planar, 201
Planet, 1, 3–6, 16, 32, 42
Planetary laws, 45
Polar coordinates, 295, 297
Poleni, G., 98
Potential energy, 96, 98, 100, 101, 105
Precession, 288, 289
Preliminary orbit, 267
Pre-telescope, 7, 9
Principia, 62, 63, 67, 93, 253
Projectile, 37, 38
Prosthaphaeresis, 14
Ptolemy, C., 3, 4
Ptolemy's model, 4–6
Pythagoras' theorem, 53, 140

R
Radius vector, 251
Range-finding, 48, 49
Reddening, 118, 119
Registax, 284
Re-orientation, 204
Reticles, 60
Right ascension (RA), 17, 19, 20, 27, 29, 30,
        121, 135, 145, 171
Rome, 32, 33, 43
Rotation matrix, 210–212
Rotations, Mars
    angles, 215–218
    apsis, 206
    centre of the Earth, 207
    coordinate systems, 207–210
    eccentric anomalies, 244
    ecliptic, 207
    elliptical orbit, 243
    fixed stars, 207
    function MMULT, 213
    Herget's method (*see* Herget's method)
    inverse matrix, 210
    Kepler's equation, 239–242
    Lagrange's f and g functions, 223–231
    major axis, 217, 219
    MMULT, Microsoft Excel, 213–215
    orbital elements, 246–255
    order of the rotations, 212
    polar coordinates, 207
    reverse direction, 208
    rotation operations, 212
    sector–triangle ratio, 231–239
    semi-major axis, 215
    software, 255
    square matrix, 208
    Sun–Mars distance, 267
    three-dimensional rotation, 207
    true anomaly, 207
    vectors, 209

S
SCT, 192
Seasons, 1
Sector–triangle ratio, 231–239
Semilatus rectum, 77
Semi-major axis, 215, 246
Series expansion of tan *x*, 143
Sextant, 16, 17, 22
Sidereal day, 122, 123
Sirius, 99
Solar day, 122, 123
Solar midday, 122

    elliptical, 121
    Greenwich Mean Time, 122
    King's Lynn, Norfolk, England, 122
    leap year, 121
    mean midday
        King's Lynn, Norfolk, England, 127, 128
    meridian, 120
    physical parameters, 124
    sidereal day, 122
    solar day *vs.* sidereal day, 123
    source of error, 123
    sunrise and sunset data, 122, 123
    sunrise/sunset, 120
    tilt, 121
    time, 122
    time zone meridian, 121
Solar system, 52, 94, 99, 233
Solar year, 145
Sphere, 10
Spherical polar coordinates, 220
Spherical triangle SNQ, 140, 141
Spherical trigonometry, 10, 137, 140, 202
Stars, 162
Stray lights, 193
Sun, 2, 5, 6, 16, 18, 42, 96
    anomaly, 139
    Earth, 139
    ecliptic, 139
    ellipse, 203
    leap year, 121
    right ascension, 135
    sunrise and sunsets (*see* Sunrise and sunsets)
    true anomaly, 139
    Vernal Equinox, 145
Sun–planet distances, 232
Sunrise and sunsets
    atmospheric conditions, 146, 147
    centre of the Sun, 147
    cloud patterns, 115
    data, 147
    dusk, 119
    Earth, 149
    ecliptic, 150
    flat area, 114
    GPS-based satellite navigation
        devices, 115
    hilly areas, 114
    indirect methods, 115
    King's Lynn, Norfolk, England, 113
    north, 118–120
    north-west, 115, 116
    reddening, 119
    south-east, 117, 118
    the Sun, 146

Superior planet, 4, 5
Supernova, 15–17
Sydney Observatory, 186

T
Telescope, 31–34, 38, 40–42, 154
Tilt, 121, 139
Time evolution
    anomaly, 78–86, 88, 91
Time of perihelion passage, 254, 267
Time zone meridian, 121
Topaz Sharpen AI, 284
Trigonometric addition formulae, 303
Trigonometric methods, 223
True anomaly, 58, 79, 81, 104, 128, 135, 139,
    207, 227, 233, 234, 252
TurboCAD, 157, 158, 161
Two New Sciences, 43
Tycho, B., 7, 10, 12, 13, 15, 17, 19, 22, 26, 28,
    29, 31
    data, 288, 289, 291, 292

Tycho Brahe Museum, 52
Tycho2, 160

U
Universal gravitation, 63, 64, 66, 67, 98
Uraniborg, 15

V
Vector eccentricity, 252
Vector language, 72
Vectors, 203, 204, 210, 220, 225, 226, 231,
    247, 250, 253, 299, 301
Venice, 33, 42
Venus, 41, 42
Victorian mathematical methods, 202
Vis viva equation, 95, 97, 98, 246

W
Wren, C., 60